A Volterra Approach to Digital Predistortion

IEEE Press
445 Hoes Lane
Piscataway, NJ 08854

A Volterra Approach to Digital Predistortion

Sparse Identification and Estimation

Carlos Crespo-Cadenas
Universidad de Sevilla
Sevilla

María José Madero-Ayora
Universidad de Sevilla
Sevilla

Juan A. Becerra
Universidad de Sevilla
Sevilla

IEEE Press
WILEY

Published by John Wiley & Sons, Inc., Hoboken, New Jersey.

Published simultaneously in Canada.

Library of Congress Cataloging-in-Publication Data Applied for:

Hardback ISBN: 9781394248124

Cover Image: Wiley
Cover Design: © Volodymyr Baleha/Shutterstock

Set in 9.5/12.5pt STIXTwoText by Straive, Chennai, India

To our families.

Contents

About the Authors

Carlos Crespo-Cadenas was born in Madrid in December 1949 and he received an MSc degree in Physics at the Universidad de La Habana, Cuba, in 1974. In 1978, he performed R&D projects on topics such as piezoelectric quartz devices and RF engineering design in the *Laboratorio Central de Telecomunicaciones* leading several development projects of radio communications equipment. His work as an assistant researcher was positively valued, and he was appointed auxiliary researcher in 1988. In 1991, he started a four-year stay in the Universidad Politécnica de Madrid, thanks to an award from the Spanish National Board of Scientific and Technological Research for the return of doctors and technologists to the Spanish research system. His research was focused on the design of microwave monolithic integrated circuits (MMIC) ending with a PhD in 1995. In 1994, he became University Assistant in the Universidad Politécnica de Madrid and Associate Professor in the Universidad de Sevilla in 1995. There, he created the research Group of Radiocommunication Systems and led numerous research projects. In 1998, he was appointed *Professor Titular de Universidad* and in 2018, he was appointed Full Professor.

His research lines are nonlinear analysis of radio frequency and microwave devices, modeling and compensation of nonlinear impairments, and measurement techniques for nonlinear communication systems. He is author or coauthor of more than 80 papers published in refereed journals or international conference proceedings. He has participated in 12 research projects funded by competitive calls and in 6 research contracts with private companies of topics related to his research lines. Furthermore, he has supervised 4 PhD thesis, 2 of which were awarded the *Premio Extraordinario de Doctorado* (Outstanding PhD Award) of the Universidad de Sevilla.

He is a member of the Institute of Electrical and Electronics Engineers (IEEE) and the Microwave Theory and Techniques (MTT) Society. He has served as a reviewer for several research journals, such as *IEEE Transactions on Microwave Theory and Techniques, IEEE Transactions on Signal Processing*, and *IEEE*

Transactions on Circuits and Systems. He participated as opponent in PhD thesis in Chalmers University of Technology, Sweden, and in University College Dublin, Ireland.

María José Madero-Ayora was born in Seville, Spain, in 1978. She received Telecommunications Engineering and a PhD from the Universidad de Sevilla, Spain, in 2002 and 2008, respectively. Since 2003, she has been with the Department of Signal Theory and Communications of the Universidad de Sevilla, where she has been an associate professor since 2012.

Her research activities are mainly focused on nonlinear analysis of RF and microwave devices, modeling and compensation of impairments in modulators and power amplifiers, and measurement techniques for nonlinear communications systems. This activity resulted in the development of novel Volterra-based behavioral models and the application of advanced signal processing techniques for model-order reduction in the design of digital predistorters, which have been presented in major conferences and published in international journals.

She is coauthor of more than 60 papers published in refereed journals or international conference proceedings. She is a member of the Institute of Electrical and Electronics Engineers (IEEE) and the Microwave Theory and Techniques (MTT) Society.

Juan A. Becerra was born in Seville, Spain, in 1986. He received the BS and MSc degrees in Telecommunication Engineering from the Universidad de Sevilla, Seville, Spain, in 2009 and 2012, respectively, a PhD in Electrical and Computer Engineering from the University of Delaware, Newark, DE, USA, in 2017, and a PhD in Telecommunication Engineering from the Universidad de Sevilla in 2019. Since 2017, he has been with the Department of Signal Theory and Communications, Universidad de Sevilla, and he has been an associate professor since 2023.

His main research areas include behavioral modeling and linearization of power amplifiers. He is specialized in compressed-sensing signal processing applied to the regression of Volterra series models, and the results of such activities have been presented at major conferences and published in international journals.

He is a member of the Institute of Electrical and Electronics Engineers (IEEE) and the Microwave Theory and Techniques (MTT) Society. He is a member of the Steering Committee of the Radio and Wireless Week (RWW) and the Technical Program Committee (TPC) of the IEEE Topical Conference on RF/Microwave Power Amplifiers for Radio and Wireless Applications (PAWR).

Preface

During the last decades, we have witnessed outstanding developments in wireless communications that have brought improved network capacity and coverage, increased energy efficiency, and reduced costs. Massive MIMO, millimeter-wave technology, and hybrid beamforming are being exploited with the deployment of 5G and 6G, and major breakthroughs are to come with future systems.

Energy-efficient base stations have been possible, thanks to the continuous development of transistor technologies and novel proposals of power amplifier schemes driven by highly varying envelope signals. On the other hand, increment of system capacity is related to the ability of the transmitter to strictly comply with the spectrum emission masks set by standardization and regulatory authorities. To meet both requirements, the use of linearization techniques is convenient, in particular those implemented with a digital predistorter in the baseband modules of the transmitter equipment to guarantee the linearity of the emitted radiofrequency signal.

The analysis and optimization of highly efficient power amplifiers have been addressed traditionally by several books. One of the most renowned, perhaps because it is tightly attached to practical objectives of the engineers, is the Steve C. Cripps book "*RF Power Amplifiers for Wireless Communications.*" Other recent texts, like "*Behavioral Modeling and Predistortion of Wideband Wireless Transmitters*" authored by Fadhel Ghannouchi, Oualid Hammi, and Mohamed Helaoui or "*Behavioral Modeling and Linearization of RF Power Amplifiers*" written by John Wood, have been devoted to transmitters linearization and have received a great deal of attention from the research community. Whereas the development of power amplifier architectures is feasible autonomously, it is reasonable to think that some knowledge on RF power amplifier schemes, although not strictly necessary, can be useful in efficient proposals of digital predistorter designs.

When we intended to compile this book, covering the field of behavioral modeling and linearization, our first thought was how we can provide something valuable and at the same time different from other publications. In second place, we also considered that fundamental concepts on digital predistortion can help RF engineers to redefine the design of a power amplifier that is to be used in any linearization system. More than bringing behavioral modeling and digital predistortion subjects up to date, one of the principal aspirations of this text is to expose our particular point of view following the Volterra series as the common thread on power amplifier and digital predistorter modeling, on one hand, and the use of pursuit techniques to determine a sparse set of regressors in transmitter linearization, on the other hand.

The book is organized in two parts with the hope that both can be regarded as closely related but also can be read independently. The first part, Chapters 1–4, establishes a connection between fundamental concepts of the microwave power amplifiers field and digital signal processing techniques, so that it can serve those not habituated engineers to familiarize with radio frequency amplifiers notions. The second part of the text, Chapters 5–7, examines statistical analysis for model identification and predistortion of communication transmitters and, although tightly related to the first part, can also be read separately.

In Chapter 2, we discuss how the RF power amplifier can be modeled under the perspective of Volterra series and how this view can explain theoretically classical nonlinear behavior of power amplifiers. Chapter 3 focuses on discrete-time Volterra models and how their complexity can be reduced. Some singular aspects in our approach are the consideration of the double Volterra, Volterra-Parafac, and complex-valued Volterra models. Chapter 4 is devoted to the discussion of model pruning based on the knowledge of the power amplifier equivalent circuit, relating the kernels structure to internal mechanisms associated with physical properties of transistors. Remarkable results of this chapter are model proposals based on power amplifier circuit knowledge and electrothermal or charge-trapping effects, in particular the bivariate model which includes the popular GMP model as a particular case. Following the spirit of this book, the same approach has been also extended to the proposal of an I/Q modulator model. Assuming that a predistorter is an bounded input-bounded output nonlinear system, we conclude that it can be also analyzed with the general Volterra structure used for power amplifier models.

The second part is more pragmatic and devoted to practical issues of regression, sparse models, and transmitter linearization with digital predistorters. Chapter 5 introduces the regression concept and the main statistics that allow to obtain an estimator of the power amplifier model. Chapter 6 explores the application of sparse signal processing to attain a reduced set of active coefficients in the model, including the doubly orthogonal matching pursuit (DOMP) and the sparse Bayesian pursuit (SBP) as techniques within this group. Finally, Chapter 7 reviews the most commonly used digital predistortion architectures and provides an extensive set of digital predistortion results based on the foundations set in previous chapters.

Sevilla, 27th November 2024

Acknowledgments

We want to manifest our gratitude to many people and institutions that supported us to carry out this task. Two researchers have had a remarkable influence on our work, and we owe them a special tribute. The first mention goes to Javier Reina, whose collaboration has been unique since the origin of our research team and throughout the time he belonged to the group. In the case of Sergio Cruces, his insightful knowledge on statistics has been decisive in the creation of our sparse model perspective.

Finally, we also acknowledge the financial and institutional support provided by the Spanish Ministerio de Ciencia, Innovación y Universidades, Junta de Andalucía, and Universidad de Sevilla.

Notation Conventions

- Time signals are denoted by lower case letters and the Fourier transform of a time function is denoted by the same capital letter.
- Vectors are denoted in boldface lower case letters $\mathbf{x}$, matrices in boldface upper-case letters $\mathbf{X}$, and $\mathbb{X}_n$ is a n-way array (or nth-order tensor). By default, $\mathbf{x}$ is a column vector.
- $\mathbf{x}^T$ is the transpose of $\mathbf{x}$ and $\mathbf{x}^H$ is the conjugate transpose of $\mathbf{x}$.
- Normally, $\mathbf{x}$ is reserved for the signal applied to the input of a system, $\mathbf{y}$ for the output signal and $\mathbf{h}$ for the coefficients vector in linear regression models.
- When necessary, $\tilde{\mathbf{x}}$ is used to represent the RF real-valued input signal and $\mathbf{x}$ its complex envelope.
- Index n is kept to indicate the nonlinear order of a term, k the sample index in a discrete-time signal.
- In a linear regression model, the basis function or regressor is denoted as $\phi(k)$. In particular, $\xi_{\mathbf{q}_n}(k)$ is used to represent a nth-order regressor of a Volterra model.
- The vector of delays of the nth-order kernel is denoted as $\mathbf{q}_n = [q_1, q_2, \dots, q_n]$.
- In compact form, we write

$$\sum_{\mathbf{q}_n=\mathbf{0}}^{\mathbf{Q}_n} h(\mathbf{q}_n) \triangleq \sum_{q_1=0}^{Q_1} \cdots \sum_{q_n=0}^{Q_n} h(q_1, \dots, q_n)$$

  to indicate multiple sums.
- Equally, we use

$$\int_{-\infty}^{\infty} (\cdot)\, d\boldsymbol{\tau}_n \triangleq \int_{-\infty}^{\infty} \int_{-\infty}^{\infty} \cdots \int_{-\infty}^{\infty} (\cdot)\, d\tau_1 d\tau_2 \cdots d\tau_n$$

- $\Re(x)$ and $\Im(x)$ denote the real and imaginary parts of x, respectively.
- The abbreviations AC and DC are used to mean simply alternating and direct, as when they modify current or voltage.
- The ordinary frequency is denoted as f and the angular frequency as $\omega = 2\pi f$.

- The symmetrized nonlinear transfer function will be explicitly designated as $\overline{H_n(\boldsymbol{\omega}_n)}$ only when necessary.
- To simplify, we use the notation

$$\prod_{r=a}^{b} x_r = x_a\, x_{a+1} \cdots x_b$$

and the product $\prod_{r=a}^{b}$ is understood to have the value 1 when $b < a$.

Acronyms

4G	fourth generation
5G	fifth generation
6G	sixth generation
AC	alternating current
ACEPR	adjacent channel error power ratio
ACPR	adjacent channel power ratio
ADC	analog-to-digital converter
AM	amplitude modulation
ARVTDNN	augmented real-valued time-delay neural network
BB	baseband
BIBO	bounded-input bounded-output
BIC	Bayesian information criterion
Bi-NL	bivariate nonlinear
BJT	bipolar junction transistor
BOTDNN	block-oriented time-delay neural network
CAD	computer-aided design
CKV	circuit knowledge Volterra
DAC	digital-to-analog converter
DC	direct current
DDR	dynamic deviation reduction
DLA	direct learning architecture
DOMP	doubly orthogonal matching pursuit
DPD	digital predistortion, digital predistorter
DSB-SC	double-sideband suppressed-carrier
DUT	device under test
DVB-T2	digital video broadcasting – terrestrial 2
ET	envelope tracking
EVBW	extended behavioral model for wideband amplifiers
EVM	error vector magnitude

FET	field-effect transistor
FFT	fast Fourier transform
FLOP	floating point operation
FV	full Volterra
GMP	generalized memory polynomial
GPIB	general-purpose instrumentation bus
HB	harmonic balance
HEMT	high-electron-mobility transistor
IF	intermediate frequency
IFFT	inverse fast Fourier transform
ILA	indirect learning architecture
IM3	third-order intermodulation
IMD	intermodulation distortion
IP_3	third-order intercept point
IP_5	fifth-order intercept point
IP_7	seventh-order intercept point
I/Q	in-phase/quadrature
LASSO	least absolute shrinkage and selection operator
LDMOS	laterally-diffused metal-oxide semiconductor
LMS	least mean squares
LNA	low noise amplifier
LO	local oscillator
LS	least squares
LSI	linear shift-invariant
MESFET	metal-semiconductor field-effect transistor
MGS	modified Gram-Schmidt
MIMO	multiple-input multiple-output
ML	memoryless
MP	memory polynomial
NL	nonlinear, nonlinearity
NMSE	normalized mean squared error
NN	neural network
NPR	noise power ratio
NQS	non-quasi-static
NVNA	nonlinear vector network analyzer
OFDM	orthogonal frequency division multiplexing
OLS	orthogonal least squares
OMP	orthogonal matching pursuit
$P_{1\,dB}$	1 dB compression point
PA	power amplifier
PAE	power added efficiency

PAPR	peak-to-average power ratio
Parafac	parallel factor decomposition
PCA	principal component analysis
PM	phase modulation
PRSS	penalized residual sum of squares
RC	resistor-capacitor
RC-DOMP	reduced complexity doubly orthogonal matching pursuit
RF	radio frequency
RPV	radially pruned Volterra
RSS	residual sum of squares
RVM	relevance vector machine
RVTDNN	real-valued time-delay neural network
Rx	receiver
SBL	sparse Bayesian learning
SBP	sparse Bayesian pursuit
SCPI	standard commands for programmable instruments
SFDR	spurious free dynamic range
SRPV	simplified radially pruned Volterra
SVD	singular value decomposition
Tx	transmitter
UWB	ultra wideband
VBW	Volterra behavioral model for wideband amplifiers
VDTDNN	vector-decomposition time-delay neural network
VSA	vector signal analyzer
VSG	vector signal generator

1

Overview of Nonlinear Effects in Wireless Communication Systems

This book is about the behavioral modeling of nonlinear communication circuits and their linearization based on a theoretical Volterra series approach. Nonlinear behavior is an inherent characteristic of electronic elements and devices, associated with the function they perform in a radio frequency (RF) communication system. To be successfully captured by a remote observer, the information-carrying signal must be strongly amplified at the cost of giving rise to nonlinear distortions. In the same way, the generation of carriers or the process of incorporating information into the carrier signal is realizable, thanks to the nonlinear operation of the different modules in the transmitter. The price for these valuable features is the generation of nonlinear imperfections in the signal sent, with the appearance of two adverse collateral consequences: provoking misinformation in the recipient of the signal and interfering with other users of the system. The central objectives of this book are the study of these nonlinear problems in wireless communication systems and the research of techniques to compensate for nonlinear impairments in order to ensure efficient, error-free transmission without interference to other users.

1.1 Wireless Communication Systems

1.1.1 Transmitters and Receivers

The scenario given by a typical wireless communication transmitter–receiver link is shown in Figure 1.1. The binary data with the information to be transmitted is converted to a baseband analog signal used to modulate an RF carrier and then amplified and radiated using the transmitter antenna. On the receiver side, the signal is captured by the receiver antenna and demodulated before its conversion to the discrete-time received sequence.

A Volterra Approach to Digital Predistortion: Sparse Identification and Estimation, First Edition.
Carlos Crespo-Cadenas, María José Madero-Ayora, and Juan A. Becerra.

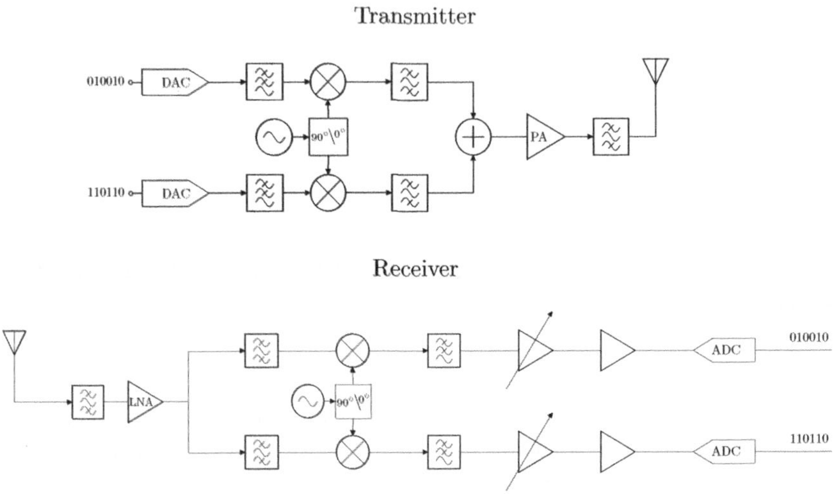

Figure 1.1 Block diagram of a typical wireless communication transmitter–receiver link.

The theoretical assumption of linear operation in wireless networks is only an approximation because the transmitters are built with several blocks, namely, modulators, mixers, power amplifiers, etc., whose electronic circuits are essentially nonlinear. In practice, undesired nonlinear effects produced in transmitters, mainly in the power amplifier, degrade the system performance and cause difficulties in meeting the stringent requirements set by the standardization entities, such as spectral masks and dispersion in the constellation.

Modern wireless communication systems are designed to operate with digitally modulated signals that have large bandwidths and high peak-to-average power ratios. Since the nonlinear behavior of the system is heavily dependent on the input signals employed, advanced knowledge of the digital world is more than advisable for the modern RF engineer.

Over the past few decades, wireless communication systems have been increasing data rates and capacity as a consequence of more sophisticated and efficient cellular networks. New generation systems employ highly spectrally efficient modulation schemes and solutions, such as orthogonal frequency division multiplexing (OFDM) and multiple-input multiple-output (MIMO). The fifth generation (5G) systems make use of both solutions, together with other technical developments that they have introduced in radio access networks. Furthermore, the trend in 5G systems to increase the use of frequency bands over 6 GHz, also forecasted for beyond 5G and 6G systems to satisfy the ever-increasing demand for connectivity, will also lead to signals more sensitive to nonlinear effects.

Noticeable nonlinear effects are generated in the circuits by the envelope variations of these signals. The nonlinear operation cannot be easily described in an analytical way; therefore, optimized designs are complex. In addition to the changes in amplitude and phase shifts typically observed in linear systems, spurious components are generated in nonlinear circuits, distorting the amplifier or mixer behavior. Among the effects of nonlinear distortions, intermodulation distortion and spectral regrowth should be taken into account since they cannot be eliminated by filtering and produce detrimental adjacent channel interference (Maas, 2003).

It is also worth noticing that, today, circuits operating in the RF range coexist with low-frequency baseband signal processing. This baseband signal processing has undergone a substantial evolution over the last few years that has led to complex modern systems. The study and prediction of the behavior of RF systems can benefit from the application of signal-processing techniques.

The present evolution toward more sustainable communications involves the search for energy-efficient transceivers, with the power amplifier being the most critical subsystem of the transmitter in terms of power consumption. In this context, the performance enhancement of a wireless communication system when its efficiency and power consumption are optimized is clear considering that these ubiquitous networks are constituted by a multitude of base stations, each one with a transmitter and a nonlinear power amplifier. However, we should recall that the power amplifier is a source of undesired nonlinear effects, more notable especially as its efficiency increases. Therefore, as RF engineers, we want to study and understand these nonlinear effects. We also want to compensate for nonlinearities and construct a linearized transmitter (predistortion). In other cases, we want to equalize the signal in the receiver (postdistortion) in the presence of a high-level noise.

1.1.2 Real-valued Continuous-time RF Signals and Complex-valued Discrete-time Baseband Signals

RF transmitters of communication systems based on modern wireless standards, like 4G, 5G, and beyond, generate real-valued continuous-time bandpass signals with a frequency response that occupies a limited bandwidth B centered around the carrier frequency f_c. The trend is to use a broad signal bandwidth, but in all cases, the condition $B \ll f_c$ is satisfied[1]. The transmitted bandpass signals are the result of modulation, i.e., the incorporation of the baseband signal information to an RF carrier.

1 Ultra wideband (UWB) signals, considered for low power systems but not for mobile communications, do not fulfil this condition. Thus, UWB is out of the scope of this text.

A common representation of a bandpass signal is[2]

$$\tilde{x}(t) = \Re\{x(t)e^{j\omega_c t}\}, \tag{1.1}$$

where $x(t)$ is the signal's complex envelope, an equivalent low-pass signal, and $\omega_c = 2\pi f_c$ is the angular frequency of the carrier. The output $\tilde{y}(t)$ of a bandpass linear system centered at ω_c, with a signal $\tilde{x}(t)$ applied at the input, is given by the convolution:

$$\tilde{y}(t) = \tilde{h}(t) * \tilde{x}(t) = \int_{-\infty}^{\infty} \tilde{h}(\tau)\tilde{x}(t-\tau)d\tau, \tag{1.2}$$

where $\tilde{h}(t)$ is the real-valued impulse response of the system. The bandpass radio communication channel is an example of a linear system described by (1.2), whereas this linear convolution is insufficient to explain the nonlinear behavior of the power amplifier, for example. The extension of the linear convolution (1.2) to the case of nonlinear systems is the Volterra series (Volterra, 1959), a major topic of discussion throughout the following chapters of this book.

Similar to the equivalent low-pass signal, the bandpass radio communication channel can be modeled with its equivalent low-pass channel impulse response $h(t)$ to obtain the output complex envelope as the complex-valued convolution $y(t) = h(t) * x(t)$, formulated analogously to equation (1.2). It should not go unnoticed that this complex-valued convolution is essentially different to the real-valued convolution (1.2), because it involves the four convolutions of the real and imaginary parts of the equivalent low-pass signal $x(t)$ and the impulse response $h(t)$.

All the interesting information is contained in the complex envelope, and as a consequence, the RF signal bandwidth can be controlled in the baseband blocks of the transmitter. This is why the low-pass equivalent representation offers not only a more convenient perspective to the analysis of a communication system, but also the possibility of counteracting in the baseband blocks for the alterations induced in the transmitted signal by the radio channel impairments. Practical techniques, e.g., OFDM, coding, equalization, and MIMO, are frequently implemented at baseband to maximize spectral efficiency in a reliable communication.

Formally, compensation of linear impairments in the radio channel can be managed with an RF linear block either in the transmitter (pre-compensation) or in the receiver (post-compensation). However, the benefits of low-frequency modules compared to high-frequency circuits suggest that a baseband filter involving a complex-valued linear convolution would be the best alternative. In this case, the pre-equalizer is the inverse module of the radio channel that introduces the necessary distortion to compensate for the transmission's linear

2 Throughout the text, the tilde ˜ is used to indicate an RF real-valued signal only when it is necessary to highlight the difference with respect to its complex envelope.

impairments. Therefore, the pre-equalizer output can be modeled with a linear convolution enunciated likewise in the baseband complex domain and in the RF real domain. The usefulness of a predistorter module to compensate for nonlinear impairments in a communication system can be substantiated similarly, giving also a justification for the baseband solution and bearing in mind an important reflection. Recalling that although convolutions for real-valued and complex-valued linear systems are strictly different because the complex-valued convolution involves four real-valued convolutions, both linear convolutions have the same form. This is not true for the case of nonlinear systems, and therefore, the reasoning behind the equivalence in the form of baseband and bandpass RF representations cannot be extended to nonlinear models. This is a crucial difference that the compensation of nonlinear impairments introduces in comparison to the linear case. Numerous publications have been devoted to the issue of baseband Volterra models, and it is also a matter of discussion in this text (Benedetto et al., 1979; Kim and Konstantinou, 2001; Morgan et al., 2006).

Digital modulation offers several advantages over analog modulation in terms of transmission capacity and robustness to channel impairments. With the same signal bandwidth, modern techniques such as OFDM allow much higher data rates as compared to other modulation schemes, and the advances in hardware and digital signal processing have led to much cheaper and more power efficient implementation of baseband modules in wireless transceivers. To profit from these aspects, modern wireless transceivers are currently built with a baseband section, where the information to be transmitted is processed digitally, converted to an analog signal, and then radiated by the RF front end. For this reason, modulation and coding as well as equalization and diversity techniques are implemented in the discrete-time digital domain, whereas frequency conversion and signal amplification in the transmitter front-end are designed within the continuous-time analog domain, creating a border between both fields marked by the digital-to-analog converter (DAC) and analog-to-digital converter (ADC).

An essential requirement for improving the performance of modern wireless systems is the cooperative development of comprehensive experimental and theoretical techniques in the two separate fields defined by RF and microwave modules involving real-valued continuous-time RF signals, on the one hand, and baseband modules where the discrete-time complex envelope is processed, on the other hand.

To begin the exposition of this book's approach to nonlinear modeling, we will briefly discuss some basic concepts about the element that contributes the most to nonlinear distortions in a wireless communications system, the power amplifier.

1.2 Modeling Power Amplifiers

The unquestionable existence of a large number of base stations in modern wireless communication systems, together with the fact that each transmitter has a power amplifier, makes this nonlinear block a determining element in terms of energy consumption and cost. Since any a priori knowledge of this block structure can be helpful in building a mathematical representation, we first discuss some power amplifier topics to support its modeling.

Design of RF power amplifiers typically starts with procedures tightly associated with the A, B, and C classification, an orthodox subject treated in traditional books and also in more recent publications. In particular, Cripps advanced in Cripps (1999) the publication of a simple theory to explain the behavior of a power amplifier, which is a type of electronic circuit with an intrinsic nonlinear complexity. It will not go unnoticed by a clever reader that some parts of our perspective on this particular topic are inspired by that idealized approach and are also intended to be developed in the same spirit.

The search for energy-efficient communications has led today to the use of other efficient transmitter architectures (Qi and He, 2019), such as the introduction of switched and continuous-mode amplifying classes; the use of approaches based on dynamic bias such as envelope tracking, where the supply voltage to the amplifier changes in accordance with the signal envelope; or the use of active load adaptation techniques, in which the impedance presented to the core amplifier changes so that the instantaneous output power can be controlled. In this third group, we can mention outphasing and Doherty power amplifiers, being the latter of primary importance in base stations due to their optimized efficiency at the back-off region. It is worth mentioning at this point that the intention of the present book is to explain the Volterra-based techniques for linearization through digital predistortion, which can be applied to all types of power amplifiers irrespective of their architecture.

A classic power amplifier is composed of a transistor with input and output matching networks and a bias circuitry that supposedly does not perturb the RF response of the device. A typical schematic of a power amplifier with a transistor as active device is shown in Figure 1.2. The great effort made to develop nonlinear modeling of transistors and characterize power amplifiers with sophisticated computer-aided design (CAD) software is now available to RF engineers, providing acceptable amplifier design methods. More than in modeling issues and flawless CAD software analysis, we are interested in the nonlinear behavior of the final amplifier design, viewing the amplifier as a block with two ports: input and output. The behavioral characterization of the power amplifier is based on a more or less complex nonlinear approach with results fitted to the measured response.

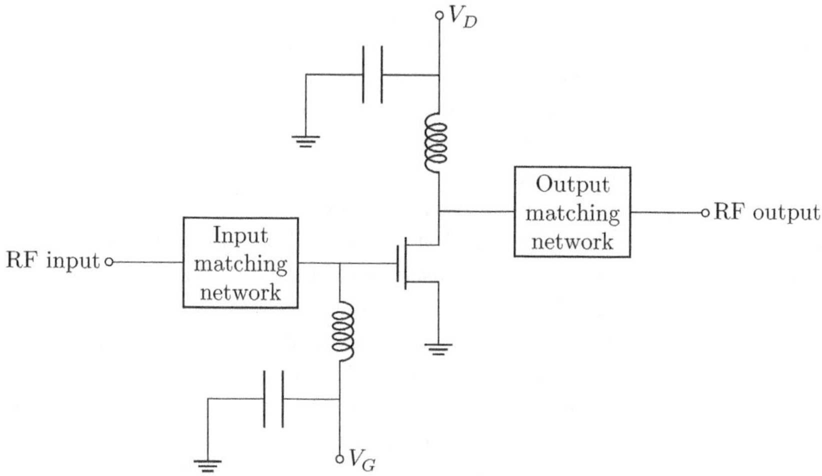

Figure 1.2 Basic circuit of an FET power amplifier.

To correctly specify a well-designed amplifier, figures of merit such as power gain, power-added efficiency, or power consumption, as well as phenomena such as harmonic and intermodulation distortions, amplitude modulation to amplitude modulation (AM–AM) and amplitude modulation to phase modulation (AM–PM) conversions, play a critical role and are most relevant to RF power amplifier designers. Assuming the reader is familiar with these significant terms, an elementary discussion of them is summarized in Section 1.5 for the case of sinusoidal or multitonal input signals.

Representation of the power amplifier with a behavioral model, without explicit knowledge of the relation of its parameters to the corresponding equivalent circuit or the transistor physics, is impeccably conceivable. However, the line of reasoning on which our research has been based is a viewpoint that pursues the connection of the nonlinear system parameters to the equivalent circuit and the physical properties of the real device. Notwithstanding that comprehensive device modeling is outside the scope of this book, this subject is briefly discussed in a separate section.

An elementary analysis of nonlinear distortions in a quasi-linear amplifier can be proposed considering the response given by a power series expansion. Referring to the amplifier represented in Figure 1.3 as a black box, when the RF signal $x(t)$ is applied to the input, the output signal can be expressed as

$$y(t) = c_1 x(t) + c_2 x^2(t) + c_3 x^3(t) + \cdots + c_n x^n(t). \tag{1.3}$$

Figure 1.3 Symbolic representation of a power amplifier.

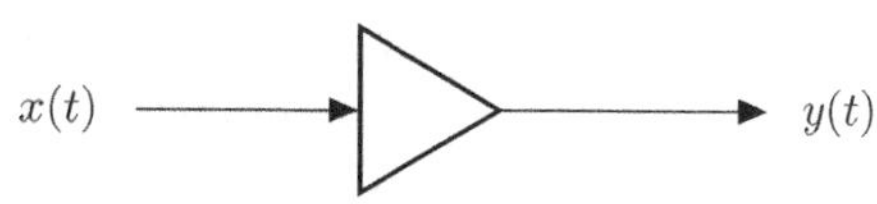

The amplifier response is a polynomial of $x(t)$ consisting of a series of nonlinear terms up to $c_n x^n(t)$ and is truncated to terms of higher orders. This straightforward result, well-grounded for the particular weakly nonlinear case and capable of demonstrating the amplitude changes produced by the input signal, e.g., the existence of AM–AM conversion in single-tone tests, is of limited accuracy in explaining nonlinear behavior in more general circumstances.

We can refer to several examples where (1.3) yields an inaccurate description of an actual amplifier behavior. The first visible weak point is its shortcoming in predicting AM–PM conversion as a consequence of the lack of phase dependence in all the terms, particularly in the linear term.

Another issue of this truncated power series model is the difficulty of accurately representing the impact of nonlinear distortions on the response of an amplifier driven by a strong input signal. Given large input signal amplitude, this happens if the number of terms of the selected polynomial is not enough for the target accuracy in a regime well into gain compression, where ideally it would be necessary $n \to \infty$. The addition of more polynomial terms with higher nonlinear degrees, introducing more complexity without improving the accuracy, is not always a practical solution. This dilemma is notably clear in the case of an amplifier driven well into compression whose output cannot be described adequately by a power series truncated to a finite number of terms. However, the treatment of the amplifier response involving an adequate analytical function, which is conceptually equivalent to its Taylor series expansion, that is, to a power series with an infinite number of terms, is mathematically possible.

Following the line of reasoning of the aforementioned Cripps' book (Cripps, 1999), which declared the idea that the performance of simple and basic models may be sufficient for a first approach to a practical power amplifier design, we first considered the simple black box power series expansion (1.3) to model the nonlinear effects in amplifiers. Unfortunately, the power series formulation beyond the compression point of the power amplifier is of limited use for the designer because nonlinear distortions become extremely strong. The ideal strongly nonlinear transfer characteristic of an field-effect transistor (FET) shown in Figure 1.4 (solid line) can illustrate the limiting behavior of the power amplifier, but this characteristic presents a singular point at $V_{gs} = 0$ where it is not differentiable and cannot be expanded in a power series. However, as depicted in the same figure, this ideal characteristic can be approximated by a smooth function (dashed line) for which the power series expansion exists and suggests, after truncation, a realistic and simple approach to power amplifier modeling between cutoff and saturation points. Nevertheless, the power series loses modeling accuracy describing the hard saturation behavior. Since this is a non-analytical function and cannot be expanded with a Taylor series for $V_{gs} = 0$, this is a good example to compare the two different approaches:

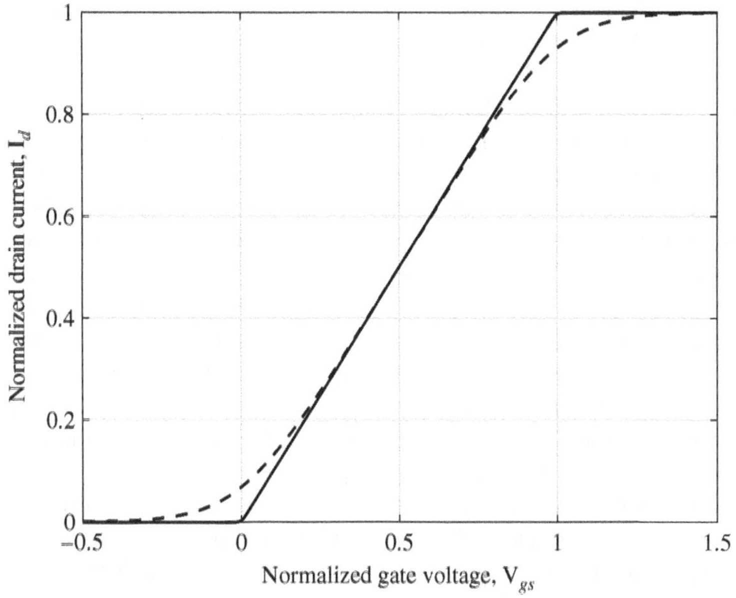

Figure 1.4 FET transfer characteristics: ideal "strongly" (solid line) and realistic (dashed line) nonlinear model.

the strongly nonlinear characteristic and the truncated smooth characteristic expansion.

The polynomial representation of the smooth transfer characteristic represents a useful contribution to analyze nonlinear distortions generated by the power amplifier and its efficiency. The truncated power series performance is easily understandable not only in the case of a class A amplifier, for which the bias point is in the middle of the quasi-linear range, about the normalized $V_{gs} = 0.5$ value in the figure. Although approximate, the analysis of other conventional modes with higher efficiency, for example, the important class B ($V_{gs} = 0$) and class AB amplifiers, is possible with this basic model as well. An adverse effect of high-efficiency power amplifiers is frequently the strong nonlinear effects they introduce in the transmitted signal.

RF engineers have available CAD software for performance analysis of possible power amplifiers' schemes, mostly based on the harmonic balance nonlinear technique, but power amplifiers' design is not in the scope of this book nevertheless. Notwithstanding, we will proceed with a basic study of the models proposed in the CAD software to evaluate the nonlinear elements of the equivalent circuit, with the purpose of exploiting any circuit-level information in the development of more advanced power amplifiers' behavioral models. An evident and necessary model improvement must include memory effects related to the reactive elements of the circuit, and the Volterra series is the natural extension to the memoryless power series.

The typical techniques to develop power amplifiers can be distinct and usually with opposed specifications since the designs are mainly conceived for a high-linearity and low-distortion functioning of the circuit, leading to a drop in efficiency. One solution is the use of methods to design highly linear amplifier architectures although with low efficiency. Another possibility is more relevant to this text and is to use a functional block to generate a distorted signal in such a way as to compensate for the nonlinear distortions introduced by a highly efficient power amplifier. This solution can be easily exemplified by considering a memoryless power amplifier with a compression region in the AM–AM conversion characteristic and a predistorter block connected at its input with a matched expansion AM–AM characteristic that compensates for nonlinearity and contributes to generate an overall linear response.

In each base station of a wireless communication system, there is a transmitter with a high power consumption block, the distorting and expensive power amplifier, causing spectral regrowth. Predistortion techniques are especially necessary to minimize the co-channel and adjacent channel nonlinear interference of the transmitted signal and, at the same time, maintain the amplifier efficiency over a widely varying envelope amplitude. Considering that implementing amplifier linearization in the baseband section of each transmitter is potentially beneficial, it is not surprising that the digital predistorter (DPD) is playing a prominent role in wireless communication engineering.

DPD design techniques are already well underway because all knowledge about power amplifiers modeling is directly applicable to digital predistorters. Moreover, a Volterra-based model is linear with respect to the coefficients and can be viewed as a regression model where the output is expressed with a linear combination of basis functions. This mathematical basis set, denoted here as regressors, provides these models with an important advantage and allows the solution to be computed straightforwardly with the least squares algorithm.

To end this section, we can remark that given the importance of microwave mixers and other RF blocks as nonlinear distortion sources, some introductory points for the analysis of these circuits are relevant and will be addressed next.

1.3 Modeling Mixers and Modulators

Mixers are fundamental circuits in radio communications. Their function is essential in superheterodyne receivers to down convert the RF signal into an intermediate frequency (IF), and they are also employed to up convert the IF

Figure 1.5 Symbol of a down-converter mixer.

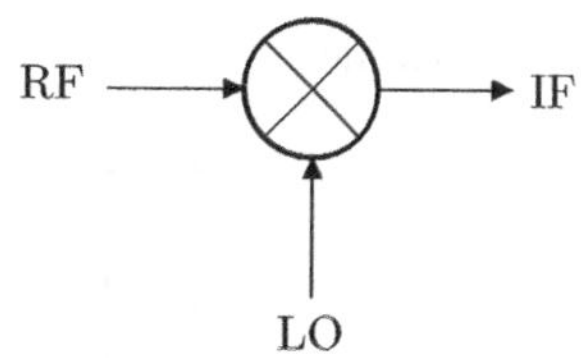

signal to RF in transmitters. This way, the signal can be digitally processed at lower frequencies, that is, modulated in the transmitter and demodulated in the receiver. From another point of view, mixers are circuits excited by two input signals: one is applied to the local oscillator (LO) port and the other to the RF or IF port. Customarily, the mixer is operated with a strong LO signal and a moderate RF (or IF) signal. This is particularly true for receivers, where the RF signal is normally weak, but also, the IF signal in transmitters is moderate to preserve the linearity of the frequency conversion operation. The mixer functioning implies an inherent multiplication of signals, and due to it, the symbol representing a mixer is the one provided in Figure 1.5, in this example a down-converter mixer.

The frequency conversion produced by mixers has its origin in the application of nonlinear devices to time-varying circuits. There is a vast variety of mixer types: based on diodes or active devices, composed of a single nonlinear device or multiple nonlinear devices, with unbalanced, singly balanced, or doubly balanced configurations (Maas, 2003).

In principle, any nonlinear device can be used as a mixer. Let us consider a nonlinear device for which the output signal can be expressed as

$$y(t) = c_0 + c_1 x(t) + c_2 x^2(t). \tag{1.4}$$

If the input signal is composed of two tones with different amplitudes and frequencies,

$$x(t) = A_1 \cos(\omega_1 t) + A_0 \cos(\omega_0 t), \tag{1.5}$$

then the output includes components at direct current (DC) and two tones at ω_1 and ω_0 produced by the linear term. In addition, the quadratic term generates the products:

$$c_2 x^2(t) = c_2 A_1^2 \cos^2(\omega_1 t) + 2c_2 A_1 A_0 \cos(\omega_1 t)\cos(\omega_0 t) + c_2 A_0^2 \cos^2(\omega_0 t). \tag{1.6}$$

The first and last terms generate signals at DC and at the second harmonic of each tone, and the second term produces an output:

$$\begin{aligned} &2c_2 A_1 A_0 \cos(\omega_1 t)\cos(\omega_0 t) \\ &\quad = c_2 A_1 A_0 \cos\left[\left(\omega_1 - \omega_0\right) t\right] + c_2 A_1 A_0 \cos\left[\left(\omega_1 + \omega_0\right) t\right], \end{aligned} \tag{1.7}$$

whose spectrum is shown in Figure 1.6. Then, if a bandpass filter is inserted at the output of the considered nonlinear device, centered for instance at $\omega_1 + \omega_0$ and

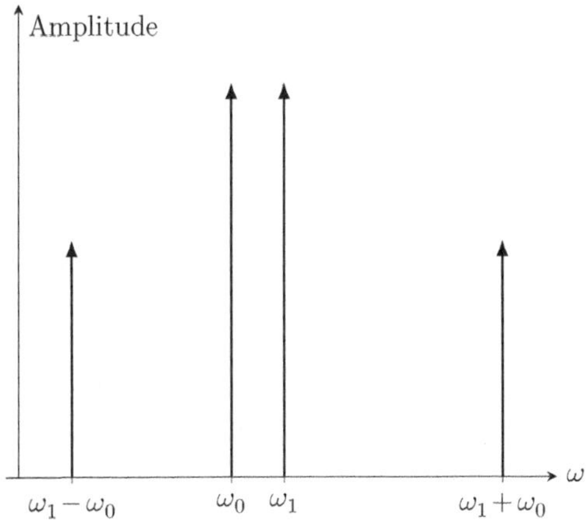

Figure 1.6 Output frequency components of a second-order nonlinear system according to (1.4).

sufficiently narrow, its output is a single tone at this frequency. Assuming constant A_0 and ω_0, the mixer has produced a linear up-conversion from the low-frequency ω_1 to the high-frequency $\omega_1 + \omega_0$.

A habitual case is a wireless transmitter where the carrier signal is strong, with A_0 large enough, and the other signal is modulated with the source message. In such case, there is an up conversion in which the new spectral component at the higher RF assimilates the same modulation as the original signal, and can be emitted by the antenna of the RF front-end with the relevant information. This RF communication signal experiences an inverted process in a superheterodyne receiver, where it is applied to the mixer together with the strong LO signal and down converted to the IF signal before the transmitted information is recovered with a demodulator. The modulation is kept in both cases, up and down-conversions; therefore, it is said that there is a linear conversion despite the mixer being an inherently nonlinear device.

It can be deduced that, in order to get a frequency conversion, it suffices that the nonlinear device presents only a quadratic term. This is not true neither for diodes nor for bipolar junction transistor (BJT) transistors whose nonlinearity is an exponential function where higher-order terms are not negligible in general. In the case of FET transistors, the output is approximately quadratic, thus presenting better characteristics as mixers.

Mixers may present conversion gain or loss that, for a downconverter mixer, is the ratio between the output power at the frequency ω_{IF} and the input power at the frequency ω_{RF}. Isolation is another important feature, measuring the attenuation undergone by the signal at one of the mixer inputs, the RF signal, or the LO, when it is measured in the other input or in the IF output. Furthermore, it must be considered that due to their inherently nonlinear principle of operation, mixers have responses at a large number of frequencies, not just the one at which they are designed to work. Namely, they present spurious responses at any frequency $\omega_{m,n} = m\omega_{\text{RF}} + n\omega_{\text{LO}}$, with m and n integers that can be both positive or negative (Maas, 2003).

Modulation consists of varying some of the properties of the sinusoidal carrier waveform—amplitude, phase, and/or frequency—according to the modulating signal that contains the information. The modulating signal is also referred to as the baseband signal. By transmitting the modulated signal, the information is transported to the other end of the communication system, where the baseband signal is extracted through the demodulation process.

As mentioned in Section 1.1.2, a modulated RF signal $\tilde{x}(t) = \Re\{x(t)e^{j\omega_c t}\}$ is a bandpass signal that can be expressed in terms of its complex envelope $x(t) = x_I(t) + jx_Q(t)$, where $x_I(t)$ is the in-phase component and $x_Q(t)$ is the quadrature component. All the carried information is contained in the complex envelope; therefore, this representation indicates how to implement the digital modulation process by means of the quadrature modulator or in-phase/quadrature (I/Q) modulator shown in Figure 1.7.

Ideally, the I/Q modulator produces an RF output signal whose complex envelope $u(t)$ is a linear transformation of the baseband signal $x(t)$. However, hardware implementations of the I/Q modulator introduce impairments in practice. Considering the schematic representation of the main blocks in Figure 1.7, differences in the gains of the in-phase path and the quadrature path produce *gain imbalance*. Carrier leakage from the LO applied to the mixers leads to the appearance of a *DC offset* in $u(t)$. The imperfection of the 90° shift needed between the in-phase and quadrature paths or a phase error in the LO generates quadrature error

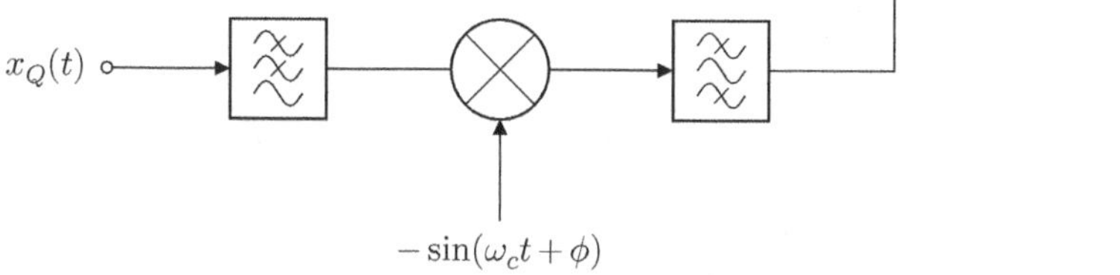

Figure 1.7 Schematic representation of an ideal I/Q modulator including the main blocks of its hardware implementation that lead to imperfections.

(Cavers and Liao, 1993; Cavers, 1997; Wisell, 2000; Ding et al., 2008). When I/Q impairments are present, the complex envelope $u(t)$ at the output of the modulator depends both on the complex-valued input signal $x(t)$ and the image signal $x^*(t)$.

Additionally, the presence of mixers and other nonlinear devices such as amplifiers in the implementation of I/Q modulators can also introduce nonlinear effects in their behavior. In the schematic representation of Figure 1.7, both RF bandpass inputs and baseband inputs are present, the latter included in the in-phase or quadrature paths and the former applied to the mixers. Therefore, the nonlinear behavior of I/Q modulators presents important differences with that of RF power amplifiers. It will be discussed in detail in Chapter 4 of this book.

1.4 Circuit Models of Nonlinear Devices

1.4.1 Nonlinear Circuit Elements Representation

In order to accomplish the nonlinear analysis of an electronic circuit, whose devices include inherent nonlinearities, it is necessary to replace each device with an equivalent model. The more usual nonlinear elements included in the models of the devices are nonlinear conductances, nonlinear capacitances, and nonlinear controlled sources. For all these cases, the current depends nonlinearly on one or more voltages, and the elements can be appropriately characterized by a power series around the bias point.

Nonlinear conductance: In Figure 1.8(a), a nonlinear conductance is shown whose voltage–current relationship can be expressed as

$$i_G = g(v_G), \tag{1.8}$$

where the explicit dependence on t has been omitted for clarity. If V_G and I_G denote the voltage and current in the operation point, respectively, the power series expansion of the function $g(\cdot)$ about the bias point can be written as

$$i_G = I_G + \sum_{n=1}^{\infty} g_n (v_G - V_G)^n, \tag{1.9}$$

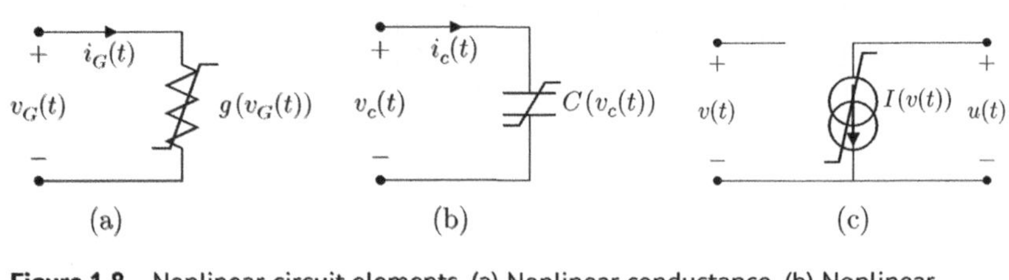

Figure 1.8 Nonlinear circuit elements. (a) Nonlinear conductance. (b) Nonlinear capacitance. (c) Nonlinear current source controlled by voltage or transfer nonlinearity.

where g_n is the nth coefficient of the power series. Making use of the definition of incremental currents and voltages $i_g = i_G - I_G$ and $v_g = v_G - V_G$, (1.9) can be reduced to

$$i_g = \sum_{n=1}^{\infty} g_n v_g^n. \tag{1.10}$$

Nonlinear capacitance: In the case of the nonlinear capacitance shown in Figure 1.8(b), we can write the current as a function of the charge by applying the chain rule:

$$i_C = \frac{dq_C}{dt} = \frac{dq_C}{dv_C} \cdot \frac{dv_C}{dt} = C(v_C)\frac{dv_C}{dt}, \tag{1.11}$$

where $C(v_C)$ is the nonlinear capacitance. Let us assume that the nonlinear capacitance has a power series expansion around the bias point; then it can be written as

$$C(v_C) = c_0 + \sum_{n=1}^{\infty} c_n (v_C - V_C)^n = \sum_{n=0}^{\infty} c_n v_c^n, \tag{1.12}$$

where $v_c = v_C - V_C$ is the incremental voltage. The current is expressed in terms of the coefficients of the nonlinear capacitance as

$$i_C = \sum_{n=0}^{\infty} c_n v_c^n \frac{dv_C}{dt}. \tag{1.13}$$

In order to write this expression as a function of the incremental voltages and currents, it is enough to rewrite (1.13) in the following equivalent form:

$$i_c - I_C = \sum_{n=0}^{\infty} c_n v_c^n \frac{d(v_c + V_C)}{dt}. \tag{1.14}$$

Taking into account that the DC current in a capacitor is $I_C = 0$ and that V_C is a constant, it can be concluded that

$$i_c = \sum_{n=0}^{\infty} c_n v_c^n \frac{dv_c}{dt}. \tag{1.15}$$

Transfer nonlinearity: Finally, it is necessary to consider the case of a transfer nonlinearity, as the one shown in Figure 1.8(c). In the case of a nonlinear current source controlled by a single voltage at a different point in the circuit, the power series expression for the current is

$$i(v) = \sum_{n=1}^{\infty} g_n v^n, \tag{1.16}$$

and in the case of a nonlinearity depending on two voltages, v and u, the power series of the current is expressed as

$$i(v,u) = \sum_{\substack{n=0 \\ n+m\geq 1}}^{\infty} \sum_{m=0}^{\infty} g_{nm} v^n u^m. \tag{1.17}$$

Regarding the obtention of the power series coefficients g_n or g_{nm}, the traditional approach is based on the Taylor series expansion around the circuit bias point.

The nonlinear device models summarized earlier are indispensable components of a nonlinear circuit and its analysis. In particular, for the nonlinear circuits based on FET devices, the main nonlinearity appears in the current that flows through the drain $i_{ds}(v_{gs}, v_{ds})$, and it has been modeled as a nonlinear controlled source expressed as shown in (1.17), due to the importance of the cross-terms g_{nm} demonstrated in several research papers (Pedro and Perez, 1994). Note that in this case, the g_{n0} coefficients are related to the nonlinear controlled source that depends on v_{gs}, while the g_{0m} are the coefficients of the nonlinear output conductance. Therefore, sometimes the following expression is preferred, equivalent to (1.17), that considers those effects separately in three different terms:

$$i(v,u) = \sum_{n=1}^{\infty} g_{n0} v^n + \sum_{m=1}^{\infty} g_{0m} u^m + \sum_{n=1}^{\infty} \sum_{m=1}^{\infty} g_{nm} v^n u^m. \tag{1.18}$$

1.4.2 Large-signal Models for FET Devices

In order to model the active FET devices, the compact equivalent three-node circuit shown in Figure 1.9 can be used. Among the elements of the equivalent

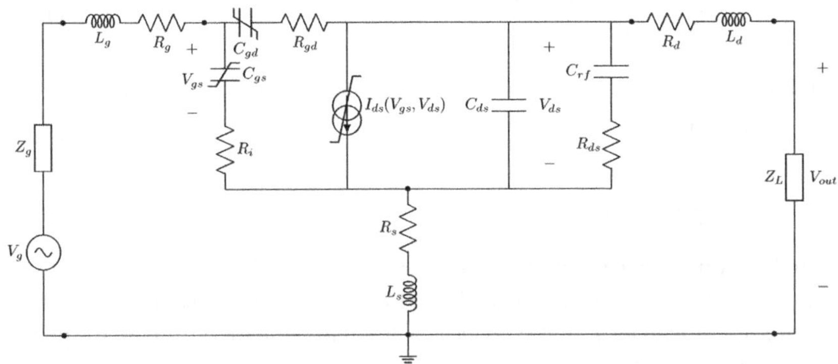

Figure 1.9 Compact equivalent three-node circuit of an FET, including the extrinsic and intrinsic large-signal circuits for gate, drain, and source nodes. The elements shown in the circuit are detailed in Section 1.4.2.

circuit, R_g, R_s, R_d, L_g, L_s, and L_d form the extrinsic circuit. Regarding the intrinsic circuit, R_i is the intrinsic resistance or the resistance of the semiconductor region under the gate, between the source and channel. C_{ds} is the drain-to-source capacitance, which is dominated by metallization capacitance, and is therefore often treated as a constant. C_{gs} and C_{gd} are the gate-to-channel capacitances. Note that the branches containing R_i–C_{gs} and the parasitic elements R_{ds}–C_{rf} introduce a filter-like frequency dependence since they account for non-quasi-static (NQS) or memory effects. V_{gs} and V_{ds} are the gate-to-source and drain-to-source voltages, respectively, and I_{ds} is the nonlinear channel current source.

Well-known examples of compact models are Curtice model (Curtice, 1980) and Statz or Raytheon model (Statz et al., 1987) for GaAs integrated circuit FETs is gallium arsenide integrated circuit field-effect transistors, Curtice cubic model (Curtice and Ettenberg, 1985) for FETs used in power amplifiers, Angelov model (Angelov et al., 1992) suitable for both metal-semiconductor field-effect transistor (MESFET) and high-electron-mobility transistor (HEMT) devices, and more general-purpose models such as Materka model (Materka and Kacprzak, 1985). Many of these compact models are extensions of the small-signal equivalent circuit model, based on the intuitive association of the circuit elements with the physical structure of the transistor, and describing the large-signal behavior by curve fitting the DC $I_d - V_{ds}$ characteristics and the capacitance–voltage relationships of the transistor.

Since it is generally accepted that the dominant source of nonlinearity in FET transistors is the drain-to-source current $I_{ds}(V_{gs}, V_{ds})$ (Brazil, 2003; Brinkhoff and Parker, 2003; Carvalho and Pedro, 2002), we will pay special attention to modeling the drain current characteristics. Furthermore, some remarks will be presented about modeling the gate capacitances, being the second source of nonlinearity.

1.4.2.1 Angelov Model for the Drain Current Characteristics

The transconductance is one of the most critical aspects for large-signal predictions. The transconductance of a HEMT device exhibits a peak value that is not present in MESFET devices. *Angelov model* (Angelov et al., 1992) takes into account this phenomenon by means of a separable expression for the drain current $I_{ds}(V_{gs}, V_{ds}) = I_{d,A}(V_{gs}) \cdot I_{d,B}(V_{ds})$. The term $I_{d,B}(V_{ds})$ is the same as the one used in Curtice model for FETs, that is

$$I_{d,B}(V_{ds}) = (1 + \lambda V_{ds}) \tanh(\alpha V_{ds}). \tag{1.19}$$

Each term emphasizes a different feature of the $I - V$ characteristic. The term $(1 + \lambda V_{ds})$ is employed to describe the finite output conductance, whose slope depends on the λ parameter. Both saturation and knee regions in the $I - V$ characteristic are modeled by means of a tanh function. The α parameter controls the sharpness of the knee region; the higher α is, the faster the saturation effects will be noticed.

On the other hand, a gate control function $I_{d,A}(V_{gs})$ is proposed wherein the first derivative has the same generic shape as the transconductance curve, yielding the following expression for the proposed nonlinear drain current:

$$I_{ds} = I_{pk}(1 + \tanh \psi) \cdot (1 + \lambda V_{ds}) \tanh(\alpha V_{ds}), \tag{1.20}$$

with

$$\psi = P_1(V_{gs} - V_{pk}) + P_2(V_{gs} - V_{pk})^2 + P_3(V_{gs} - V_{pk})^3 + \cdots \tag{1.21}$$

where I_{pk} and V_{pk} are the drain current and gate voltage corresponding to the peak transconductance, respectively. In these expressions, the various P_i coefficients are empirical polynomial fitting parameters describing the dependence on an effective gate potential, ψ. The model is sufficiently accurate even if ψ is approximated by a linear function. Note that this drain current function has well-defined derivatives with respect to the gate voltage, which enables the identification of the higher-order derivative terms with the order of the distortion components.

1.4.2.2 Models for the Gate Capacitances

Although many of the remaining components in the large-signal equivalent circuit exhibit little or no voltage dependence, the input capacitances C_{gs} and C_{dg} can vary significantly with bias in a nonlinear fashion. Early large-signal MESFET models, such as Curtice model, incorporate a voltage dependence for the model capacitors based on that of the ideal metal–semiconductor junction. Newly proposed models are functions of both V_{gs} and V_{ds}, as well as they are fully charge-conservative (Aaen et al., 2007). For instance, in Angelov model, the same tanh functions employed in the drain current expression are used to model gate-to-source and gate-to-drain capacitances, and a certain cross-coupling of both V_{gs} and V_{ds} is taken into account in C_{gs}.

1.4.2.3 Simplified Nonlinear Models for FET Amplifiers

The equivalent three-node circuit for FET devices shown in Figure 1.9 can be simplified as in the study by Minasian (1980). Parasitic elements can be neglected, as well as the feedback capacitance C_{gd}, thus making the circuit unilateral. Recalling that the dominant source of nonlinearity in FET transistors is the drain-to-source current, we neglect the nonlinear behavior of C_{gs} and C_{ds}. Furthermore, the linear effect of C_{ds} can be included in the output impedance Z_0. The resulting elementary model for an amplifier with a unilateral FET is shown in Figure 1.10. The drain current, denoted here as I_d, is actually dependent on $V_{gs} = V_{GS} + v_g$ and $V_{ds} = V_{DS} + v_d$, where V_{GS} and V_{DS} are DC values and v_g and v_d are the corresponding gate and drain AC waveforms. However, assuming fixed DC bias voltages, the explicit dependence can be concentrated to v_g and v_d. Furthermore, we have included the dependence on an additional internal variable z to facilitate

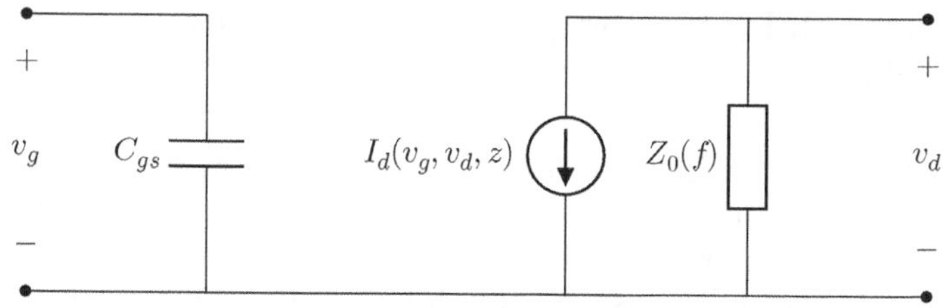

Figure 1.10 Simplified unilateral nonlinear model for an FET amplifier.

models with novel structures. This internal variable can be, for example, the junction temperature rise when considering an electrothermal subcircuit, as the distributed resistor-capacitor (RC) network with n stages provided in Figure 1.11 (Dai et al., 2003). Another case is the internal mechanism of charge trapping, which is the cause of drain-lag effects in FET devices (Jardel et al., 2007). The simplified diagram of a one-trap model shown in Figure 1.12 is mounted in series between the gate resistance and the current source, modifying the control voltage of the current source by adding transients that are related to the capture or emission of charges by the traps. Resistance R_{fill} together with the capacitance C provide the capture time constant, while the emission time constant is given by R_{empty} and C. Therefore, the charge-trapping subcircuit generates the new internal variable z that additionally controls the current source of the transistor.

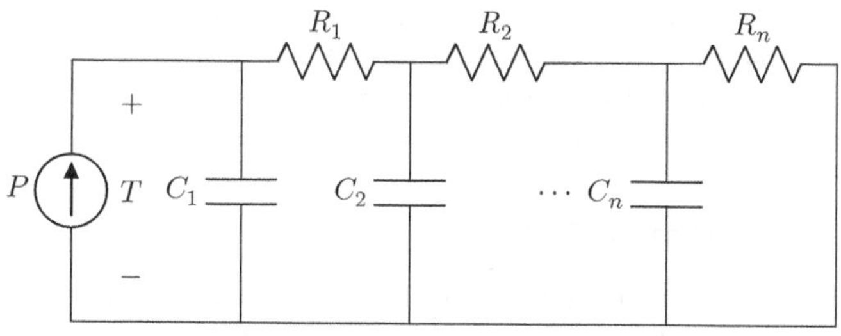

Figure 1.11 Distributed RC subcircuit with n stages to model thermal effects.

Figure 1.12 Simplified layout of a charge-trapping subcircuit.

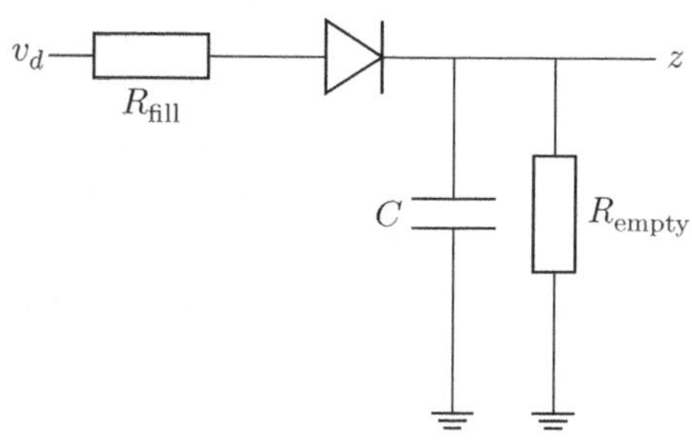

1.5 Experimental Evaluation of Nonlinear Circuits: Classical Methods

The observable properties of electronic devices are represented by some figures of merit obtained by experimental characterization. For a power amplifier, for instance, widely used figures of merit are its gain, efficiency, and some quantitative measurements of its nonlinear distortions such as the 1-dB compression point and third-order intercept point.

Considering that nonlinear devices do not comply with the superposition principle, their response to a certain input may vary depending on the kind of input employed (Pedro and Carvalho, 2003). This fact makes advisable to use communication signals as test inputs because they are similar to the excitation expected in real operation. In an attempt to approximate communication signals, which present a band-limited power spectral density containing a large number of spectral lines, classical methods to evaluate nonlinearity have employed sinusoidal signals or tones.

In this way, single-tone tests can be viewed as the simplest approximation for the recommended probing signals, in which all the power is concentrated in a single spectral line. Although the one-tone test is suited for evaluating distortion in the fundamental frequency, it is a limited characterization tool for nonlinear systems since it can only produce output spectral components that are harmonically related to the input frequency. The advantage of two-tone tests, where the input signal is composed of two tones of equal amplitudes and certain frequency separation related to the channel of interest, is that a large number of mixing products generate in-band distortion. More sophisticated multitone signals are adopted when targeting to measure co-channel distortion.

1.5.1 One-tone Characterization Tests

The frequency-domain transfer function $H(\omega)$, which is a common way to identify linear devices, can be obtained by applying a signal $x(t) = A\cos(\omega_c t)$ at the input of the device under test (DUT) and measuring its output $y_{\mathrm{LIN}}(t)$ at the same input frequency ω_c, referred to as the fundamental frequency. By sweeping the frequency of the input signal ω_c, we can track changes in the amplitude and phase of the output, $y_{LIN}(t) = A_o(\omega_c)\cos[\omega_c t + \theta(\omega_c)]$, and construct the complex-valued transfer function $H(\omega) = A_o(\omega)\angle\theta(\omega)$. However, when the sinusoidal tone is applied to a nonlinear device, in particular an amplifier, new frequency components located at the harmonic frequencies $m\omega_c$, $m = 2, 3, \ldots$, will be generated having amplitudes and phases that vary nonlinearly with the stimulus level. The response can be written as:

$$y_{\mathrm{NL}}(t) = \sum_{m=0}^{\infty} A_o(m\omega_c, A)\cos[m\omega_c t + \theta(m\omega_c, A)]. \tag{1.22}$$

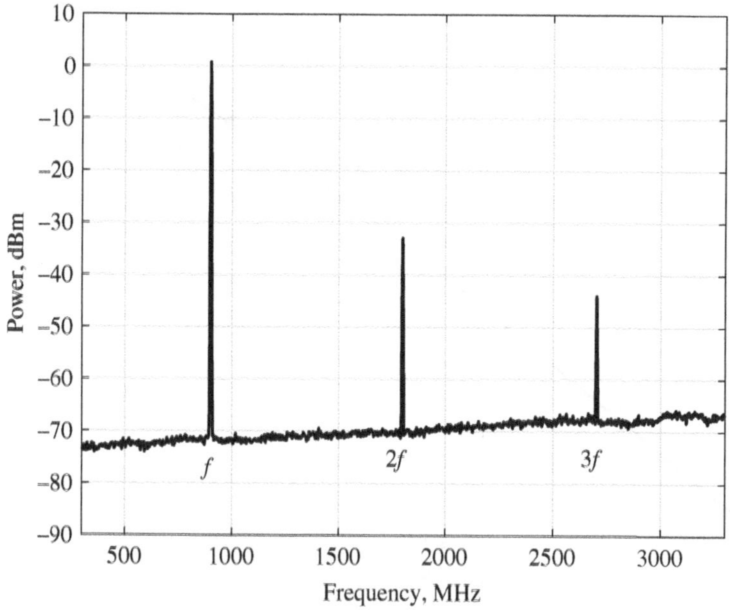

Figure 1.13 Spectral components at the output of a power amplifier, a nonlinear device, for a one-tone test: fundamental frequency and harmonics.

Figure 1.13 illustrates those output spectral components at the fundamental and harmonic frequencies.

The main figures of merit associated with one-tone tests are the following:

- **AM–AM conversion**: Considering closely the practical case of the response at the fundamental frequency ω_c, the output amplitude and phase will depend on the input amplitude A. However, whereas in a linear system, an amplitude sweep of the input tone produces an output amplitude proportional to A, in nonlinear systems, the changes of the output amplitude deviate from the linear behavior. This property is referred to as *AM–AM conversion* and describes the relation between the output amplitude at the fundamental frequency and the input amplitude at a fixed input frequency. In particular, it characterizes gain compression of a nonlinear device versus the input level.
- **1-dB compression point**: In the AM–AM conversion characteristic of Figure 1.14, the level at which the output power is compressed by 1 dB has been considered as a standard reference point to set the limit for a "linear" behavior of the amplifier. The *1-dB compression point*, P_{1dB}, is defined as the power level at which the signal output is 1 dB below the output that would be obtained by extrapolating the linear small-signal characteristic of the system. The 1-dB compression point may be referred to the input or the output power levels. A gain plot provides an immediate way for evaluating the 1-dB compression point

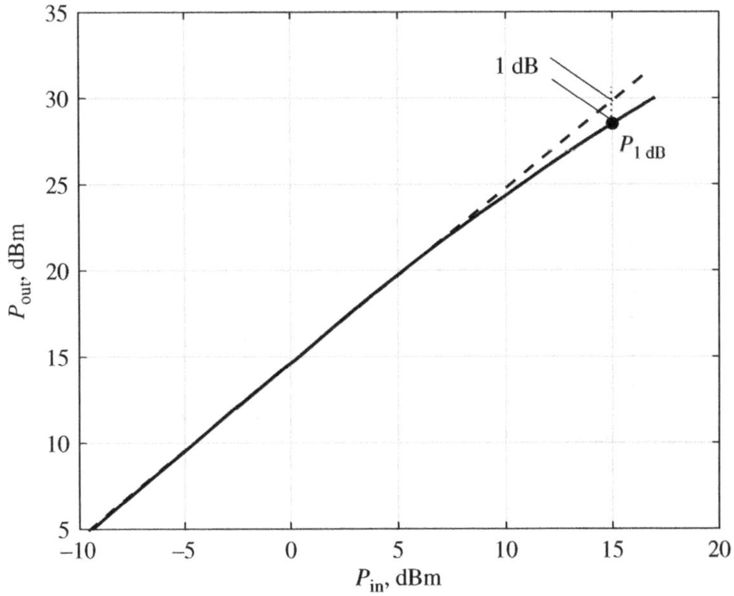

Figure 1.14 AM–AM characteristic of a power amplifier and 1-dB compression point, $P_{1\ dB}$.

since it is simply the power at which the gain has already tailed off 1 dB from its small-signal value.

- **AM–PM conversion**: Analogously to AM–AM conversion, the output phase at the fundamental frequency depends on the input amplitude A in such a way that an amplitude sweep of the input tone produces a certain behavior of the output phase with respect to A. In nonlinear systems, vector addition of the output fundamental with distortion components determines a phase variation of the resultant output when the input level varies. It is important to mention that AM–AM nonlinear behavior can be observed independently of the possible memory effects exhibited by the system. However, only dynamic nonlinear systems or nonlinearities with memory exhibit AM–PM conversion, including those usually called quasi-memoryless systems (Ding and Sano, 2004) (See Chapter 2 for details).

1.5.1.1 Implementation of One-tone Tests

One-tone tests can be accomplished with a vector network analyzer. Both AM–AM and AM–PM conversions can be characterized in a single power sweep since the vector network analyzer can simultaneously measure the magnitude and phase of the DUT's gain. If only AM–AM conversion is needed, a spectrum analyzer can also be employed. The setup based on a spectrum analyzer shows the advantage that it is also possible to measure the power of the harmonics.

There are other characterizations that construct the AM–AM conversion by measuring the input and output power of the device with a power meter and use a calibrated phase shifter and a spectrum analyzer for AM–PM conversion. In the latter, a shifted sample of the input is added to the output of the DUT trying to cancel the output fundamental signal (Pedro and Carvalho, 2003).

Another option with more sophisticated laboratory equipment is based on Nonlinear Vector Network Analyzers (NVNAs). These instruments are able to measure and display both the amplitude and phase of the full output spectra—fundamental, harmonics, and cross-frequency products—in time, frequency, power, or user-defined custom domains. They extend the concept of linear scattering parameters to the nonlinear field by means of X-parameters or parameters of the polyharmonic distortion model. However, they are not yet available in most laboratories.

1.5.2 Two-tone Characterization Tests

The use of two-tone stimulus enables the identification of new mixing components close to the fundamentals which constitute distortion components. If a two-tone excitation $x(t) = A_1 \cos(\omega_1 t) + A_2 \cos(\omega_2 t)$ is applied to the amplifier, the output is given by a very large number of mixing terms. Referring to a usual narrowband RF subsystem, two types of information can be extracted from a two-tone test:

1.5.2.1 In-band Intermodulation Distortion

In-band distortion products are the mixing components falling in the output fundamental frequencies zone. For example, in-band measurements would include the fundamental frequencies, ω_1 and ω_2 and also the third-order components at $2\omega_1 - \omega_2$ and $2\omega_2 - \omega_1$. These distortion products are referred to as third-order intermodulation (IM3) products and contribute to the intermodulation distortion (IMD). They form a group of lower and upper sidebands, separated from the signals by the tones frequency difference, $\Delta\omega = \omega_2 - \omega_1$. Figure 1.15 shows the spectrum at the output of a power amplifier where third and fifth order intermodulation products appear together with the fundamental tones. In this case, equal amplitude tones have been applied to the power amplifier, $A_1 = A_2 = A$, and we can observe that the lower intermodulation products present the same level than their corresponding upper intermodulation products. However, there are cases in which the levels of lower and upper intermodulation products are different, leading to an asymmetrical spectrum. Those situations are often called IMD asymmetries and are associated to the presence of memory effects, as it will be detailed in Section 1.5.3.

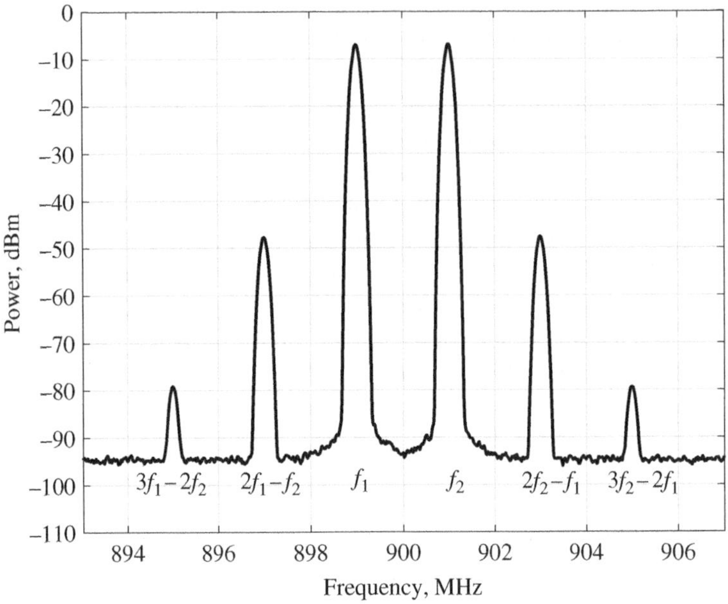

Figure 1.15 Spectral components at the output of a power amplifier for a two-tone test: tones and intermodulation products.

The fundamental output power at ω_1 and the IMD power level of the distortion response at $2\omega_1 - \omega_2$ are plotted in Figure 1.16 for the particular case of equal amplitude tones $A_1 = A_2 = A$ against the power of the input tones. For every increment of 1 dB in the input power, the output power of the fundamental tones is increased 1 dB until it starts saturating and departs from its linear behavior. On the other hand, the output power of the IM3 products shows an approximate increment of 3 dB when the input power is increased 1 dB before also entering into saturation. Another interesting observation is that there are some situations, referred to as sweet-spots, where the IMD drops for a certain input power due to particular conditions of biasing and load impedance of the power amplifier.

If we extrapolate the line with slope of 1 dB/dB that approximates the fundamental output power and the line with 3 dB/dB slope that approximates the IM3 power, the point where both intersect is called the *third-order intercept point*, IP_3. The intercept point can be referred to the input or the output of the DUT. It is commonly employed as a reference for the degree of nonlinear distortions that we can expect from a device. The higher the IP_3 is in a certain device, the weaker the nonlinear distortions it produces. Observe that IP_3 is a concept valid for an ideal device with a 3 dB/dB slope in the IM3 characteristic, which is not the case in an actual device. In the example provided in Figure 1.16, the definition is convenient

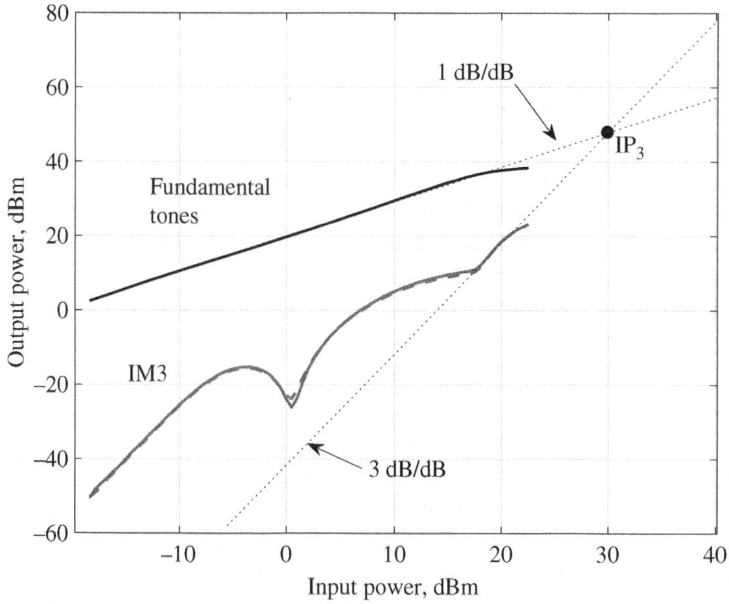

Figure 1.16 Power sweep with a two-tone test: third-order intercept point IP_3.

only for operating points around 20 dBm of input power. Accordingly, there is no rigorous mathematical relation between the 1-dB compression point and the intercept point IP_3, although design engineers use a difference of 10 dB as a common rule of thumb.

Considering that IMD generally increases with increasing signal levels, IP_3 may also be used to establish the *dynamic range* of a system. The signal level at which the IMD level meets the noise floor is employed to define the *spurious free dynamic range* (SFDR), which is the ratio of the output power level at the fundamental frequency to the noise power level that equals the IMD power level.

We can also mention that some other intercept figures of merit can be defined for the fifth-order (IP_5) or seventh-order (IP_7) distortion. However, they are infrequent.

All the previously indicated about intermodulation distortion has been explained from the point of view of a power amplifier, where the output fundamental frequencies of the tones are the same as those of the input tones. However, intermodulation distortion is present for any nonlinear device, including mixers. The only difference in the particular case of mixers is due to frequency conversion. If two tones at the fundamental frequencies $\omega_{RF,1}$ and $\omega_{RF,2}$ are applied at the input of a down-converter mixer together with a local oscillator signal at a frequency ω_{LO}, the output fundamental frequencies are converted to

$\omega_{IF,1} = \omega_{RF,1} - \omega_{LO}$ and $\omega_{IF,2} = \omega_{RF,2} - \omega_{LO}$ and, therefore, the IM3 products are placed at $2\omega_{IF,2} - \omega_{IF,1}$ and $2\omega_{IF,1} - \omega_{IF,2}$.

1.5.2.2 Out-of-band Components

Out-of-band distortion components are the harmonics of each of the tones, but also the mixing products that fall close to the various harmonics, for example, the products at $\omega_1 + \omega_2$ near the second harmonic, at $2\omega_1 + \omega_2$ and $\omega_1 + 2\omega_2$ near the third harmonic, and so on. As their name indicates, out-of-band distortion components appear at zones of the output spectrum quite far from the fundamental signals; therefore, they are simple to be filtered in narrowband systems. There are also mixing products located at DC that describe the bias point shift from the quiescent point, when the input driving level increases.

1.5.2.3 Implementation of Two-tone Tests

The usual equipment to accomplish two-tone tests is a spectrum analyzer since mixing products involving a combination of both ω_1 and ω_2 have different frequencies from either of the inputs. The two-tone stimulus is usually produced with two signal generators, each one providing an input tone that is applied to the DUT by means of a power combiner or a directional coupler. Phases of the two tones generated in this way are uncorrelated (Pedro and Carvalho, 2003).

A two-tone signal can be also implemented with only one generator including arbitrary waveform generator by modulating the amplitude of a single carrier at frequency $\omega_c = (\omega_1 + \omega_2)/2$ using a double-sideband suppressed-carrier (DSB-SC) format where the envelope varies according to the sinusoidal signal

$$x_m(t) = 2A\cos(\omega_m t) \tag{1.23}$$

with the modulation frequency $\omega_m = \Delta\omega/2$, dependent on the separation between the two tones. The result is a signal given by

$$x(t) = 2A\cos(\omega_m t)\cos(\omega_c t) = A\cos(\omega_1 t) + A\cos(\omega_2 t), \tag{1.24}$$

where $\omega_1 = \omega_c - \omega_m$ and $\omega_2 = \omega_c + \omega_m$ are the frequencies of the two sinusoidal tones.

The existing relationship between the two tones separation and the envelope or baseband frequencies makes two-tone IMD measurements appropriate for nonlinear characterization taking into account memory effects, as it will be detailed in Section 1.5.3.

However, a complete two-tone IMD characterization makes advisable the measurement of not only magnitude using a spectrum analyzer but also phase. Varying the frequency separation $\Delta\omega$ between the two tones, as shown in Figure 1.17, provides interesting information about memory effects. Different sophisticated methods have been proposed for the measurement of intermodulation distortion

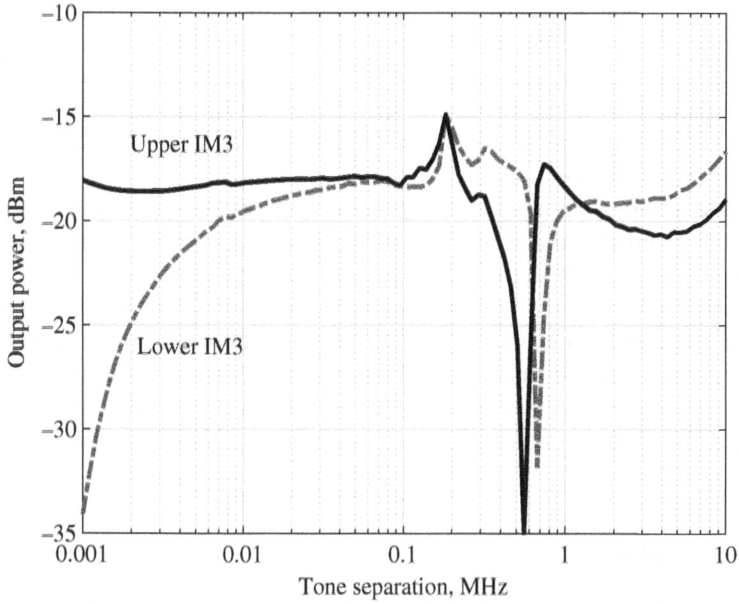

Figure 1.17 Power of the third-order intermodulation products versus the tone separation. Their variations with the tone separation evidence memory effects.

phase using synchronized generators and several vector network analyzers (Vuolevi et al., 2001). The use of a nonlinear vector network analyzer allows the accurate measurement of the magnitude and phase of all distortion products at the input and output of a device with a calibrated instrument. Other researchers have proposed signal processing techniques to obtain the IMD phase information employing standard uncorrelated two-tone and multitone excitations (Martins and Carvalho, 2005). Probably the most important methods are those using signal generators with arbitrary modulation capability in order to generate a two-tone signal, as described by equations (1.23) and (1.24), and time-domain operation for its measurement, either by means of the acquisition of samples with a digital oscilloscope or a vector signal analyzer (Draxler et al., 2003).

Based on the use of arbitrary modulation in the signal generator, in Crespo-Cadenas et al. (2005) a simplified method to measure the phase of intermodulation products relative to the tones using non-sophisticated communications equipment was presented. It considers a two-tone signal that is created by using a DSB-SC signal modulated by a sinusoidal baseband waveform with frequency ω_m, which produces two coherent tones with the same level and an exactly constant separation $\Delta\omega = 2\omega_m$. With this method, the relative phase between tones can be controlled by software definition of the modulating signal. The baseband signal at the output of the nonlinear device can be acquired with a vector signal analyzer using an appropriate sampling rate. The available samples contain all the information, i.e., magnitude and phase, of the original modulating signal and also of the generated intermodulation products. Therefore, after a Fourier transform, the relative phases corresponding to each intermodulation product with respect to the input tones can be evaluated.

It is necessary to correct the recovered phase of each intermodulation product to take into account that the acquired signal experiences a certain delay while propagating from the DUT to the measurement instrument. This phenomenon produces a different shift in phase for each of the frequency components of the output signal, i.e., the two tones and the intermodulation products, which is added to the phase that needs to be measured. Therefore, it is necessary to estimate the delay in order to correct the measured phases. In this way, it is possible to obtain the relative phases of the intermodulation products with respect to the tones' phases.

1.5.3 Memory Effects

In the context of nonlinear systems, the term *memory* was proposed by Chua et al. (1987) to describe the influence on the output of a system at a time t of the input signal not only at that time instant, but also spanning a finite history of the input signal, to some time in the past. Since the influence of the input signals deep in the past fades to zero, we use the term *fading memory* and the largest considered time interval determines the *memory length* of the system. As we will discuss in the next chapter, a strictly memoryless amplifier may only cause an AM–AM conversion, never an AM–PM conversion, so that an amplifier with some phase distortion must possess a certain amount of memory.

Essentially, by memory effects we are describing the dynamical behavior of the system. Thus, amplifiers with reactive components introducing phase shifts have memory, with characteristic times that are generally either of the same timescale as the signal frequency—*short-term memory*—or at much slower rates—*long-term memory*. Because of that, nonlinear systems that introduce a time delay equivalent to a small fraction of the RF cycle exhibit only short-term memory effects and are denoted by many authors as *quasi-memoryless systems* (Bosch and Gatti, 1989), for which the amount of amplitude and phase distortions are modeled by *static* AM–AM and AM–PM conversions or, equivalently, complex-valued but constant nonlinear transfer functions, as it will be explained in Chapter 2. By contrast, nonlinear systems with time delays in the order of the symbol duration or larger, exhibit short- and long-term memory effects, and are usually considered to present *dynamic* AM–AM and AM–PM characteristics which are modeled by frequency-dependent nonlinear transfer functions.

Memory effects can be experimentally observed as shifts in the amplitude and phase of IMD components caused by changes in the modulation frequency that controls the tone frequency separation in two-tone tests (Bosch and Gatti, 1989), or also by hysteresis in the AM–AM and AM–PM plots (Cabral et al., 2005). In the literature, these phenomena are also referred to as *bandwidth-dependent IMD behavior* (Brinkhoff, 2004), *dynamic system effects* (Pedro et al., 2003), *rate-dependent effects* (Parker and Rathmell, 2003) or *NQS effects* (Brazil, 2003; Aaen et al., 2007). The variations with the tone frequency separation in the output power of the IM3 products shown in Figure 1.17 evidence that the power amplifier under test exhibited memory effects. A difference between the upper and lower intermodulation products for a two-tone test or between the upper and lower adjacent channels for a modulated input is referred to as an *asymmetry*, and it is another indication of memory effects. IMD asymmetries can be also observed in Figure 1.17, where the output power of the upper IM3 product differs from that of the lower IM3 product for almost all the values of the tones separations.

When we talk about memory effects in RF power amplifiers, the major sources of these effects are (Vuolevi et al., 2001; Aaen et al., 2007):

- **Short-term memory effects**: The high-frequency dynamics of the amplifiers are determined by the reactances associated with the transistor. In the usual description of a transistor model, like the one discussed in Section 1.4.2, these reactances comprise the capacitances and inductances associated with the parasitic elements of the extrinsic model, and also the nonlinear charge storage within the transistor's active region, in the intrinsic model.

  For small-signal characterization, the short-term memory effects are simply the frequency response of the transistor, which is bias-dependent and requires that the capacitances describing the linearized charge storage behavior in the transistor are also bias-dependent. Under large-signal conditions, the voltage- or current-dependence of the charge storage functions becomes important. The changing dynamical behavior with signal drive manifests when measuring the AM–AM and AM–PM characteristics of the system. The AM–PM conversion is essentially the nonlinear behavior that is often referred to as short-term memory effects.

  The matching networks employed in power amplifiers are also a source of short-term memory effects, since they are built from reactive components or transmission lines whose frequency dependence contribute to the short-term dynamics.
- **Long-term memory effects**: The three main causes of long-term memory effects are:
  - **Thermal effects**: The transistor channel can heat up non-uniformly when driven by modulated signals. Because of this local change of temperature,

some of the parameters of the transistor will be slightly different, for example, a reduction of the gain from the equilibrium-temperature value is common. The time constants associated with thermal transients—as modeled with the subcircuit of Figure 1.11—are generally in the order of milliseconds, which is close to the timescale of low modulation frequencies, in the range of 100 kHz and below. These long-term memory effects can be seen in the AM–AM characteristics of RF power amplifiers as a spread around the mean gain compression curve.

- **Charge trapping**: Imperfections and defects in the semiconductor occur in several locations at the internal structure of the transistors. These imperfections often manifest themselves as available states that can capture and release electrons and holes, a mechanism governed by local potentials and temperature. The action of trapping or releasing an electron can effectively change the charge density in the channel of the transistor, and its rate is on a timescale of kilohertz through megahertz, depending on the nature of the trapping center. In Figure 1.12, a subcircuit providing capture and emission time constants to model these charge trapping effects was presented. Therefore, charge trapping is a mechanism causing long-term memory effects. GaAs and GaN FETs display several trap-related phenomena, while other transistors like laterally-diffused metal-oxide semiconductor (LDMOS) do not suffer from them.
- **DC bias networks**: The DC bias networks of the power amplifier provide a low impedance path for the DC bias connections that is simultaneously a high impedance for the RF signal. As shown in Figure 1.18, this path has inductances and capacitances that control the frequency response from the DC to a few tens of MHz. Therefore, any signal component in this frequency range will experience memory effects. The signal components appearing in this baseband frequency range will mix with the RF components as a result of the even-order nonlinearities in the active device. This is the most usual cause of asymmetries in IMD responses. As an example of memory effects produced by DC bias networks, Figure 1.19 shows the relative phase of the IM3 products with respect to the tones, and their phase difference for a tones separation sweep. In this case, a bias network designed with a resonance near 130 kHz was employed, and the memory effects it produced are clearly observable in the IMD phase measurement.

As mentioned above, the presence of long-term memory effects can be revealed by measurements consisting in two-tone tests where a sweep of the tone spacing is performed. However, the identification of the origins of the long-term memory is more difficult. Thermal effects can be observed by making pulsed DC measurements and varying the pulse width and duty cycle.

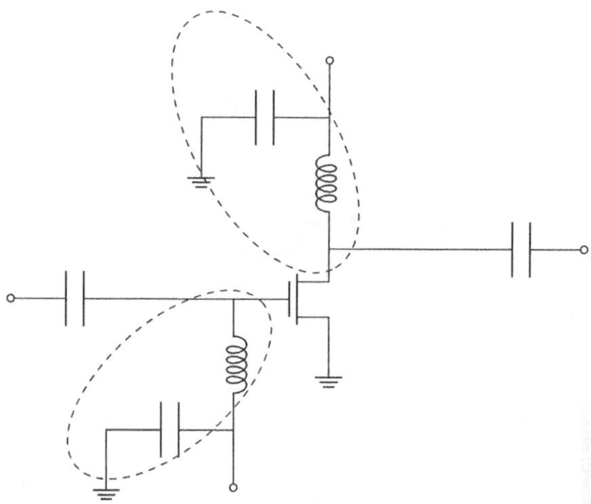

Figure 1.18 DC bias networks in a power amplifier responsible for memory effects.

Traditionally, technical datasheets of commercial RF and microwave circuits show the behavior of the different forms of nonlinear performance using an arbitrary modulation bandwidth or a fixed tone separation in two-tone tests. However, intermodulation distortion levels can experience significant variations when excitations with different bandwidths are used. This phenomenon is the result of complex interactions among the active devices and the rest of the circuit, which make distortion strongly dependent on the characteristics of the modulation signal.

One-tone power sweep measurements are very helpful during the design of power amplifiers, but they do not provide enough information about the memory of the system. On the contrary, two-tone intermodulation characterizations with varying frequency separation between the tones are common measurements in nonlinear characterization to provide memory information. The comprehension of its peculiarities has attracted research interest. For example, the classification of amplifiers regarding their memory effects (Bosch and Gatti, 1989), the study of asymmetries in IM3 products revealing that terminating impedances at baseband or difference frequencies are the main cause for distortion sideband asymmetries (Carvalho and Pedro, 2002; Brinkhoff and Parker, 2003), the proposal of metrics to quantify memory effects (Martins et al., 2006), and the use of intermodulation distortion profiles produced under multitone excitations to quantify the impact of dynamic effects on RF systems (Figueiredo et al., 2021).

However, the concepts of memory effects and asymmetries are misleadingly exchanged sometimes, thus a clarification on their relation is not out of place here. Recall that, in the context of two-tone and multitone tests, memory effects can be detected as changes in the amplitude and phase of intermodulation components with tones spacing, whereas a difference between the upper and lower

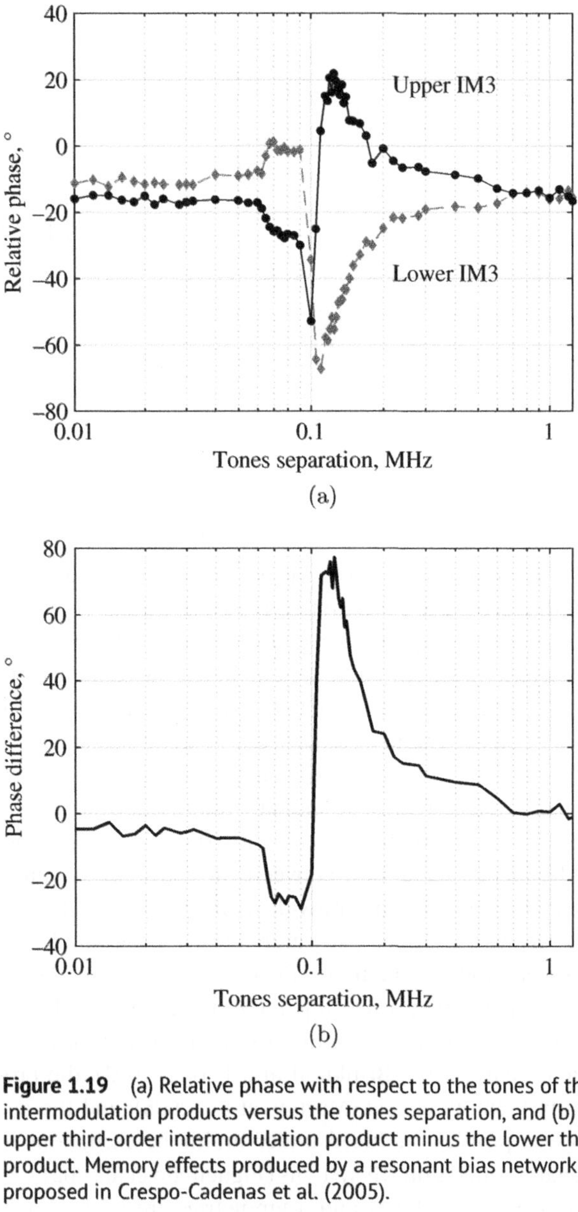

Figure 1.19 (a) Relative phase with respect to the tones of the third-order intermodulation products versus the tones separation, and (b) phase difference of the upper third-order intermodulation product minus the lower third-order intermodulation product. Memory effects produced by a resonant bias network measured with the method proposed in Crespo-Cadenas et al. (2005).

intermodulation products is referred to as an asymmetry. Therefore, any system that presents a frequency-dependent baseband load impedance is a nonlinear system with memory. Despite asymmetry being another indication of memory effects, not all nonlinear systems with memory exhibit spectral asymmetries, as we will illustrate in Chapter 2. Notice that if the characteristic observed in a spectrum analyzer has (magnitude) asymmetry, then it is possible to say that the amplifier has memory, since the condition for a magnitude asymmetry can only be met for those bias and matching networks in which the baseband load impedance is also frequency-dependent (Crespo-Cadenas et al., 2006). Even in the case of a spectrum with symmetric magnitude, it is possible to observe memory effects in an amplifier if they are related to phase asymmetry, or if they are hidden by dominant third-order nonlinearities.

1.6 Behavioral Modeling and Linearization of Nonlinear Systems

One of the most important drawbacks of nonlinear circuits is that there are neither universal analysis methods nor universal models. Therefore, it is difficult to establish a unified approach valid for all nonlinear systems, and every model and simulation technique will perform well in a specific application only (Kundert et al., 1990).

Different approaches to nonlinear systems analysis, such as large-signal scattering parameters, time-domain differential equations, harmonic-balance analysis, neural networks, to mention some of them, have been elaborated to study a given class of nonlinear systems (Maas, 2003). Let us take a look at some of the nonlinear techniques more relevant to the objective of this text.

1.6.1 Harmonic Balance

Harmonic balance (HB) is a frequency-domain technique suitable to the analysis of strongly nonlinear circuits, for example, a power amplifier driven by a continuous wave sinusoidal signal. HB is based on the equivalent circuit of the power amplifier, which can contain lumped elements as well as distributed transmission lines. The response is a periodic signal that can be expressed as a linear combination of sinusoidal signals with related frequencies and, therefore, simple equations are obtained for the solution at each harmonic. Referring to Figure 1.20, the HB technique first solves the linear circuit assuming an estimated value for the Fourier components of the voltages V_1 and V_2 at the nonlinear nodes. These values allow to compute the nonlinear current across the corresponding nonlinear element in the time domain $i_1(v_1)$ and $i_2(v_2)$, and after Fourier transformation, the

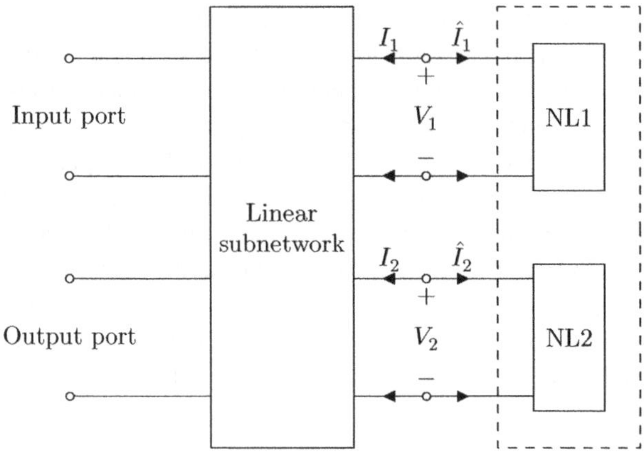

Figure 1.20 Harmonic-balance technique. The nonlinear circuit is conceived with the two nonlinear current sources separated from the linear subnetwork.

frequency components of the nonlinear currents $\hat{I}_1$ and $\hat{I}_2$. New values of V_1 and V_2 can be computed by solving again the linear subcircuit but now with the nonlinear currents $\hat{I}_1$ and $\hat{I}_2$. The final solution is given by the values when Kirchoff's law is satisfied, that is, when the sum of the linear current and the nonlinear current at each harmonic is zero, $I_1 + \hat{I}_1 = 0$ and $I_2 + \hat{I}_2 = 0$. Newton's method is the most common algorithm used to solve the optimization problem of the HB technique.

The excellent capability of the HB technique in the analysis of nonlinear circuits allows accurate predictions not only of the harmonic components generated at the output of a power amplifier when a single sinusoidal tone is applied at its input, as in the experiment of Figure 1.13, or the AM–AM conversion, as in Figure 1.14. HB is also a useful tool to analyze the output of a power amplifier with a multitone excitation, for example, a two-tone input and the corresponding intermodulation products represented in Figures 1.15–1.17.

Notwithstanding the remarkable features of the HB technique in the case of strongly nonlinear circuit analysis with multitone excitations, it has a limited value when predicting the nonlinear distortions produced to a modulated signal by the blocks of a wireless communications system.

1.6.2 Volterra Series

A second way to study the behavior of a nonlinear system is the Volterra series approach. Bedrosian and Rice wrote a significant paper demonstrating the help of Volterra series to evaluate distortion in nonlinear systems with memory, in which the authors express their honest view:

> In practice it appears that Volterra series do not enable us to do anything that cannot be done otherwise.
>
> (Bedrosian and Rice, 1971)

This is not a very encouraging statement to initiate a text written with the perspective of the Volterra series technique, but the authors immediately point out an important clue about its convenience in the analysis of nonlinear systems:

> However, a direct attack on modulation problems often leads to a morass of algebra. The Volterra series approach has the virtue that many such problems can be treated in an orderly way by first computing the [transfer functions] *G*, and then substituting them in the appropriate general formulas.
>
> (Bedrosian and Rice, 1971)

Surely, the study of the nonlinear effects in power amplifiers is demanding, and perhaps the first reason to consider the Volterra-based approach a useful procedure is its fair grade of simplicity to alleviate the "morass of algebra" in nonlinear analysis.

There are however more reasons that make the Volterra series a viewpoint widely accepted by RF engineers to develop frequency-domain analysis of nonlinear circuits (Bussgang et al., 1974; Benedetto et al., 1979; Mirri et al., 2002). It is worth mentioning the application of this excitation-independent representation to the analysis and design of circuits, or its ability to express the response of some nonlinear systems by means of closed-form expressions.

In particular, the important help this mathematical tool gives to the development of models describing a physical system, for example, a power amplifier, deserves special attention and will be at the center of our discussion. This way, a Volterra model may help to study and explain a stable time-invariant nonlinear system with memory, and to make predictions about its nonlinear behavior.

Although the details of the Volterra formulation are discussed in following chapters, let us advance that the output of an amplifier can be expressed with a Volterra series written as

$$
\begin{aligned}
y(t) = &\int_{-\infty}^{\infty} h_1(\tau_1)x(t-\tau_i)d\tau_1 \\
&+ \int_{-\infty}^{\infty}\int_{-\infty}^{\infty} h_2(\tau_1,\tau_2)x(t-\tau_1)x(t-\tau_2)d\tau_1 d\tau_2 \\
&+ \int_{-\infty}^{\infty}\int_{-\infty}^{\infty}\int_{-\infty}^{\infty} h_3(\tau_1,\tau_2,\tau_3)x(t-\tau_1)x(t-\tau_2)x(t-\tau_3)d\tau_1 d\tau_2 d\tau_3 + \cdots,
\end{aligned}
\tag{1.25}
$$

a representation that can be considered as a power series expansion like (1.3) with memory. The power amplifier is completely described by the signal-independent

Volterra kernels $h_n(\tau_1, \tau_2, \dots, \tau_n)$, a set of parameters that can be estimated using direct experimental data. This parameters estimation is one step forward toward obtaining predictions of nonlinear effects.

As a modeling technique linear with respect to the unknown parameters, the statistical properties of the resulting Volterra model estimators are easier to determine, compared to models which are nonlinearly related to their parameters. For example, the Saleh model (Saleh, 1981) represents the amplifier output as nonlinearly dependent on two parameters. However, the Volterra series offer the ability to more easily fit experimental data, and this is perhaps the most outstanding aspect of the fascination that this approach has produced in researchers. There are other reasons that make the modeling of the Volterra series so attractive. Within this perspective, RF system designers have representations with a well-grounded theory and capable of providing highly accurate modeling results compared to measured data. In addition, all the device information available at the physical or circuit level can be exploited in the deduction of Volterra-based behavioral models, and we can track the relation of the model kernels to the physical or the equivalent circuit parameters.

Unfortunately, a major drawback of Volterra series pointed out by experts resides in its inability to handle large-signal distortion problems and to accurately represent an amplifier. The lack of ability in the case of a large input signal is explained by mathematicians by cryptically saying that the limited range of convergence of the Volterra series is exceeded. A typical example is the compression presented by the AM–AM characteristic of an amplifier, a nonlinear behavior not explained adequately by a Volterra model without memory: the polynomial model (1.3).

To overcome this inconvenience, an alternative representation of the nonlinear distortions in the case of a memoryless power amplifier is a model described by closed-form expressions with parameters adjusted to measured data. The AM–AM characteristic of the Saleh model is plotted with a thick solid line in Figure 1.21, clearly showing the typical power amplifier behavior with separated quasi-linear and compression zones. To illustrate the aforementioned discussion about the limited range of convergence of the Volterra series, the adjusted characteristic of a 3rd-degree polynomial model demonstrates a poor accuracy in the quasi-linear zone and a visible divergence in the compression zone. An increment of the polynomial model terms to 5th-degree allows a better adjustment in the quasi-linear zone but an accentuated divergence in the compression zone. Adding more terms do not substantially improve the model results because, although there is a better fit in a limited range of the compression zone, the divergence gets worse, even in the case of a polynomial extended to 9th-degree.

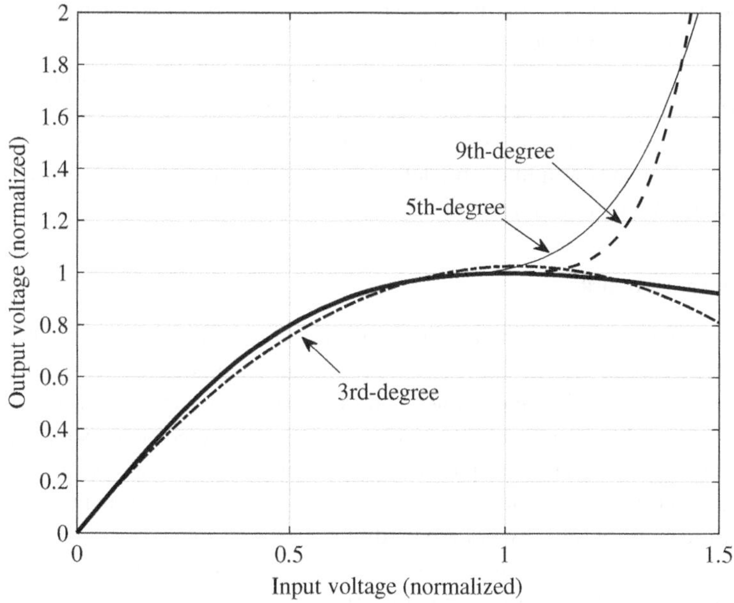

Figure 1.21 Normalized AM–AM characteristic of a memoryless power amplifier.

In this discussion, we can remark that the primary reason of divergence lies in the practical impossibility of computationally implementing a series with an infinite number of terms, unfeasibility naturally associated with the use of a truncated Volterra series. However, management of an infinite Volterra series in a mathematical development is not necessarily a theoretical limitation. To be precise, consider the example of a nonlinear function with a convergent Taylor series expansion and recall that it is a polynomial series composed by an infinite number of terms, a representation mathematically equivalent to the original function.

This section can be concluded with a summary of some statements about a Volterra approach to modeling and its relevance to the objectives of this book. Having in mind its limitations, we can agree that experts have widely considered the important role that the Volterra series has played in the field of distortion studies. Among all nonlinear circuit analysis methods, the Volterra series stand out for their ability to explain memory effects, for their natural connection with physical or circuit models, and for their condition of linearity in the parameters. This linear dependence on the unknown parameters makes the Volterra models easier to fit than other models having a nonlinear relation with the parameters. At this point, we can mention the incorporation of statistics in addition to the RF power amplifiers and baseband digital signal processing disciplines. The ease of determining the estimators is perhaps the main feature that makes linear regression such an attractive statistical approach for modeling the relationship between the response of a nonlinear power amplifier and its input signal.

According to the ideas discussed above, and taking into consideration mainly the special attention Volterra approach deserves, our analysis will be focused on a set of additional considerations to deepen the perception of the usefulness of the Volterra series in power amplifiers modeling and linearization, discussions to be addressed in the following chapters of this book.

1.6.3 Neural Networks

In addition to Volterra-based modeling, neural networks (NNs) (Haykin, 1999) have attracted a certain interest in the research community applied to linearization through digital predistortion due to their nonlinear modeling capabilities (Zhang et al., 2003). Neural networks are massively parallel distributed processors that are composed of a number of simple processors units, the artificial *neurons*. To achieve a good performance, they need to undergo a process of learning in which the interconnections of the network are modified.

The basic model of a neuron is provided in Figure 1.22, where three main elements must be stressed. First, the interconnecting links of the inputs x_j to neuron k by means of *synaptic weights* w_{kj}, so that the input is multiplied by the synaptic weight. Second, the linear combiner that sums the weighted inputs to neuron k together with an external *bias* b_k. That is, at the output of the linear combiner, we have

$$v_k = \sum_{k=1}^{m} w_{kj} x_j + b_k. \tag{1.26}$$

The third element is the *activation function* $f_{act}(\cdot)$ that limits the amplitude of the output of the neuron y_k and is usually defined in a nonlinear way. Although there

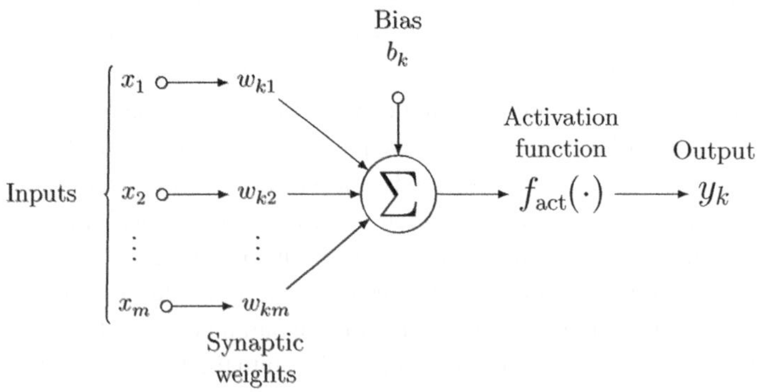

Figure 1.22 Model of a neuron.

is a wide variety of activation functions, basic types are the threshold function

$$y_k = \begin{cases} 1, & \text{if } \upsilon_k \geq 0 \\ 0, & \text{if } \upsilon_k < 0 \end{cases} \tag{1.27}$$

and the sigmoid function, for which the hyperbolic tangent function can be employed showing good performance when applied to the compensation of distortion and impairments of transmitters (Wang et al., 2019)

$$f_{\text{act}}(\upsilon_k) = \tanh(\upsilon_k). \tag{1.28}$$

The neurons of a neural network are organized in layers, as it is shown in Figure 1.23. There is an *input layer* containing the source nodes associated with the input signals. Afterward, there are one or more *hidden layers*, not seen directly from the input or the output of the network, that enable the interconnections between the neurons according to the structure of each type of neural network. The output signals of the neurons in the final layer, referred to as the *output layer*, provide the overall response of the neural network. Neural networks modeling capabilities are based on the universal approximation theorem (Cybenko, 1989) that states that a single-layer neural network can approximate any continuous nonlinear function. The employed network architecture determines the type of learning algorithm that is necessary to train the network, that is, to find the synaptic weights that produce the best input–output mapping.

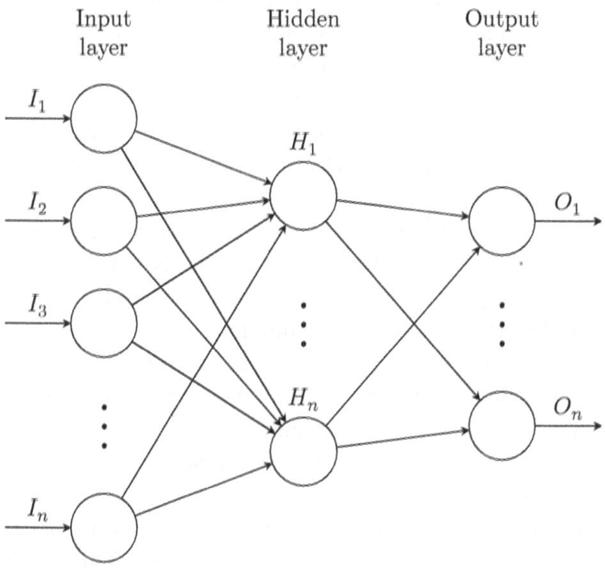

Figure 1.23 Feedforward network with the input layer containing the source nodes and two layers of neurons: a hidden layer and the output layer.

Examples of neural network-based architectural designs applied to modeling and linearization of power amplifiers are the real-valued time-delay neural network (RVTDNN) (Liu et al., 2004; Rawat et al., 2010), the augmented real-valued time-delay neural network (ARVTDNN) (Wang et al., 2019), the vector-decomposition time-delay neural network (VDTDNN) (Zhang et al., 2019) and the block-oriented time-delay neural network (BOTDNN) (Jiang et al., 2022). Neural networks were also applied to crosstalk, I/Q imbalance and nonlinearity in MIMO transmitters (Jaraut et al., 2018). The concept of sparsity was explicitly applied to neural networks in (Tanio et al., 2020).

Although neural networks theoretically have great advantages in modeling and linearization performance, they pose three drawbacks for their use: (i) a long training time, specially problematic in real-time scenarios such as the DPD forward path, (ii) high complexity, (iii) loss of track of the physical meaning of the basis functions, since the nonlinearity is defined in the training of the network and not by its architecture as it is the case of Volterra series.

1.7 Regression

One of the convenient characteristics that Volterra series feature is that their coefficients can be estimated through linear regression. Although Volterra series represent a nonlinear function with memory relating their input to their output, the relation between the kernels and the output signal is linear. This differentiated property is opposed to other nonlinear structures that require optimization algorithms and nonlinear solving methods to estimate the model coefficients. The objective of regression is to calculate an estimator of the coefficients so that the error between the output of the model and the real output signal is minimized. The direct mathematical minimization of the error leads to a very standard approach known as least squares. The least squares method allows to obtain an estimator following a simple analytic closed-form expression, therefore its convenience. This way, by using the least squares method as a basis, machine learning techniques can be directly applied in the estimation.

The main drawback of Volterra series models is that their number of coefficients rapidly grows with the nonlinearity order and the memory depth. A high number of coefficients heavily affects the performance of the estimator after the regression, therefore not only the computational complexity needs to be carefully addressed considering that the computation capabilities are limited. In this scenario, the concept of coefficients selection—also known as model order reduction or *a posteriori* pruning—plays a fundamental role. This set of techniques aims at exploiting the sparsity of the model. There are different ways to understand sparsity. The most intuitive approach is to think that the same information is distributed among several model components, and by deactivating certain parts of the model, the performance remains the same. Thus, if the most relevant model coefficients are selected and the rest are discarded, the model exhibits an equivalent level of performance but the number of coefficients is reduced, enhancing the regression performance and reducing the computational complexity. A special emphasis on the use of regression in Volterra-based models and making use of their sparsity is laid in the last group of chapters of this book.

1.8 Structure of the Book

This book contains seven chapters, including this initial chapter, which serves as an introduction to the research objectives and sets the scheme for our studies. The rest of the chapters can be approximately divided in three groups, according to the research field in which they are focused.

Chapter 2 belongs to the first group, which can be roughly framed within a Volterra series approach to model classical continuous-time and real-valued nonlinear systems, with particular emphasis on the power amplifiers of a wireless communications network. Although the focus here is not on the design of high-frequency electronic circuits, this chapter is near to RF-designers interest and the practical relevance is concentrated on the response of the nonlinear system driven by periodic inputs, such as a continuous wave single sinusoidal signal or a multitone signal. To describe the dynamic behavior of a power amplifier following this Volterra series perspective, the nonlinear transfer functions are derived based on a simple and conventional equivalent circuit. The harmonic input and the nonlinear currents methods are discussed and applied to a power amplifier, illustrating the benefits of the Volterra series approach as a forceful tool in the analysis of intermodulation distortion in systems with memory. To complete the Volterra perspective, the chapter includes the equivalent representation of an amplifier with a Volterra series in the frequency domain. Since the power amplifier is driven by a single input signal, the conventional input–output Volterra relationship is implicitly a univariate model. However, modern RF communications hardware is also made up of electronic circuits with more than one input port: mixers and modulators are two of the most obvious examples. Due to the importance of these nonlinear bivariate systems, Chapter 2 introduces the discussion of a double Volterra series proposal to describe their behavior and advances a new modeling approach. To illustrate the Volterra series effectiveness researching the dependence of intermodulation on frequency in nonlinear systems with memory, the analysis of an amplifier model and the deduction of the nonlinear transfer functions are next shown. Finally, an extension of the conventional Volterra series representation of an RF bandpass system to its univariate low-pass equivalent involving complex envelope signals is presented. This modeling approach allows evaluating and controlling the nonlinear performance of digital communication systems operating within the discrete-time baseband modules of a wireless transmitter.

A second group consists of Chapters 3 and 4. Chapter 3 describes the method to relate the continuous-time RF Volterra model and the corresponding discrete-time Volterra model. Discrete-time baseband Volterra models are indispensable to analyze the nonlinear performance of the system and are employed to manage the complex-valued signals in the digital processing modules of wireless communications transmitters. Despite the premise of a high complexity imputed to the Volterra models, one unquestionable feature of this approach is the explicit relationship of the discrete-time Volterra kernels and the corresponding nonlinear transfer functions, a result that enables the proposal of heuristic behavioral models with simple structures. The univariate types of the memory polynomial and the generalized memory polynomial models are two examples of the heuristic structures shown in this chapter. Because some of the bivariate systems represented by a double Volterra series are driven by one baseband signal, at least, use of the corresponding discrete-time double Volterra model disclosed in the chapter is of utmost importance. Furthermore, due to the natural link between discrete-time Volterra kernels viewed as tensors, and a multidimensional arrays approach, a brief discussion on Volterra–Parafac models is advanced (Favier et al., 2012; Bouilloc and Favier, 2012). Issues like the convenience of a Volterra model in the frequency domain, the discrete-frequency transfer functions representation and the example of a power amplifier linearization with a digital predistorter based on a Volterra–Parafac model, exploiting the direct dependence of the output spectrum on the input spectrum, are also included. After that, the general problem of a complex-valued nonlinear system is formally addressed under a Volterra series perspective and a widely nonlinear transformation is demonstrated. To conclude the chapter, some figures of merit used to characterize nonlinear distortions in communication systems are reviewed.

Chapter 4 deals with Volterra models derived by relying on the corresponding power amplifier circuit model. Beginning with a circuit level view, these models were deduced based on the available information and with the objective of pruning the huge number of parameters of a general Volterra model. Examples mentioned in the chapter are the Volterra model for wideband systems and the (univariate) circuit-knowledge Volterra model. The bivariate Volterra models extend the circuit-knowledge approach by including a new variable generated by an internal subnetwork of the power amplifier equivalent circuit, in particular the charge trapping subnetwork. Under this perspective, the memory polynomial and the generalized memory polynomial are particular cases of a power amplifier bivariate Volterra model. To close the topic of power amplifiers modeling, the proposal of a Volterra model customized to the particular case of a dual band signal is demonstrated. Finally, because of the importance of I/Q modulators in wireless transmitters, a model is introduced including the formal deduction based on a double Volterra series approach.

The third group of chapters of the book comprises Chapters 5, 6, and 7. Chapter 5 is devoted to linear regression of Volterra-based models. The matrix formulation of a power amplifier modeling framework is developed, paying special attention to the statistical properties of the variables involved in the process. The least-squares method is visited and its formulation is derived, providing insights on the numerical performance of the estimators and the consequences of having a high number of coefficients in the model. Next, ℓ_2 regularization is proposed as a means of reducing the variance of the estimator and ℓ_1 regularization is presented as the link to sparse machine learning techniques, which are covered in the next chapter. Finally, adaptive optimization is introduced to serve as the foundation of iterative regression.

Sparse signal processing is addressed in Chapter 6, where the basics of attaining models in which not all the components are active are studied. First, the concept of thresholding is analyzed as the basics of coefficients selection and the local search techniques are presented through the hill climbing algorithms. Next, the group of greedy pursuits, that are characterized by selecting components following a local decision in each iteration, are reviewed. The group of techniques that rely on Bayesian statistics are covered next, providing pruning techniques based on knowledge of events. Rules for selecting the optimum number of coefficients are analyzed, where the Bayesian information criterion (BIC) is covered in depth. Finally, a practical sparse regression exercise is provided, in which a set of exercises are performed paying attention to the particularities of sparse regression.

Chapter 7 aims to provide an overview of linearization strategies. The concept of digital predistortion is formulated along with its repercussions for the underlying digital predistortion architectures. The most commonly used architectures, namely the direct learning architecture (DLA) and the indirect learning architecture (ILA) are discussed and the regression characteristics in these schemes are covered. To wrap up the chapter, the regression in both architectures is combined with the use of sparse machine learning techniques, exemplifying their application in an extensive set of predistortion scenarios.

Bibliography

P. Aaen, J.A. Plá, and J. Wood. *Modeling and Characterization of RF and Microwave Power FETs*. The Cambridge RF and Microwave Engineering Series. Cambridge University Press, Cambridge, UK, 2007. ISBN 9780521870665.

I. Angelov, H. Zirath, and N. Rosman. A new empirical nonlinear model for HEMT and MESFET devices. *IEEE Transactions on Microwave Theory and Techniques*, 40(12):2258–2266, 1992. doi: 10.1109/22.179888.

E. Bedrosian and S.O. Rice. The output properties of Volterra systems (nonlinear systems with memory) driven by harmonic and gaussian inputs. *Proceedings of the IEEE*, 59(12):1688–1707, 1971. doi: 10.1109/PROC.1971.8525.

S. Benedetto, E. Biglieri, and R. Daffara. Modeling and performance evaluation of nonlinear satellite links — A Volterra series approach. *IEEE Transactions on Aerospace and Electronic Systems*, AES-15(4):494–507, 1979. doi: 10.1109/TAES.1979.308734.

W. Bosch and G. Gatti. Measurement and simulation of memory effects in predistortion linearizers. *IEEE Transactions on Microwave Theory and Techniques*, 37(12):1885–1890, 1989. doi: 10.1109/22.44098.

T. Bouilloc and G. Favier. Nonlinear channel modeling and identification using baseband Volterra–Parafac models. *Signal Processing*, 92(6):1492–1498, 2012. doi: 10.1016/j.sigpro.2011.12.007.

T.J. Brazil. Simulating circuits and devices. *IEEE Microwave Magazine*, 4(1): 42–50, 2003. doi: 10.1109/MMW.2003.1188235.

J. Brinkhoff. *Bandwidth-Dependent Intermodulation Distortion in FET Amplifiers*. PhD thesis, Macquarie University, Sydney, Australia, 2004.

J. Brinkhoff and A.E. Parker. Effect of baseband impedance on FET intermodulation. *IEEE Transactions on Microwave Theory and Techniques*, 51(3):1045–1051, 2003. doi: 10.1109/TMTT.2003.808704.

J.J. Bussgang, L. Ehrman, and J.W. Graham. Analysis of nonlinear systems with multiple inputs. *Proceedings of the IEEE*, 62(8):1088–1119, 1974. doi: 10.1109/PROC.1974.9572.

P.M. Cabral, J.C. Pedro, and N.B. Carvalho. Dynamic AM-AM and AM-PM behavior in microwave PA circuits. In *2005 Asia-Pacific Microwave Conference Proceedings*, volume 4, pages 4, 2005. doi: 10.1109/APMC.2005.1606809.

N.B. Carvalho and J.C. Pedro. A comprehensive explanation of distortion sideband asymmetries. *IEEE Transactions on Microwave Theory and Techniques*, 50(9):2090–2101, 2002. doi: 10.1109/TMTT.2002.802321.

J.K. Cavers. The effect of quadrature modulator and demodulator errors on adaptive digital predistorters for amplifier linearization. *IEEE Transactions on Vehicular Technology*, 46(2):456–466, 1997. doi: 10.1109/25.580784.

J.K. Cavers and M.W. Liao. Adaptive compensation for imbalance and offset losses in direct conversion transceivers. *IEEE Transactions on Vehicular Technology*, 42(4):581–588, 1993. doi: 10.1109/25.260752.

L.O. Chua, C.A. Desoer, and E.S. Kuh. *Linear and Nonlinear Circuits*. Electrical & electronic engineering. McGraw-Hill, New York, 1987. ISBN 9780071001670.

C. Crespo-Cadenas, J. Reina-Tosina, and M.J. Madero-Ayora. Phase characterization of two-tone intermodulation distortion. In *IEEE MTT-S International Microwave Symposium Digest, 2005*, pages 1505–1508, 2005. doi: 10.1109/MWSYM.2005.1516979.

C. Crespo-Cadenas, J. Reina-Tosina, and M.J. Madero-Ayora. IM3 and IM5 phase characterization and analysis based on a simplified Newton approach. *IEEE Transactions on Microwave Theory and Techniques*, 54(1):321–328, 2006. doi: 10.1109/TMTT.2005.861659.

S.C. Cripps. *RF Power Amplifiers for Wireless Communications*. Artech House Microwave Library. Artech House, Boston, 1999. ISBN 9780890069899.

W.R. Curtice. A MESFET model for use in the design of GaAs integrated circuits. *IEEE Transactions on Microwave Theory and Techniques*, 28(5): 448–456, 1980. doi: 10.1109/TMTT.1980.1130099.

W.R. Curtice and M. Ettenberg. A nonlinear GaAs FET model for use in the design of output circuits for power amplifiers. *IEEE Transactions on Microwave Theory and Techniques*, 33(12):1383–1394, 1985. doi: 10.1109/TMTT.1985.1133229.

G.V. Cybenko. Approximation by superpositions of a sigmoidal function. *Mathematics of Control, Signals and Systems*, 2:303–314, 1989. doi: 10.1007/BF02551274.

W. Dai, P. Roblin, and M. Frei. Distributed and multiple time-constant electro-thermal modeling and its impact on ACPR in RF predistortion. In *Conference, 2003. Fall 2003. 62nd ARFTG Microwave Measurements*, pages 89–98, 2003. doi: 10.1109/ARFTGF.2003.1459759.

L. Ding, Z. Ma, D.R. Morgan, M. Zierdt, and G.T. Zhou. Compensation of frequency-dependent gain/phase imbalance in predistortion linearization systems. *IEEE Transactions on Circuits and Systems I: Regular Papers*, 55 (1):390–397, 2008. doi: 10.1109/TCSI.2007.910545.

Y. Ding and A. Sano. Time-domain adaptive predistortion for nonlinear amplifiers. In *2004 IEEE International Conference on Acoustics, Speech, and Signal Processing*, volume 2, pages ii–865, 2004. doi: 10.1109/ICASSP.2004.1326395.

P. Draxler, I. Langmore, T.P. Hung, and P.M. Asbeck. Time domain characterization of power amplifiers with memory effects. In *IEEE MTT-S International Microwave Symposium Digest, 2003*, volume 2, pages 803–806, 2003. doi: 10.1109/MWSYM.2003.1212492.

G. Favier, A.Y. Kibangou, and T. Bouilloc. Nonlinear system modeling and identification using Volterra-PARAFAC models. *International Journal of Adaptive Control and Signal Processing*, 26(1):30–53, 2012. doi: 10.1002/acs.1272.

R. Figueiredo, N.B. Carvalho, A. Piacibello, and V. Camarchia. Nonlinear dynamic RF system characterization: Envelope intermodulation distortion profiles: A noise power ratio-based approach. *IEEE Transactions on Microwave Theory and Techniques*, 69(9):4256–4271, 2021. doi: 10.1109/TMTT.2021.3092398.

S.S. Haykin. *Neural Networks: A Comprehensive Foundation*. International edition. Prentice Hall, Upper Saddle River, NJ, 1999. ISBN 9780132733502.

P. Jaraut, M. Rawat, and F.M. Ghannouchi. Composite neural network digital predistortion model for joint mitigation of crosstalk, I/Q imbalance, nonlinearity in MIMO transmitters. *IEEE Transactions on Microwave Theory and Techniques*, 66(11):5011–5020, 2018. doi: 10.1109/TMTT.2018.2869602.

O. Jardel, F. De Groote, T. Reveyrand, J.-C. Jacquet, C. Charbonniaud, J.-P. Teyssier, D. Floriot, and R. Quere. An electrothermal model for AlGaN/GaN power HEMTs including trapping effects to improve large-signal simulation results on high VSWR. *IEEE Transactions on Microwave Theory and Techniques*, 55(12):2660–2669, 2007. doi: 10.1109/TMTT.2007.907141.

C. Jiang, H. Li, W. Qiao, G. Yang, Q. Liu, G. Wang, and F. Liu. Block-oriented time-delay neural network behavioral model for digital predistortion of RF power amplifiers. *IEEE Transactions on Microwave Theory and Techniques*, **70**(3):1461–1473, 2022. doi: 10.1109/TMTT.2021.3124211.

J. Kim and K. Konstantinou. Digital predistortion of wideband signals based on power amplifier model with memory. *Electronics Letters*, 37(23):1417–1418, 2001. doi: 10.1049/el:20010940.

K. Kundert, J. K. White, and A. Sangiovanni-Vincentelli. *Steady-State Methods for Simulating Analog and Microwave Circuits*. Kluwer Academic Publishers, Boston, 1990. ISBN 0792390695.

T. Liu, S. Boumaiza, and F.M. Ghannouchi. Dynamic behavioral modeling of 3G power amplifiers using real-valued time-delay neural networks. *IEEE Transactions on Microwave Theory and Techniques*, 52(3):1025–1033, 2004. doi: 10.1109/TMTT.2004.823583.

S.A. Maas. *Nonlinear Microwave and RF Circuits*. Artech House, Boston, 2nd edition, 2003. ISBN 9781580536110.

J.P. Martins and N.B. Carvalho. Multitone phase and amplitude measurement for nonlinear device characterization. *IEEE Transactions on Microwave Theory and Techniques*, 53(6):1982–1989, 2005. doi: 10.1109/TMTT.2005.848841.

J.P. Martins, P.M. Cabral, N.B. Carvalho, and J.C. Pedro. A metric for the quantification of memory effects in power amplifiers. *IEEE Transactions on Microwave Theory and Techniques*, 54(12):4432–4439, 2006. doi: 10.1109/TMTT.2006.882871.

A. Materka and T. Kacprzak. Computer calculation of large-signal GaAs FET amplifier characteristics. *IEEE Transactions on Microwave Theory and Techniques*, 33(2):129–135, 1985. doi: 10.1109/TMTT.1985.1132960.

R.A. Minasian. Intermodulation distortion analysis of MESFET amplifiers using the Volterra series representation. *IEEE Transactions on Microwave Theory and Techniques*, 28(1):1–8, 1980. doi: 10.1109/TMTT.1980.1129998.

D. Mirri, G. Luculano, F. Filicori, G. Pasini, G. Vannini, and G.P. Gabriella. A modified Volterra series approach for nonlinear dynamic systems modeling.

IEEE Transactions on Circuits and Systems I: Fundamental Theory and Applications, 49(8):1118–1128, 2002. doi: 10.1109/TCSI.2002.801239.

D.R. Morgan, Z. Ma, J. Kim, M.G. Zierdt, and J. Pastalan. A generalized memory polynomial model for digital predistortion of RF power amplifiers. *IEEE Transactions on Signal Processing*, 54(10):3852–3860, 2006. doi: 10.1109/TSP.2006.879264.

A.E. Parker and J.G. Rathmell. Bias and frequency dependence of FET characteristics. *IEEE Transactions on Microwave Theory and Techniques*, 51(2):588–592, 2003. doi: 10.1109/TMTT.2002.807819.

J.C. Pedro and N.B. Carvalho. *Intermodulation Distortion in Microwave and Wireless Circuits*. Artech House Microwave Library. Artech House, Boston, 2003. ISBN 9781580533560.

J.C. Pedro and J. Perez. Accurate simulation of GaAs MESFET's intermodulation distortion using a new drain-source current model. *IEEE Transactions on Microwave Theory and Techniques*, 42(1):25–33, 1994. doi: 10.1109/22.265524.

J.C. Pedro, N.B. Carvalho, and P.M. Lavrador. Modeling nonlinear behavior of band-pass memoryless and dynamic systems. In *IEEE MTT-S International Microwave Symposium Digest, 2003*, volume 3, pages 2133–2136, 2003. doi: 10.1109/MWSYM.2003.1210584.

T. Qi and S. He. Power up potential power amplifier technologies for 5G applications. *IEEE Microwave Magazine*, 20(6):89–101, 2019. doi: 10.1109/MMM.2019.2904409.

M. Rawat, K. Rawat, and F.M. Ghannouchi. Adaptive digital predistortion of wireless power amplifiers/transmitters using dynamic real-valued focused time-delay line neural networks. *IEEE Transactions on Microwave Theory and Techniques*, 58(1):95–104, 2010. doi: 10.1109/TMTT.2009.2036334.

A.A.M. Saleh. Frequency-independent and frequency-dependent nonlinear models of TWT amplifiers. *IEEE Transactions on Communications*, 29(11): 1715–1720, 1981. doi: 10.1109/TCOM.1981.1094911.

H. Statz, P. Newman, I.W. Smith, R.A. Pucel, and H.A. Haus. GaAs FET device and circuit simulation in SPICE. *IEEE Transactions on Electron Devices*, 34(2):160–169, 1987. doi: 10.1109/T-ED.1987.22902.

M. Tanio, N. Ishii, and N. Kamiya. Efficient digital predistortion using sparse neural network. *IEEE Access*, 8:117841–117852, 2020. doi: 10.1109/ACCESS.2020.3005146.

V. Volterra. *Theory of Functionals and of Integral and Integro-Differential Equations*. Dover Books on Intermediate and Advanced Mathematics. Dover Publications, New York, 1959.

J.H.K. Vuolevi, T. Rahkonen, and J.P.A. Manninen. Measurement technique for characterizing memory effects in RF power amplifiers. *IEEE Transactions on Microwave Theory and Techniques*, 49(8):1383–1389, 2001. doi: 10.1109/22.939917.

D. Wang, M. Aziz, M. Helaoui, and F.M. Ghannouchi. Augmented real-valued time-delay neural network for compensation of distortions and impairments in wireless transmitters. *IEEE Transactions on Neural Networks and Learning Systems*, 30(1):242–254, 2019. doi: 10.1109/TNNLS.2018.2838039.

D. Wisell. Identification and measurement of transmitter nonlinearities. In *56th ARFTG Conference Digest-Fall*, volume 38, pages 1–6, Boulder, CO, 2000.

Q. J. Zhang, K.C. Gupta, and V.K. Devabhaktuni. Artificial neural networks for RF and microwave design - from theory to practice. *IEEE Transactions on Microwave Theory and Techniques*, 51(4):1339–1350, 2003. doi: 10.1109/TMTT.2003.809179.

Y. Zhang, Y. Li, F. Liu, and A. Zhu. Vector decomposition based time-delay neural network behavioral model for digital predistortion of RF power amplifiers. *IEEE Access*, 7:91559–91568, 2019. doi: 10.1109/ACCESS.2019.2927875.

2

Volterra Series Approach

2.1 Introduction

In the first chapter, we outlined the subjects of particular interest in this text, focused mainly on the nonlinear behavior of power amplifiers and on all other aspects related to their linearization, but not only those. The nonlinear phenomena revealed in mixers, frequency converters, and modulators were included in the objectives of our analysis.

The problems derived by the presence of dispersive elements, such as transmission lines, the circumstance of large time constants compared to the period of the fundamental frequency of the excitation signal, or the very likely high number of reactive elements in the analyzed circuit, have been successfully overcome by the harmonic balance technique in the case of a periodically excited nonlinear network (Kundert et al., 1990; Pedro et al., 2018). Most strongly nonlinear circuits can be analyzed by harmonic balance without difficulty, but in the analysis of nonlinear systems driven by wireless communication signals, the harmonic balance usefulness is limited.

Many respectable researchers have been of the conviction that Volterra-series analysis is applicable to weakly nonlinear circuits driven by signals with moderate strength. In consequence, the Volterra approach was used to provide a significant insight into the distortions generation mechanisms in power amplifiers, mixers, and other microwave circuits and became an invaluable tool for the radio frequency (RF) design engineer (Maas, 2003; Pedro and Carvalho, 2003).

As well as being a useful tool for evaluating the 1-dB compression point, third-order intercept point, the intermodulation products, or the AM–AM and AM–PM characteristics, in a conventional analysis of nonlinear circuits, Volterra series is a valid approach to behavioral modeling of amplifiers or mixers in a scenario of wireless communication systems.

The difficulty of behavioral analysis in the case of nonlinear systems is unavoidable, but Volterra series is without doubt the simplest methodology we have at

A Volterra Approach to Digital Predistortion: Sparse Identification and Estimation, First Edition.
Carlos Crespo-Cadenas, María José Madero-Ayora, and Juan A. Becerra.

hand to base a system design, a good reason to inspire the conception of this book and devote this chapter to its preliminary treatment.

In the following section, we deal with the introductory concepts of Volterra series and comment on some considerations about convergence, homogeneity of the terms, linearity with respect to the kernels, memory, and the need for a careful application of the Volterra series in complex-valued systems. Once these formal aspects have been reviewed, in the next section, we exemplify the application of the Volterra approach to the analysis of a simplified field-effect transistor (FET) amplifier driven by a sinusoidal input signal, allowing comparison with a conventional analysis of class-A and class-B amplifiers. We also review some procedures to determine the nonlinear transfer functions in a more realistic amplifier and discuss the Volterra series in the frequency domain. Two-block models are also discussed and a double Volterra series approach is introduced to analyze mixers. Next, we exemplify the Volterra series procedure to compute intermodulation distortion. Since the amplifier baseband equivalent representation is the main objective of this chapter, its formal deduction based on the classical continuous-time-domain Volterra series is finally considered.

2.2 Volterra Series

A real-valued continuous-time system can be described by an *operator* that transforms the *input signal* $x(t)$ into the *output signal* $y(t)$. The signals $x(t)$ and $y(t)$ are continuous real functions of time t. If the system is *causal*, the output at any instant does not depend on the values of the input at future instants, but only at past instants, and it is said that the system has *memory*. A system with fading memory for which the output is dependent on instants within a delimited time interval is called a *finite memory* system, and if the dependence is only on the input at the present instant it is called a *memoryless* system. In this text, we assume causal and finite memory systems.

Within some restrictions, the nonlinear system output is represented by a Volterra series expansion (Bussgang et al., 1974) and can be written as

$$y(t) = h_0 + \sum_{n=1}^{\infty} y_n(t), \tag{2.1}$$

for which the output component of order n is[1]

$$y_n(t) = \int_{-\infty}^{\infty} h_n(\boldsymbol{\tau}_n) \prod_{i=1}^{n} x(t-\tau_i) d\boldsymbol{\tau}_n \tag{2.2}$$

---

1 We have used the compact notation

$$\int_{-\infty}^{\infty} h_n(\boldsymbol{\tau}_n) \prod_{i=1}^{n} x(\tau_i) d\boldsymbol{\tau}_n \triangleq \int_{-\infty}^{\infty}\int_{-\infty}^{\infty} \cdots \int_{-\infty}^{\infty} h_n(\tau_1, \tau_2, \dots, \tau_n)$$
$$\times\, x(t-\tau_1)x(t-\tau_2)\cdots x(t-\tau_n) d\tau_1 d\tau_2 \cdots d\tau_n$$

where $h_n(\boldsymbol{\tau}_n)$ is the nth-order Volterra *kernel* of the system and $\boldsymbol{\tau}_\mathrm{n}$ is a vector of delays. In the following, we will assume systems with no output if the input signal is zero $x(t) = 0$, hence $h_0 = 0$.

For causal systems, the lower integration limit is zero, and for finite memory systems, the upper limit is bounded. If the system is memoryless, $\boldsymbol{\tau}_n = \mathbf{0}$ for all n and (2.1) reduces to a Taylor series expansion.

The integral (2.2) can be viewed as the extension of the response of a linear system to the case of nonlinear systems, and the kernel $h_n(\boldsymbol{\tau}_n)$ can be named the *nonlinear impulse response* of order n. Since the kernels $h_n(\boldsymbol{\tau}_n)$ are not dependent on time t, the Volterra series is a good choice to describe *nonlinear time-invariant* systems.

The n-dimensional Fourier transform of the multidimensional nth-order kernel, $H_n(\omega_1, \omega_2, \ldots, \omega_n)$, is referred to as the *nonlinear transfer function* of order n, and using a compact notation similar to (2.2), can be expressed as the multidimensional Fourier transform of the Volterra kernel, given by

$$H_n(\boldsymbol{\omega}_n) = \int_{-\infty}^{\infty} h_n(\boldsymbol{\tau}_n) e^{-j\boldsymbol{\omega}_n^T \boldsymbol{\tau}_n} d\boldsymbol{\tau}_n, \tag{2.3}$$

with $\boldsymbol{\omega}_n^T \boldsymbol{\tau}_n = \omega_1\tau_1 + \omega_2\tau_2 + \cdots + \omega_n\tau_n$. The vector of angular frequencies $\boldsymbol{\omega}_\mathrm{n}$ is indicated in bold. For a memoryless system, the nonlinear transfer function is frequency independent $H_n(\boldsymbol{\omega}_n) = H_{0n}$.

Making use of the inverse Fourier transform, the impulse response of order n can be written as

$$h_n(\boldsymbol{\tau}_n) = \frac{1}{(2\pi)^n} \int_{-\infty}^{\infty} H_n(\boldsymbol{\omega}_n) e^{j\boldsymbol{\omega}_n^T \boldsymbol{\tau}_n} d\boldsymbol{\omega}_n. \tag{2.4}$$

Substituting (2.4) into (2.2) and executing the multiple integrals on $\boldsymbol{\tau}_n$, we obtain the nth-order term of the Volterra series now expressed by means of products of the input spectrum $X(\omega)$ in the integrand,

$$y_n(t) = \frac{1}{(2\pi)^n} \int_{-\infty}^{\infty} H_n(\boldsymbol{\omega}_n) \prod_{i=1}^{n} X(\omega_i) e^{j\omega_i t} d\omega_i. \tag{2.5}$$

Considering that the term (2.5) is the same for any permutation of the ω_i, from now on, we assume that the nonlinear transfer functions are symmetric with respect to their arguments, that is,

$$H_n(\omega_1, \ldots, \omega_r, \ldots, \omega_s, \ldots, \omega_n) = H_n(\omega_1, \ldots, \omega_s, \ldots, \omega_r, \ldots, \omega_n), \tag{2.6}$$

under any permutation of their arguments ω_r and ω_s. In case that the transfer function needs to be explicitly symmetrized, it will be designated as $\overline{H_n(\boldsymbol{\omega}_n)}$ (Bussgang et al., 1974).

The Volterra series was first proposed for the analysis of nonlinear circuits in a pioneering paper of Wiener and some years later Narayanan applied this approach to evaluate the distortions in a nonlinear transistor model (Wiener, 1942; Narayanan, 1967). More work was additionally dedicated to Volterra series analysis (Schetzen, 1981; Boyd and Chua, 1985). Assuming a mild excitation, the analysis was based on the nonlinear equivalent circuit of the common-emitter amplifier with analytic nonlinearities and demonstrated the usefulness of the Volterra series for investigating the frequency-dependent nonlinear behavior of the device under study.

In line with a neo-traditionalist position, we believe that power amplifiers can still be adequately modeled in a wide range of practical scenarios using the Volterra series.

2.2.1 Properties of the Volterra Series

It is opportune to review the Volterra series properties, so we next briefly discuss some remarkable characteristics like convergence, homogeneity, linearity, and memory, including also a short consideration of its application to complex-valued systems.

2.2.1.1 Convergence

Focusing on nonlinear devices and circuits for wireless communications, the main contribution of the Volterra series to the practical problem of distortions analysis is its ability to describe the dynamic behavior of the nonlinear system at higher frequencies. The application of this fundamental tool is only possible if the conditions for its existence are satisfied. In the case of devices with nonlinearities that are not infinitely differentiable (non-analytic), we know that the Volterra series do not converge.

Keeping to the subject of memoryless systems, three examples of real-valued nonlinearities with a discontinuous derivative are represented with solid lines in Figure 2.1. The non-analytic function in Figure 2.1(a) represents an ideal saturator with a discontinuity at the origin. The second example of Figure 2.1(b) represents a simple approximation to the I–V characteristic of a FET and shows a function with two singular points where the first derivative is discontinuous. The third example, Figure 2.1(c), stands for an envelope detector with input RF signal x and output $|x|$, which is a non-analytic function with a first derivative discontinuous at the origin. Therefore, a system with any of these nonlinearities can not be described by a polynomial series.

It is widely considered that the major drawback of Volterra series consists in its inability to manage distortions analysis in systems of this type. Nevertheless, we can circumvent this difficulty by approximating the nonlinearity with an infinitely differentiable analytic function. To illustrate this point, let us consider the envelope detector of Figure 2.1(c) with the non-analytic characteristic $|x| = \sqrt{x^2}$, differentiable elsewhere except in the isolated singularity at $x = 0$. Including a negligible parameter $\epsilon > 0$, it is possible approximate the envelope with the differentiable nonlinearity $\sqrt{x^2 + \epsilon}$, which is an analytic function of

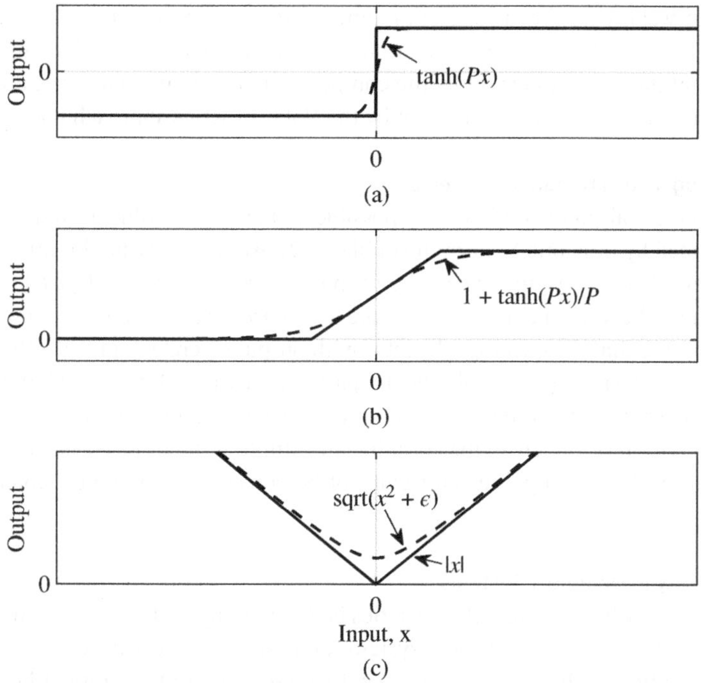

Figure 2.1 Nonlinearities with discontinuous derivatives (solid line) and smooth approximations (dashed lines). (a) Ideal saturator, (b) I–V FET characteristic, (c) Envelope detector (real-valued x).

x^2. In that case, the new nonlinearity can be approximated by a Taylor series with even order terms $x^2, x^4, x^6, \ldots$. Therefore, the envelope $|x|$ can be formally approximated by a Volterra series operator in the limit when $\epsilon \to 0$.

The above example reveals the second weakness of the Volterra series: even though the envelope detector output can be expressed as a polynomial series, the number of necessary terms is excessively high for an acceptable description accuracy. Moreover, for this particular case, computing the approximated Taylor series exceeds the numerical cost of calculating $|x|$ from x by simply changing the sign for $x < 0$.

Although in a simple memoryless approach the piecewise-linear characteristic of the I–V characteristic of Figure 2.1(b) can be helpful to deal with nonlinear systems, the implicit discontinuity discards its implementation in a Volterra series analysis. Fortunately, the nonlinearities of real-world devices do not present that discontinuous properties, making a theoretical Volterra series approach possible. After replacement of the piecewise linear I–V characteristic of Figure 2.1(b) with a hyperbolic tangent function, the Taylor series around $x = 0$, truncated to a reasonable number of terms, provides a satisfactory approximation. We expect only a few terms in weakly nonlinear operation, and an increasing number of terms as the system is driven toward the stronger nonlinear saturation region. Here, we can remark that the limitation is set by the computational inability to manage an excessively large number of terms, and not by the Volterra series approach itself.

2.2.1.2 Homogeneous Nonlinear Systems

From a mathematical point of view, it is possible to realize a nonlinear system $\hat{S}_n[x(t)]$ described by only one Volterra term like (2.2). After modifying the input by $Ax(t)$, where A is a constant scalar, the output is given by $y(t) = \hat{S}_n[Ax(t)] = A^n\hat{S}_n[x(t)]$. Then, the system $\hat{S}_n[x(t)]$ is denoted a *homogeneous* nonlinear system of order n. Observe that the envelope detector with output given by $|x|$ fulfills the condition $y(t) = |Ax(t)| = A|x(t)|$ only for the particular case of A being real and positive. Moreover, recalling the discussion in the previous paragraph, the Taylor series approximation of the envelope contains an infinite number of even order terms. Therefore, the envelope detector can not be considered a homogeneous system.

2.2.1.3 Linearity in Nonlinear Systems

A very attractive attribute of the Volterra series is the linearity with respect to the kernels, though it is a tool for nonlinear systems analysis. The linearity is evident from the convolution multiple integral (2.2), which gives rise to the simple filter interpretation in the case of the linear term, $n = 1$. The straightforward generalization of this view to a nth-order nonlinear term is clear considering a linear filter with a nonlinear impulse response $h_n(\tau_n)$ and an input given by the product of the delayed values $x(t-\tau_1)x(t-\tau_2)\cdots x(t-\tau_n)$. Linearity is a crucial property in the wide popularity that the Volterra series have in the implementation of power amplifiers behavioral models and digital predistorters, as the reader will have the opportunity to see in the following chapters.

2.2.1.4 Memory and Memoryless Systems

The Volterra series is a beneficial tool for the analysis of a nonlinear power amplifier based on its equivalent circuit model. As it was commented previously, if the memory length of a system is zero, $\boldsymbol{\tau_n} = \mathbf{0}$ for all n and (2.1) reduces to the Taylor series expansion

$$y(t) = c_1x(t) + c_2x^2(t) + c_3x^3(t) + \cdots. \tag{2.7}$$

This memoryless expression serves to model an amplifier only in the case that its equivalent circuit does not include memory components like capacitances or inductances, so that the coefficients c_k are real-valued. Consequently, a *strictly-memoryless* amplifier driven by a single tone has no AM–PM conversion, as it will be deduced below.

Actually, the strictly-memoryless model is an abstract concept because the equivalent circuit of a real power amplifier does include, at least, capacitances and inductances. Even at low frequencies, these linear components with memory introduce a phase deviation in the output, and the amplifier with a single tone input can only be described by a series like (2.7) if phase shifts θ_n, dependent on the order n, are included in the terms $x^n(t)$.

The first observation that we can highlight in the model of our amplifier is that the output signal at a certain instant does not depend on the input signal at past time instants. A second consideration is that the phase distortion of the output of this *quasi-memoryless* model leads to an AM–PM characteristic with phase conversion. Observe that the memory of this model is fully revealed as the amplifier is driven by signals with higher frequencies and larger bandwidths, evidencing the need of a complete Volterra series model.

The presence of thermal and charge-trapping effects in the transistors of the amplifier as well as the time constants of the components in the biasing networks, produce additional memory effects referred to as *long-term memory*, to differentiate from the *short-term memory* that was discussed above.

2.2.1.5 Volterra Series for Complex-valued Systems

The Volterra series (2.1)–(2.2) is a theoretical tool limited to the analysis of real-valued systems and cannot be applied directly to a system with complex-valued kernels or input signals. However, that is precisely the case present in many problems we face in the analysis of wireless communication systems using the baseband equivalent model. To overcome that difficulty, some papers have been published to demonstrate how the Volterra series can be adapted to analyze complex-valued systems, a topic that we will discuss in the last section of this chapter.

2.3 Volterra Series Applied to RF Amplifier Modeling

The object of this section is to apply the Volterra series to the analyses of systems driven by sine wave signals, highlighting the convenience of this mathematical tool. In a communications system, the input is typically amplified and some nonlinear distortions are generated in the process. As it was mentioned in Chapter 1, measurements of the output signal magnitude and phase for a one-tone input power sweep or intermodulation distortion for two-tone inputs are customary tests that can be compared with the theoretical results. Although these are relatively simple problems, the discussion will aid to familiarize the reader with the implementation of Volterra series to situations of more complexity to be studied in the rest of the book.

The Volterra series approach has the merit that these problems are treated following a procedure in which the nonlinear transfer functions $H_n(\boldsymbol{\omega}_n)$ are first computed and then utilized in the standard expressions to evaluate AM–AM and AM–PM characteristics, two-tone intermodulation, harmonic levels, etc. Observe that this feature of a Volterra analysis gives the freedom to establish a connection between the transfer functions and the parameters of the circuit under study, an option unlikely with other techniques, as the neural networks, for instance. For now, let us assume an amplifier, with known transfer functions $H_n(\boldsymbol{\omega}_n)$ and driven by an input signal $x(t)$ given by a single sine wave.

2.3.1 Response of an Amplifier to a Single Sine Wave

The first case we will examine is the basic amplifier of Figure 2.2(a). The study assumes a FET with the simplified equivalent circuit shown in Figure 2.2(b), composed of a current source I_d controlled by the gate-to-source voltage V_{gs}. Figure 2.2(c) displays the normalized "ideal strongly nonlinear" characteristic using a continuous broken line. In this piecewise linear I–V characteristic, the drain current I_d is zero for input voltages V_{gs} below the cutoff point $V_t = 0$, and it is perfectly linear for higher voltages until an abrupt saturation of $I_{\max} = 1$ is reached at $V_{gs} = 1$. This broken characteristic does not seem appropriate to analyze the circuit in a Volterra series approach, but has been successfully employed in textbooks to deduce some practical formulas as a guide to evaluate amplifiers performance. We address here a more realistic evaluation of the amplifier output to exemplify the alternative of an elementary Volterra approach to base an amplifier design on.

To comply with the customary Volterra series notation, in this part, we employ $x(t)$ and $y(t)$ to denote the amplifier input and output AC voltages, respectively. Assuming a load impedance $Z_L(\omega) = R_L$, and referring to the quiescent bias gate-to-source voltage as V_q and quiescent drain current as I_q, we obtain the expressions $x(t) = V_{gs} - V_q$ and $y(t) = (I_d - I_q)R_L$ for the amplifier input and output, respectively.

If we suppose an input given by a single tone input signal $x(t) = A\cos(\omega_c t + \varphi)$, the Fourier transform is a sum of two delta functions $X(\omega) = \pi\left[\mathcal{A}\delta(\omega - \omega_c) + \mathcal{A}^*\delta(\omega + \omega_c)\right]$, where we have used the phasor notation $\mathcal{A} = Ae^{j\varphi}$. Substituting in (2.5), the nonlinear term is

$$y_n(t) = \frac{1}{2^n}\int_{-\infty}^{\infty} H_n(\boldsymbol{\omega}_n)e^{j\overline{\omega}_n t}\prod_{i=1}^{n}\left[\mathcal{A}\delta(\omega_i - \omega_c) + \mathcal{A}^*\delta(\omega_i + \omega_c)\right] d\omega_i, \tag{2.8}$$

with $\overline{\omega}_n = \omega_1 + \omega_2 + \cdots + \omega_n$. When the integrals with the delta functions are carried out, the frequencies are restricted to be $\omega_i = \omega_c$ or $\omega_i = -\omega_c$ so that for a given

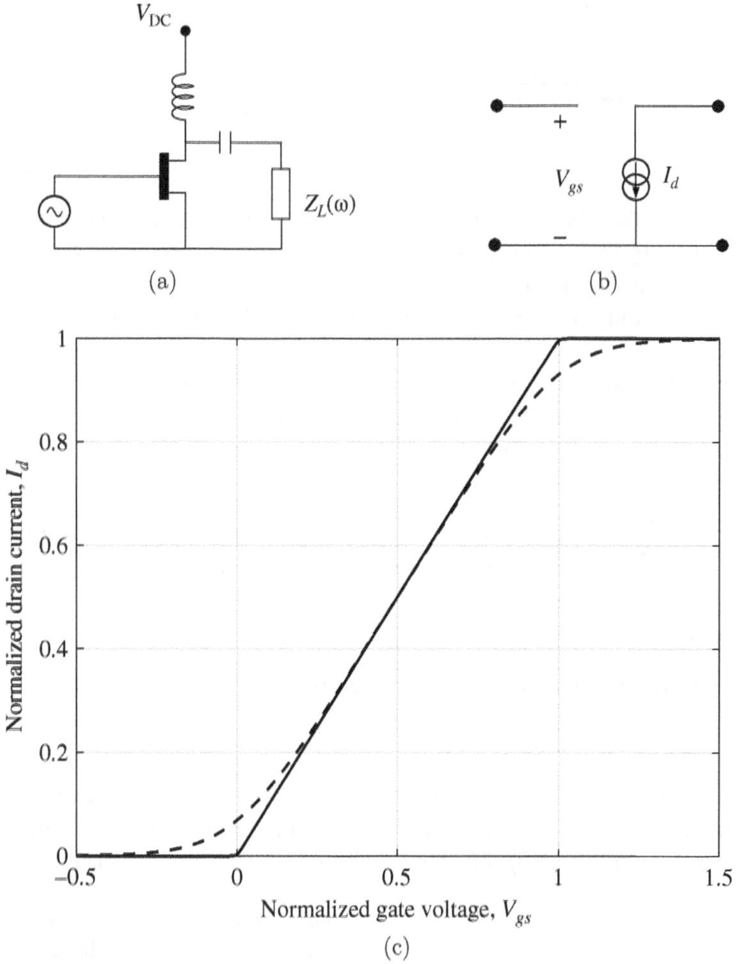

Figure 2.2 (a) Elementary power amplifier circuit, (b) Simple FET model, (c) Normalized FET transfer characteristics: ideal "strongly" nonlinear model (solid line) and realistic nonlinear model (dashed line).

combination in which p frequencies are equal to ω_c and the remaining $n - p$ equal to $-\omega_c$, the result is

$$\frac{1}{2^n}\mathcal{A}^p(\mathcal{A}^*)^{n-p}H_n(\boldsymbol{\omega}_n)e^{j\omega t}, \tag{2.9}$$

with a frequency $\omega = (n - 2p)\omega_c$ ranging from $n\omega_c$ for $p = 0$ to $-n\omega_c$ for $p = n$. Assuming $H_n(\boldsymbol{\omega}_n)$ symmetrical, there are $n!/[p!(n-p)!]$ permutations of the same set of frequencies ω_i with identical values that can be grouped together allowing to write the nth-order output as

$$y_n(t) = \left(\frac{A}{2}\right)^n \sum_{p=0}^{n} \frac{n!}{p!(n-p)!} H_n(\underbrace{\omega_c, \dots, \omega_c}_{n-p \text{ times}}, \underbrace{-\omega_c, \dots, -\omega_c}_{p \text{ times}}) e^{j(n-2p)(\omega_c t+\varphi)}. \tag{2.10}$$

It becomes evident that the nonlinear term $y_n(t)$ is real-valued because the expression is constituted by the sum of pairs of conjugate components producing the output sine wave signals at new harmonic frequencies $2\omega_c, 3\omega_c, \dots, n\omega_c$. Figure 2.3 sketches the amplitude response of an amplifier at fundamental and harmonic frequencies.

Focusing on the response at the fundamental frequency ω_c of a (quasi-) memoryless amplifier with constant transfer functions $H_n(\omega_n) = |H_n|e^{j\theta_n}$, we observe that only odd-order terms $n = 2p - 1$ contribute to the output

$$y(t)|_{\omega_c} = \sum_{\substack{n=1 \\ (n \text{ odd})}}^{\infty} \Re\left\{\frac{1}{2^{n-1}} \frac{n!}{\left(\frac{n+1}{2}\right)!\left(\frac{n-1}{2}\right)!} A^n |H_n| e^{j(\omega_c t+\varphi+\theta_n)}\right\}. \tag{2.11}$$

Without loss of generality, we can set $\varphi = 0$ and the leading terms give

$$\begin{aligned} y(t)|_{\omega_c} &= |H_1|A\cos(\omega_c t + \theta_1) + \frac{3}{4}|H_3|A^3\cos(\omega_c t + \theta_3) \\ &\quad + \frac{5}{8}|H_5|A^5\cos(\omega_c t + \theta_5) + \cdots . \end{aligned} \tag{2.12}$$

The amplitude and phase of $y(t)|_{\omega_c}$ are determined by nonlinear relations to the input amplitude A. For small signal amplitude, the nonlinear terms in (2.12) can be neglected and the amplifier is operated in linear mode. As the input level becomes

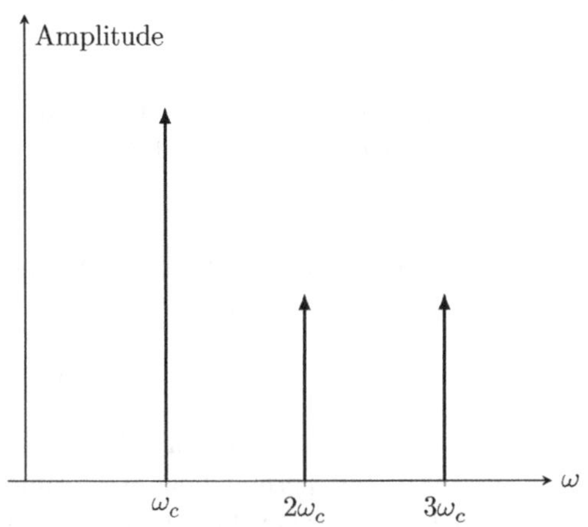

Figure 2.3 Amplitude response of an amplifier at fundamental and harmonic frequencies.

greater, the nonlinear terms get larger significance, and the increasing distortion of the response deviates the amplifier from the linear operation, so that the output amplitude is no longer proportional to A and the output phase experiences a shift with respect to the input phase (Maas, 2003).

The response (2.12) explains the experimental amplitude and phase behavior observed in the output of actual amplifiers, in particular, the AM–AM and AM–PM conversion characteristics described in Chapter 1. On the other hand, in a fictitious amplifier with real-valued transfer functions $H_n(\omega_n)$, the parameters $\theta_n = 0$ for all n, and $y(t)|_{\omega_c}$ is strictly memoryless with no phase shift with respect to $x(t)$, that is, there is no AM–PM conversion. As we will see in the following chapter, the amplifier memory is completely revealed when the input signal is modulated and has a large bandwidth.

The amplitude and phase characterization measured in an experimental setup, where a single-tone input signal is swept in amplitude, provides a series of practical data that serves to construct the best fitted AM–AM and AM–PM curves. Although the nonlinear relationship (2.12) can be modeled by adjusting the parameters of an appropriate nonlinear function, we can exploit the fact that the involved problem is linear with respect to the unknown coefficients and the curve fitting can be addressed using a polynomial regression. Another virtue of the Volterra series approach is its inherent nexus with the amplifier circuit-model, making feasible to deduce first the transfer functions $H_n(\omega_n)$ and then substitute in (2.12).

2.3.1.1 Volterra-based, Yet Simple, Analysis of Conventional Amplifier Modes

In revisiting, the conventional analysis of amplifier modes, like class A or class B configurations, based on an ideal strongly nonlinear I–V FET characteristic (Cripps, 1999), we address a Volterra series perspective with a FET modeled as a voltage-controlled current source in which the ideal piecewise-linear characteristic is approximated by

$$I_d = \frac{I_{\max}}{2}\left[1 + \frac{1}{P}\ln\left(\frac{\cosh(\alpha P V_{gs})}{\cosh[\alpha P(V_{gs} - V_{\text{sat}})]}\right)\right], \tag{2.13}$$

where I_d and V_{gs} are the drain current and the gate-to-source voltage, as previously defined.

In the analysis, the drain current is normalized with $I_{\max} = 1$ mA, $V_{\text{sat}} = 1$ V and $\alpha = 1\,\text{V}^{-1}$ as it is plotted in Figure 2.2(c) with a dashed line for a finite P. The current shows a quasi-perfect linear behavior in the middle range with a gradual cutoff and a similar gradual saturation. Notice that (2.13) is an absolutely differentiable function from which the ideal cutoff and hard saturation nonlinear characteristic is obtained in the limit as $P \to \infty$. In a way similar to Cripps (1999), we can define the classical modes of operation according to Table 2.1.

Table 2.1 Amplifier modes of operation (normalized model).

Mode	DC bias point V_q	Signal amplitude A	Load impedance R_L
Class-A	0.5	0.5	1
Class-B	0	1	1
Class-C	<0	>1	R_{opt}

2.3.1.2 Class-A Mode

In class-A mode operation, the amplifier is biased with a DC voltage V_q in the midpoint of the *quasi-linear* region between cutoff and saturation. Considering the realistic transfer characteristic of the FET, the drain current (2.13) can be approximated with a Taylor series as

$$I_d = I_q + g_1 x + g_2 x^2 + g_3 x^3 + \cdots, \tag{2.14}$$

where I_q and g_n are given by the current and its derivatives evaluated at the quiescent voltage V_q, and $x = V_{gs} - V_q$. At this point, the reason for having selected the expression (2.13) for the drain current is evident because the computation of the ln [cosh(x)] function derivatives is immediate with closed-form expressions. For example, the first derivative is tanh(x), the second derivative is $1 - \tanh^2(x)$, and so on.

Figure 2.4 shows the dependence of the normalized drain current and g_n parameters on the input voltage for $P = 10$. Observe that, if $x = 0$, the input voltage $V_{gs} = V_q$ is equal to the quiescent bias voltage. In particular, the dependence is on the quiescent voltage V_q when $x(t) = 0$. We observe that the selected bias voltage $V_q = 0.5$ of this class-A amplifier has the distinct property that $g_1 = 1$, small values for all g_n with odd $n > 1$, and $g_n = 0$ for coefficients with n even. Compared to the linear response of the strongly nonlinear characteristic of Figure 2.2(c) to a sinusoidal drive signal $x(t)$ with an amplitude inside the range between cutoff and saturation, this model will generate an output current without even-harmonics and minor, but appreciable, odd-harmonics content. Substituting in (2.12) and focusing on the fundamental frequency ω_c, we obtain for the output signal $y(t)$ the general expression[2]

$$\begin{aligned} y(t) &= g_1 R_L A \cos(\omega_c t) + \frac{3}{4} g_3 R_L A^3 \cos(\omega_c t) \\ &\quad + \frac{5}{8} g_5 R_L A^5 \cos(\omega_c t) + \cdots, \end{aligned} \tag{2.15}$$

where a minus sign due to the 180° phase change has been omitted.

2 Recall that in this strictly-memoryless amplifier, $\theta_n = 0$ for all n.

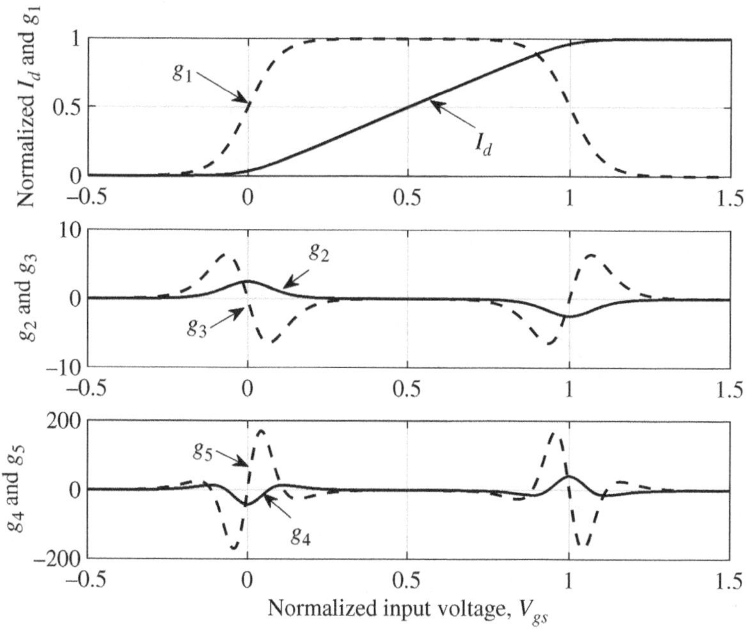

Figure 2.4 Dependence of the drain current I_d (solid line) and parameters g_n on the normalized input voltage V_{gs} ($P = 10$) (g_1, g_3, and g_5 are represented with dashed lines). With no AC input voltage, $x = 0$ and the figure shows the dependence on the normalized quiescent voltage $V_{gs} = V_q$.

Since $g_1 R_L = 1$, an input tone with small amplitude $A < 0.5$ will produce a nearly sinusoidal output with the same amplitude. As the signal level is increased, the nonlinear distortions of the response at ω_c, as well as the signal strength at other harmonics, become more significant. Focusing on the fundamental frequency, the Volterra-based AM–AM conversion characteristic of the class-A amplifier has been plotted in Figure 2.5. To compare, the dashed line corresponding to the idealized piecewise linear transfer expression is also shown clearly demonstrating the 1-dB compression point at an input of about 0 dB.

2.3.1.3 Class-B Mode

Another particular operation mode is for the amplifier biased with a quiescent voltage $V_q = 0$ near cutoff, denoted class-B mode. This bias condition coincides with a point with $g_1 = 0.5$ and negligible g_3 and g_5 (see the corresponding dashed lines in Figure 2.4), suggesting a highly linear behavior if we assume an impedance load which is a conceptual short at even harmonics. For this class-B mode, $g_1 R_L = 0.5$ and the gain is −6 dB with respect to the class-A amplifier

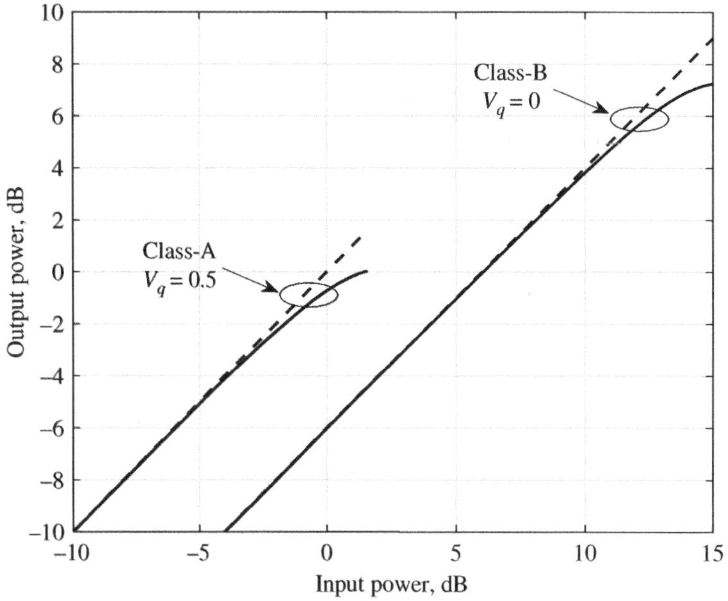

Figure 2.5 Intput and output RF power for amplifiers operating in class-A and class-B modes.

gain. On the other hand, the linear operation range is extended beyond the 1-dB compression point of the class-A amplifier.

2.3.2 Determining Nonlinear Transfer Functions

Equation (2.5) enables the direct evaluation of the nth-order Volterra terms as a function of the input spectrum $X(\omega)$ once the nonlinear transfer functions $H_n(\omega_n)$ are known. Then, the first issue we face is how these transfer functions can be deduced. To answer this question, we address the *nonlinear currents* and the *harmonic input* methods (Bussgang et al., 1974).

2.3.2.1 Nonlinear Currents Method

The *nonlinear currents* method serves to calculate the response of a circuit with a nonlinear current source given by a Taylor series, for example, (2.14). We consider first the application of this method to an amplifier with the elementary circuit model shown in Figure 2.6(a), for which the gate capacitance C_g has been included. This equivalent circuit is formed by the gate and the drain subcircuits with two useful particularities. Firstly, the subcircuits are decoupled and can be analyzed separately, and secondly, the gate subcircuit is linear.

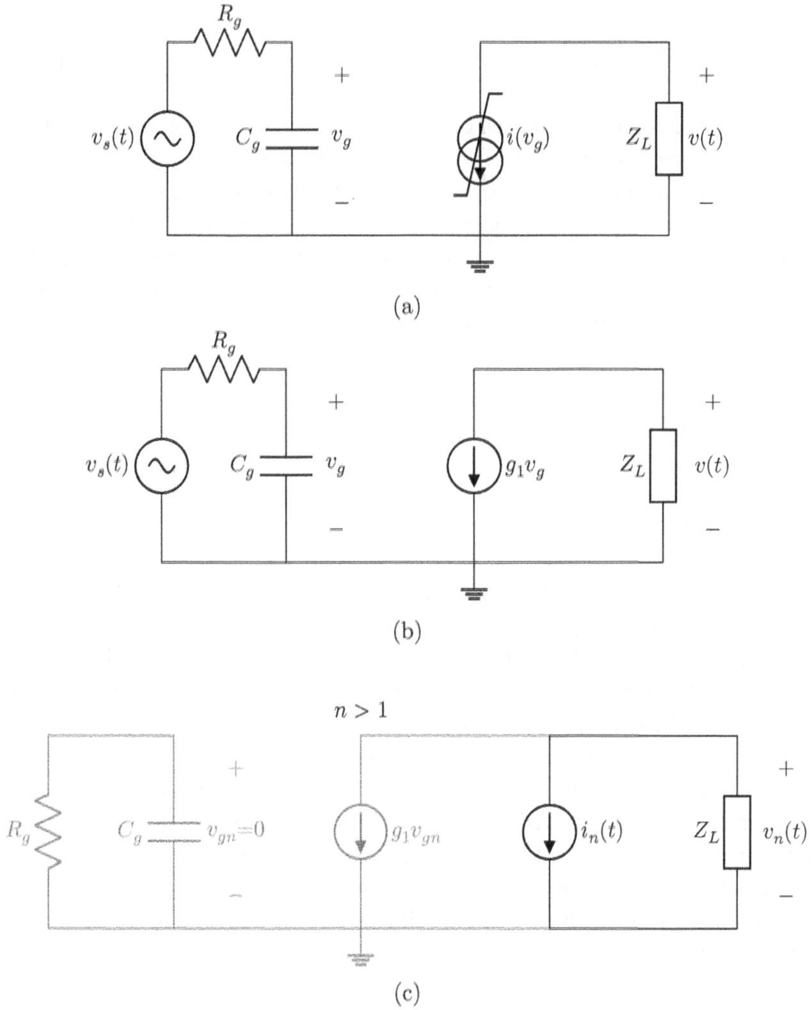

Figure 2.6 (a) Equivalent nonlinear circuit for an amplifier, (b) the associated linear circuit, valid to compute v_1, and (c) the associated linear circuit to compute v_n.

The network analysis is initiated by recalling that the drain nonlinear current source is a transfer nonlinearity—see Figure 1.8(c)—dependent on the voltage across the gate capacitor v_g as the only nonlinear element of the amplifier. In that case, the transfer nonlinearity expression (1.16) is directly applied, and the AC drain current $i(v_g)$ in Figure 2.6(a) can be written as:

$$i(v_g) = \sum_{n=1}^{\infty} g_n v_g^n(t) = g_1 v_g(t) + i_{NL}(t), \tag{2.16}$$

where

$$i_{NL}(t) = \sum_{n=2}^{\infty} g_n v_g^n(t). \tag{2.17}$$

Observe that the voltage-controlled current source $i(v_g)$ can be seen as constituted by the linear current source $g_1 v_g(t)$ and a nonlinear component $i_{NL}(t)$. To form the so-called *associated linear circuit* of Figure 2.6(b), we assume the nonlinear component is removed, i.e., $i_{NL}(t) = 0$, and the actual linear elements of Figure 2.6(a) are extended with the linear current source controlled by the gate voltage $g_1 v_g(t)$.

If the input is given by the voltage source $v_s(t)$, the nonlinear currents method demonstrates that the node voltages $v_g(t)$ and $v(t)$ can be expressed, respectively as

$$v_g(t) = \sum_{n=1}^{\infty} v_{gn}(t), \tag{2.18}$$

and

$$v(t) = \sum_{n=1}^{\infty} v_n(t), \tag{2.19}$$

where $v_{gn}(t)$ and $v_n(t)$ are the nth-order components of $v_g(t)$ and $v(t)$, respectively. Here we summarize the procedure to consecutively compute these components following a network analysis of the two-node associated linear circuit. For the interested reader, a more detailed exposition of the nonlinear currents method can be consulted in Bussgang et al. (1974).

At first, we compute the voltages $v_g = v_{g1}$ and $v = v_1$, which are the responses of the associated linear circuit of Figure 2.6(b) as if the nonlinear source $i_{NL}(t)$ were "disconnected". The analysis is straightforward thanks to the fact that the gate and drain branches are decoupled and can be considered separately.

In the next step, the second-order nonlinear terms $v_{g2}(t)$ and $v_2(t)$ are calculated. Once the linear voltages have been specified, (2.18) is substituted in (2.17) and the terms with the same order n are grouped in a current through the transfer nonlinearity denoted as $i_n(t)$, resulting in

$$i_{NL}(t) = \sum_{n=2}^{\infty} i_n(t). \tag{2.20}$$

Since in this elementary example the gate subcircuit is linear and $v_{gn} = 0$ for $n \geq 2$, equation (2.18) simplifies to $v_g = v_{g1}$, and the second-order nonlinear current $i_2(t)$ can be written as

$$i_2(t) = g_2(v_{g1}(t) + v_{g2}(t) + \cdots)^2 = g_2 v_{g1}^2(t). \tag{2.21}$$

Now, the second-order terms $v_{g2}(t)$ and $v_2(t)$ can be calculated by short-circuiting the input source $v_s(t)$ and connecting the current source $i_2(t)$ to the nonlinear port, i.e., $i_n(t) = i_2(t)$ in Figure 2.6(c).

Following the procedure, the third-order nonlinear term $v_3(t)$ is deduced by repeating the analysis of the associated linear circuit driven now by the third-order nonlinear current

$$i_3(t) = 2g_2 v_{g1}(t) v_{g2}(t) + g_3 v_{g1}^3(t) = g_3 v_{g1}^3(t). \tag{2.22}$$

Note that $v_{g1}(t)v_{g2}(t)$ is a third-order term given by the product of linear and second-order terms. Applying recursively the method, the successive nonlinear terms are computed in a similar fashion.

A helpful recurrent formula valid for a general nonlinear conductance and a transfer nonlinearity with a dependence $i[v(t)]$ is

$$i_n(t) = \sum_{m=2}^{n} g_m v_{m,n}, \tag{2.23}$$

where

$$v_{m,n} = \sum_{i=1}^{n-m+1} v_i(t) v_{m-i,n-1}, \tag{2.24}$$

and

$$v_{m,1} = v_m(t). \tag{2.25}$$

Nonlinear currents i_n for $n = 2, 3, 4$, and 5 are shown in Table 2.2.

2.3.2.2 Harmonic Input Method

The nonlinear transfer functions still remain to be derived. Fortunately, the *harmonic input* method is a procedure to find the solution to this problem (Bussgang et al., 1974).

To calculate the transfer functions with the harmonic input method, the input signal in (2.2) is supposed to be a sum of exponentials

$$x(t) = e^{j\omega_1 t} + e^{j\omega_2 t} + \cdots + e^{j\omega_n t}, \tag{2.26}$$

Table 2.2 Nonlinear currents $i_n(t)$.

$i_n(t)$	g_2	g_3	g_4	g_5
$i_2(t)$	v_1^2			
$i_3(t)$	$2v_1v_2$	v_1^3		
$i_4(t)$	$2v_1v_3 + v_2^2$	$3v_1^2v_2$	v_1^4	
$i_5(t)$	$2v_1v_4 + 2v_2v_3$	$3v_1^2v_3 + 3v_1v_2^2$	$4v_1^3v_2$	v_1^5

where, the ω_i are incommensurate.[3] Notice that no actual measurements are possible with this analytic (not real) signal. After substitution in (2.5) and (2.1), the method demonstrates that there is a term of order n in the output $y(t)$ given by

$$n!\overline{H_n(\boldsymbol{\omega}_n)}e^{j(\omega_1+\omega_2+\cdots+\omega_n)t}, \tag{2.27}$$

where $\overline{H_n(\boldsymbol{\omega}_n)}$ is the symmetrized nonlinear transfer function. In other words, it means that the symmetrized nonlinear transfer function is the coefficient of the system output term $n!e^{j(\omega_1+\omega_2+\cdots+\omega_n)t}$. In the following reasoning, $H_n(\boldsymbol{\omega}_n)$ is assumed symmetrized even if not explicitly indicated.

This result suggests a procedure in which the input signal of the nonlinear system is first a single exponential to determine $H_1(\omega)$, then a sum of two exponentials is applied and $H_2(\boldsymbol{\omega}_2)$ is deduced as a function of $H_1(\omega)$. If this recursion procedure is generalized to the input of n exponentials (2.26), $H_n(\boldsymbol{\omega}_n)$ can be formulated in terms of the lower-order transfer functions $H_1(\omega), H_2(\boldsymbol{\omega}_2), \ldots, H_{n-1}(\boldsymbol{\omega}_{n-1})$.

In order to exemplify the procedure, we apply this method to our simple amplifier circuit of Figure 2.6(a). If we relate $v_s(t)$ and $v(t)$ with the input and output signals, $x(t)$ and $y(t)$, respectively, the method in the frequency domain can be summarized as follows. Denoting $F[\cdot]$ the Fourier transform operation, $V_g(\omega) = F[v_g(t)]$ and $V(\omega) = F[v(t)]$ are the Fourier transforms of the gate and output voltages, respectively, and the Fourier transform of the generic input signal (2.26) is the sum of n delta functions

$$V_s(\omega) = F[v_s(t)] = 2\pi\delta(\omega - \omega_1) + 2\pi\delta(\omega - \omega_2) + \cdots + 2\pi\delta(\omega - \omega_n). \tag{2.28}$$

In the frequency domain, the resulting equations for the equivalent nonlinear circuit are

$$\begin{aligned} (V_g - V_s)/R_g + j\omega C_g V_g &= 0 \\ F[i(t)] + V/Z_L(\omega) &= 0. \end{aligned} \tag{2.29}$$

The first probing input is a single exponential signal and the linear transfer functions are just the responses of the associated linear circuit of Figure 2.6(b). A straightforward analysis of this two-node linear circuit allows us to derive the node admittance matrix, given by:

$$\mathcal{Y}(\omega) = \begin{bmatrix} 1/R_g + j\omega C_g & 0 \\ g_1 & 1/Z_L(\omega) \end{bmatrix}. \tag{2.30}$$

The first-order transfer functions $H_{g1}(\omega_1)$ and $H_1(\omega_1)$ are the coefficients of $2\pi\delta(\omega - \omega_1)$ in V_g and V when $V_s = e^{j\omega_1 t}$. In that case, the matrix equation to be solved is

$$\begin{bmatrix} H_{g1}(\omega_1) \\ H_1(\omega_1) \end{bmatrix} = \mathcal{Y}^{-1}(\omega_1) R_g^{-1} \begin{bmatrix} 1 \\ 0 \end{bmatrix}. \tag{2.31}$$

3 Frequencies ω_r and ω_s are incommensurate if their ratio ω_r/ω_s is an irrational number.

Thus, the first-order transfer functions can be expressed as:

$$H_{g1}(\omega_1) = \frac{1}{1 + j\omega_1 C_g R_g}$$
$$H_1(\omega_1) = -g_1 H_{g1}(\omega_1) Z_L(\omega_1). \tag{2.32}$$

The gate capacitance and the load impedance $Z_L(\omega)$ introduce a phase component in the linear responses indicating that the system is not strictly memoryless. This is a very noteworthy difference between the real-valued polynomial model of the strictly-memoryless amplifier in Figure 2.2 and the more realistic response of this simple model.

Once the linear term is obtained, we can calculate the nonlinear current $i_2(t)$ and solve the equations of the associated linear circuit with the input source short-circuited, as in Figure 2.6(c). Repeating successively the procedure, the response of any order can be found with the same linear subcircuit. The nth-order nonlinear transfer functions $H_{gn}(\boldsymbol{\omega}_n)$ and $H_n(\boldsymbol{\omega}_n)$ are the coefficients of $n!\,2\pi\delta(\omega - \omega_1 \cdots - \omega_n)$ in $V_g(\omega)$ and $V(\omega)$ when the input source $v_s(t)$ is short-circuited and the associated linear circuit is driven by the nth-order nonlinear current $i_n(t)$, with the source voltage given by (2.28). In the case of $n \leq 5$, we can operate with the results of $i_n(t)$ shown in Table 2.2.

The spectral term of interest in $F[i_n(t)]$ is the coefficient of $n!\,2\pi\delta(\omega - \omega_1 \cdots - \omega_n)$, which is denoted by $F_n[i_n(t)]$. Therefore, the general equation is

$$\begin{bmatrix} H_{gn}(\boldsymbol{\omega}_n) \\ H_n(\boldsymbol{\omega}_n) \end{bmatrix} = \mathcal{Y}^{-1}(\omega_1 + \omega_2) \begin{bmatrix} 0 \\ -F_n[i_n(t)] \end{bmatrix}, \tag{2.33}$$

where

$$\mathcal{Y}^{-1}(\omega) = \frac{Z_L(\omega)}{1/R_g + j\omega C_g} \begin{bmatrix} 1/Z_L(\omega) & 0 \\ -g_1 & 1/R_g + j\omega C_g \end{bmatrix}. \tag{2.34}$$

In this particular power amplifier example, the drain subcircuit is decoupled to the input linear subcircuit, in which no nth-order responses are generated for $n \geq 2$, and this is reflected by the following result deduced from equation (2.33)

$$H_{gn}(\boldsymbol{\omega}_n) = 0, \qquad \text{for} \quad n \geq 2. \tag{2.35}$$

That being the case, we can concentrate on the output node transfer functions $H_n(\boldsymbol{\omega}_n)$ and look first into the Fourier transform of the elementary second-order nonlinear current $i_2 = g_2 v_{g1}^2$. Since the excitation is given by two exponentials, the spectrum is composed of three terms of which the relevant one is at $\omega_1 + \omega_2$, and $F_2[i_2] = g_2 \overline{H_{g1}(\omega_1) H_{g1}(\omega_2)}$ is the coefficient we are interested in. Although, we use the overline to indicate symmetrization, observe that $\overline{H_{g1}(\omega_1) H_{g1}(\omega_2)} = H_{g1}(\omega_1) H_{g1}(\omega_2)$, and it can be omitted in this and all similar cases. Substituting in equation (2.33), the second-order output transfer function is obtained in terms of the first-order function

$$H_2(\boldsymbol{\omega}_2) = -g_2 H_{g1}(\omega_1) H_{g1}(\omega_2) Z_L(\omega_1 + \omega_2). \tag{2.36}$$

The next step of the analysis repeats the procedure when the excitation is given by three exponentials, $e^{j\omega_1 t} + e^{j\omega_2 t} + e^{j\omega_3 t}$, and the associated linear circuit is driven by $i_3(t) = g_3 v_{g1}^3(t)$ computed with (2.22). The relevant coefficient of the Fourier transform is

$$F_3[i_3] = g_3 H_{g1}(\omega_1) H_{g1}(\omega_2) H_{g1}(\omega_3), \tag{2.37}$$

and the equation to be solved is

$$\begin{bmatrix} H_{g3}(\boldsymbol{\omega}_3) \\ H_3(\boldsymbol{\omega}_3) \end{bmatrix} = \mathcal{Y}^{-1}(\omega_1 + \omega_2 + \omega_3) \begin{bmatrix} 0 \\ -g_3 H_{g1}(\omega_1) H_{g1}(\omega_2) H_{g1}(\omega_3) \end{bmatrix}, \tag{2.38}$$

resulting in the third-order output nonlinear transfer function given by

$$H_3(\boldsymbol{\omega}_3) = -g_3 H_{g1}(\omega_1) H_{g1}(\omega_2) H_{g1}(\omega_3) Z_L(\omega_1 + \omega_2 + \omega_3). \tag{2.39}$$

Higher-order nonlinear transfer functions can be derived continuing the procedure, for which it is necessary to determine the spectral coefficient of interest in each step $F_n[i_n]$. A formula to calculate this coefficient for other terms of nonlinear currents was published in Bussgang et al. (1974) and some examples computed with this formula are shown in Table 2.3.

So, while the response of a strictly memoryless amplifier to a single tone excitation at ω_c is given by (2.15), the output signal of the simple amplifier with memory drawn in Figure 2.6(a) can be expressed as a Volterra series by substituting the nonlinear transfer functions $H_n(\boldsymbol{\omega}_n)$ in equation (2.11). For the single tone excitation, the series truncated to the 5th order is

$$\begin{aligned} y(t) = \Re \Big\{ & g_1 H_{g1}(\omega_c) Z_L(\omega_c) A e^{j(\omega_c t + \varphi)} + \frac{3}{4} g_3 H_{g1}^2(\omega_c) H_{g1}(-\omega_c) Z_L(\omega_c) A^3 e^{j(\omega_c t + \varphi)} \\ & + \frac{5}{8} g_5 H_{g1}^3(\omega_c) H_{g1}^2(-\omega_c) Z_L(\omega_c) A^5 e^{j(\omega_c t + \varphi)} + \cdots \Big\}. \end{aligned} \tag{2.40}$$

Table 2.3 Coefficient $F_n[i_n]$ for some terms of the nonlinear currents.

Nonlinear order	$i_n(t)$	$F_n[i_n]$
$n = 2$	$g_2 v_{g1}^2$	$g_2 \overline{H_{g1}(\omega_1) H_{g1}(\omega_2)}$
$n = 3$	$2g_3 v_{g1} v_{g2}$	$2g_3 \overline{H_{g1}(\omega_1) H_{g2}(\omega_2, \omega_3)}$
$n = 4$	$3g_4 v_{g1}^2 v_{g2}$	$3g_4 \overline{H_{g1}(\omega_1) H_{g1}(\omega_2) H_{g2}(\omega_3, \omega_4)}$
n	$g_n v_{g1}^n$	$g_n \overline{H_{g1}(\omega_1) \cdots H_{g1}(\omega_n)}$

The minus sign is omitted because it only means an overall phase change. Although this expression is only a short example of possible models with memory, it is able to explain the response phase shift and the experimental AM–PM conversion characteristic of the amplifier.

A remarkable advantage of this nonlinear representation is its inherent potential to theoretically analyze power amplifiers biased in linear class-A or class-B and even in nonlinear class-C modes. Taking into account that the I–V characteristic of the drain current source is smooth, the parameters g_n can be computed with the corresponding derivatives evaluated at the quiescent voltage. Of course, the practical series truncated to the 5th-order can be adequate to describe the behavior of a class-A, and also a class-B amplifier for a moderate level signal, but most likely it will present convergence problems if the signal drives the amplifier into the compression zone.

Equation (2.40) clearly reveals that the origin of memory in this model lies in the input and output matching networks, represented by $H_{g1}(\omega)$ and $Z_L(\omega)$, and according to the discussion in Chapter 1, they are a factor in the short-term memory effects. Observe that long-term memory effects caused by DC bias networks or by internal mechanisms, like electrothermal or charge trapping subnetworks, are not included in this representation suggesting the need for a better Volterra model with a more complex structure.

2.4 Volterra Series in the Frequency Domain

The approach developed in the previous sections is a frequency domain analysis based on the Volterra series, which is a functional expansion defined in the time domain. The technique was successfully applied to the description of the power amplifier behavior relating the time domain input and output signals and making use of the system nonlinear transfer functions, defined in the frequency domain.

Although most of the research publications on amplifier modeling with signals based on conventional digital modulations have been carried out employing essentially this traditional approach, other studies discussing Volterra series in the frequency domain have been also published (Bussgang et al., 1974; Lang and Billings, 1996, 1997). More recently, the appearance of new wireless standards with signals in orthogonal frequency division multiplexing (OFDM) format has led to think about the usefulness of an analysis biased toward the frequency domain (Allegue-Martínez et al., 2012; Brihuega et al., 2021).

The first step forward is the Fourier transform of the ordinary Volterra series equation (2.1), written as:

$$Y(\omega) = \sum_{n=1}^{\infty} Y_n(\omega), \tag{2.41}$$

where the nth-order output spectrum $Y_n(\omega)$ is expressed as a function of the input spectrum $X(\omega)$, derived by the result of Fourier transforming both sides of (2.5). Since the Fourier transform of $e^{j(\omega_1+\cdots+\omega_n)t}$ is $2\pi\delta(\omega - \overline{\omega}_n)$, where $\overline{\omega}_n = \omega_1 + \cdots + \omega_n$, the integration is constrained by the condition $\overline{\omega}_n = \omega$ and the nth-order term can be written as:

$$Y_n(\omega) = \frac{1}{(2\pi)^{n-1}} \int_{\overline{\omega}_n=\omega} H_n(\boldsymbol{\omega}_n) \prod_{i=1}^{n} X(\omega_i) d\omega_i. \tag{2.42}$$

At this point, it seems opportune to introduce the multispectral density function $Y_n(\boldsymbol{\omega}_n)$, defined as:

$$Y_n(\boldsymbol{\omega}_n) = H_n(\boldsymbol{\omega}_n) \prod_{i=1}^{n} X(\omega_i). \tag{2.43}$$

Observe that this multispectral density function can be formulated as an n-fold Fourier transform of the auxiliary multidimensional time function $y_n(\mathbf{t}_n) = y_n(t_1, t_2, \ldots, t_n)$, using the following expression

$$Y_n(\boldsymbol{\omega}_n) = \int_{-\infty}^{\infty} y_n(\mathbf{t}_n) e^{-j(\omega_1 t_1+\cdots+\omega_n t_n)} d\mathbf{t}_n, \tag{2.44}$$

so that

$$y_n(\mathbf{t}_n) = \frac{1}{(2\pi)^n} \int_{-\infty}^{\infty} Y_n(\boldsymbol{\omega}_n) e^{j(\omega_1 t_1+\cdots+\omega_n t_n)} d\boldsymbol{\omega}_n. \tag{2.45}$$

Referring again to the multispectral density function $Y_n(\boldsymbol{\omega}_n)$, and after substitution in (2.42), the nth-order output spectrum is written as:

$$Y_n(\omega) = \frac{1}{(2\pi)^{n-1}} \int_{\overline{\omega}_n=\omega} Y_n(\boldsymbol{\omega}_n) d\omega_i, \tag{2.46}$$

which can be understood as the integral of the multispectral density $Y_n(\boldsymbol{\omega}_n)$ subject to the constraint $\omega_1 + \cdots + \omega_n = \omega$.

After substitution of the aforementioned definitions, the Volterra series in the frequency domain reads:

$$Y(\omega) = \sum_{n=1}^{\infty} H_n(\boldsymbol{\omega}_n) \prod_{i=1}^{n} X(\omega_i). \tag{2.47}$$

Using the basic time-domain memoryless Taylor series (2.7) of a hypothetical strictly-memoryless amplifier, and the Volterra series (2.40), which represents the simple amplifier with memory of Figure 2.6, we can easily illustrate how the analysis in the frequency domain contributes to gain an insight into amplifiers modeling.

To exemplify, if we suppose first the memoryless amplifier with $\boldsymbol{\tau}_n = \mathbf{0}$ for all n, then the nonlinear transfer functions $H_n(\boldsymbol{\omega}_n) = H_{n0}$ are frequency independent, reducing the model (2.41) to

$$Y(\omega) = \sum_{n=1}^{\infty} Y_n^{ML}(\omega) = \sum_{n=1}^{\infty} \frac{1}{(2\pi)^{n-1}} H_{n0} \int_{\overline{\omega}_n=\omega} \prod_{i=1}^{n} X(\omega_i) d\omega_i. \tag{2.48}$$

Now, the nth-order output spectrum of a general amplifier with memory can be written as:

$$Y_n(\omega) = \mathcal{F}_n(\omega) Y_n^{ML}(\omega), \tag{2.49}$$

where $\mathcal{F}_n(\omega)$ is a nth-order memory factor defined as:

$$\mathcal{F}_n(\omega) = \frac{\displaystyle\int_{\overline{\omega}_n=\omega} H_n(\boldsymbol{\omega}_n) \prod_{i=1}^{n} X(\omega_i) d\omega_i.}{\displaystyle H_{n0} \int_{\overline{\omega}_n=\omega} \prod_{i=1}^{n} X(\omega_i) d\omega_i}. \tag{2.50}$$

The nth-order output spectrum of a general amplifier can be interpreted as the result of filtering the corresponding memoryless output spectrum $Y_n^{ML}(\omega)$ with the memory factor $\mathcal{F}_n(\omega)$.

At this point, it is opportune to recall that in the case of OFDM transmitters the symbols containing the coded information are incorporated to subcarriers in the frequency domain, and this Volterra series model, valid for any general amplifier, makes evident how this frequency-domain version is valuable.

Going further, if we return to the amplifier of Figure 2.6 and consider frequency components in a narrow band, the gate linear transfer function $H_{g1}(\omega)$ is constant. Therefore, only the output network transfer functions are frequency dependent $H_n(\boldsymbol{\omega}_n) = c_n Z_L(\omega_1 + \cdots + \omega_n)$, and all the nth-order memory factors have the same expression given by:

$$\mathcal{F}_n(\omega) = \frac{\displaystyle\int_{\overline{\omega}_n=\omega} c_n Z_L(\omega_1 + \cdots + \omega_n) \prod_{i=1}^{n} X(\omega_i) d\omega_i.}{\displaystyle c_n Z_{L0} \int_{\overline{\omega}_n=\omega} \prod_{i=1}^{n} X(\omega_i) d\omega_i} = \frac{1}{Z_{L0}} Z_L(\omega). \tag{2.51}$$

The approach is equivalent to a two-block structure modeled by a memoryless nonlinearity and a linear filter $\mathcal{F}(\omega) = \mathcal{F}_n(\omega)$ connected in cascade, a structure that can be analyzed by the following simple formulation

$$Y(\omega) = \mathcal{F}(\omega) Y^{ML}(\omega). \tag{2.52}$$

It proves the feasibility of implementing the linearization in OFDM systems with an inverted two-block scheme, predistorting first the information symbols, directly in the frequency domain, and then completing the compensation with a memoryless nonlinearity. We will return to this topic in Chapter 3 where the discrete-time baseband Volterra models are discussed.

2.5 Two-block Models: Wiener and Hammerstein

In nonlinear systems evaluation, it is sometimes desirable to analytically derive the Volterra kernels. This derivation is possible in the case of simple nonlinear systems composed of the cascade of two blocks, a memoryless nonlinearity and a linear filter, whatever the order they are connected in. Two-block models are widely considered elsewhere (Morgan et al., 2006; Schreurs et al., 2008; Wood, 2014; Ghannouchi et al., 2015), so we dedicate only a few comments to complete the chapter discussion.

Wiener Model: We first discuss the Wiener model, whose structure, depicted in Figure 2.7(a), is composed of a linear filter followed by a memoryless nonlinearity. Thus, the input–output relationship of the Wiener system is

$$y(t) = \sum_{n=1}^{\infty} c_n \left[\int_{-\infty}^{\infty} h(\tau)x(t-\tau)d\tau \right]^n, \tag{2.53}$$

where c_n are the polynomial coefficients of the memoryless nonlinearity and $h(\tau)$ is the impulse response of the linear filter.

The Wiener structure eradicates the limitation of a memoryless model by straightforwardly integrating memory effects in a nonlinear system. This two-block model has been implemented to characterize nonlinear amplifiers

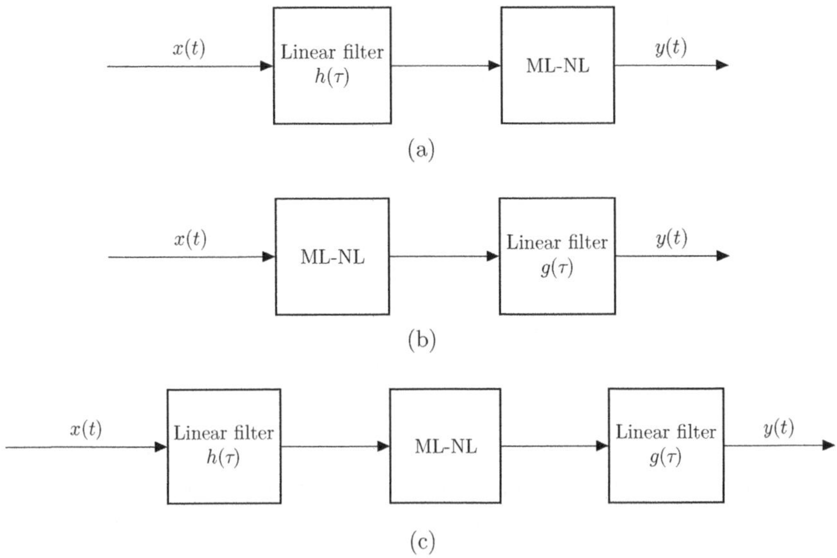

Figure 2.7 Block nonlinear models with memory: (a) Wiener, (b) Hammerstein, and (c) Wiener-Hammerstein.

with memory, as well as in transmitters predistortion, although its application is questionable as a consequence of two facts. In the first place, its modeling performance is insufficient, and on the other hand, the nonlinear dependence that the nth power introduces to the kernels produces an undesirable effect on the parameters estimation procedure.

Hammerstein Model: If the memoryless nonlinearity is followed by the linear filter in the cascade connection, as in the scheme of Figure 2.7(b), the system is denoted as Hammerstein model, for which the input–output relationship is

$$y(t) = \int_{-\infty}^{\infty} h(\tau) \left[\sum_{n=1}^{\infty} c_n x^n(t-\tau) \right] d\tau = \sum_{n=1}^{\infty} \int_{-\infty}^{\infty} h_n(\tau) x^n(t-\tau) d\tau, \quad (2.54)$$

where $h_n(\tau) = c_n h(\tau)$ is the nth-order kernel of the corresponding Volterra model. Even though the Hammerstein model has a moderate prediction efficiency, it exhibits an advantageous linear dependence with respect to the kernels $h_n(\tau)$, unlike the nonlinear kernel dependence of the Wiener model.

Other Block Models: Among all other possible block models we can imagine, now we comment briefly two examples, the **Wiener–Hammerstein** model and the **two-block feedback** model. The former type is a more sophisticated block model that can be easily proposed if we extend the two-block structure to a cascade connection of three blocks consisting of an input linear filter, a memoryless nonlinearity and an output linear filter, as in Figure 2.7(c). Identically to the Wiener model, it is also affected by the nonlinear dependence on the input filter impulse response.

The later type is a two-block structure in which the linear filter connects the system output and input. This feedback model has been derived in Narayanan (1970) for a shunt-output shunt-input feedback amplifier, as well as proposed to conceptually represent an amplifier with no need of a feedback physical path (Pedro et al., 2003).

Compared to the Volterra series, a block model is a distinct *aprioristic* approach proposed by trial and error. It means that, unlike Volterra models, these two-block representations are proposals with no need for information of the actual system internal structure.

An opportune topic is the comparison of the Volterra model derived for the simple amplifier with memory of Figure 2.6 and the Hammerstein model. Recalling the discussion of Section 2.4 and the result (2.52), we can observe that, in this particular case, the Volterra and the Hammerstein models are entirely equivalent. The point in favor of the Volterra model is the explicit relationship of the kernels and the nonlinear transfer functions with the circuit components, like capacitances, inductances and resistances as in equations (2.32), (2.36), and (2.39), to give some examples.

2.6 Double Volterra Series

Study of distortions in microwave mixers has been a topic of high interest in several publications. In the particular case of microwave amplifiers, the most important nonlinear techniques used to evaluate intermodulation distortions have been harmonic balance and Volterra series.

In the case of analysis with Volterra series applied to one input-port systems, most of the publications have followed the results of Bedrosian and Rice (1971) and Bussgang et al. (1974) summarized in this chapter. However, mixers in transmitters are two input-ports circuits: one for the large local oscillator signal and a second input for the weak low-frequency signal to be up converted to RF. The above-mentioned simple results can be generalized in a straightforward way to analyze mixers under a Volterra series perspective. This approach was presented by Rice in an early paper in which the output of a system with two input-ports is expressed as a double Volterra series (Rice, 1973), demonstrating its effectiveness in the distortion evaluation of a frequency converter.

In the same way that conventional Volterra series has proven to be a valuable technique in the study of power amplifiers behavior, the double Volterra series is also a powerful tool to analyze nonlinear distortions in microwave mixers.

Referring to the nonlinear system with two inputs $x(t)$ and $z(t)$ of Figure 2.8, the output $y(t)$ can be expressed by the double Volterra series

$$y(t) = \sum_{n=0}^{\infty} \sum_{\substack{m=0 \\ m+n\neq 0}}^{\infty} \int h_{nm}(\boldsymbol{\tau}_n; \boldsymbol{\zeta}_m) \prod_{k=1}^{n} x(t-\tau_k)d\tau_k \prod_{l=1}^{m} z(t-\zeta_l)d\zeta_l. \tag{2.55}$$

The integration is multiple and the product operator $\prod_{k=1}^{n}$ is understood to have the value 1 when $n = 0$, with no need for the corresponding integration. In equation (2.55), $\boldsymbol{\zeta}_m$ is the delay vector $[\zeta_1 \cdots \zeta_m]^T$, defined in a similar way as $\boldsymbol{\tau}_n$ in the conventional Volterra series, and the kernel $h_{nm}(\boldsymbol{\tau}_n; \boldsymbol{\zeta}_m)$ is the nonlinear impulse response of order $n + m$.

Denoting the $(n + m)$th-order nonlinear transfer function $H_{nm}(\boldsymbol{\omega}_n; \boldsymbol{\xi}_m)$ to be the multidimensional Fourier transform of the kernel $h_{nm}(\boldsymbol{\tau}_n; \boldsymbol{\zeta}_m)$, equation (2.55) can

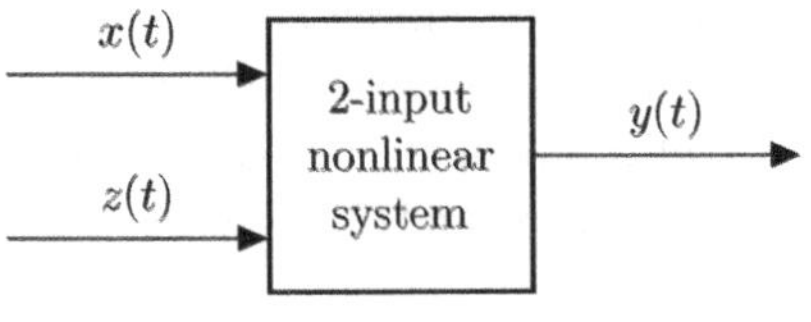

Figure 2.8 Schematic of a two-input nonlinear system.

be rewritten as:

$$y(t)=\sum_{\substack{n=0\\n+m\geq 1}}^{\infty}\sum_{m=0}^{\infty}\frac{1}{(2\pi)^{n+m}}\int H_{nm}(\boldsymbol{\omega}_n;\boldsymbol{\xi}_m)$$
$$\times\prod_{k=1}^{n}X(\omega_k)e^{j\omega_k t}d\omega_k\prod_{l=1}^{m}Z(\xi_l)e^{j\xi_l t}d\xi_l, \tag{2.56}$$

where $\boldsymbol{\omega}_n$ and $\boldsymbol{\xi}_m$ are the vectors $[\omega_1\cdots\omega_n]^T$ and $[\xi_1\cdots\xi_m]^T$, respectively. Equation (2.56) allows to express the $(n+m)$th-order term of the output as a function of the spectra $X(\omega)$ and $Z(\omega)$ of the two input signals.

Following a development resembling the conventional Volterra series analysis discussed above, the nonlinear current and harmonic input methods can be extended and applied to this double Volterra series for determining the nonlinear transfer functions $H_{nm}(\boldsymbol{\omega}_n;\boldsymbol{\xi}_m)$.

When we consider the mixer in a transmitter, one of the inputs is usually a low-frequency signal $z(t)$, and the local oscillator is applied to the second input $x(t)$. That being so, the mixer response can be expressed as the sum of three kinds of terms.

The first part is the contribution of the local oscillator, which is simply the output of a one-input nonlinear system given by equation (2.56) with $m=0$. Then, without loss of generality, it is possible to consider a single sinusoidal signal $x(t)=A\cos\omega_c t$ as local oscillator, so the output at ω_c can be computed following the same procedure demonstrated in Section 2.3.1 to obtain the amplifier response given by equation (2.11).

Another part is made up of the terms with $n=0$ and represents the contribution to the response from the low-frequency signal alone. Since the frequency band of $z(t)$ is much lower than the local oscillator frequency ω_c, this part of the response is easily filtered out and can be ignored.

Finally, the output of interest in the band around ω_c deals with the spectra of the two signals, $x(t)$ and $z(t)$, expressed as

$$y(t)=\Re\left\{\sum_{m=0}^{\infty}\int\hat{H}_m(t;\boldsymbol{\xi}_m)\prod_{l=1}^{m}Z(\xi_l)e^{j\xi_l t}\frac{d\xi_l}{2\pi}\right\}, \tag{2.57}$$

where

$$\hat{H}_m(t;\boldsymbol{\xi}_m)=e^{j\omega_c t}\sum_{\substack{n=1\\(n\,\text{odd})}}^{\infty}\frac{A^n}{2^{n-1}}\frac{n!}{\left(\frac{n+1}{2}\right)!\left(\frac{n-1}{2}\right)!}$$
$$\times H_{nm}(\underbrace{\omega_c,\ldots,\omega_c}_{\frac{n+1}{2}\text{ times}},\underbrace{-\omega_c,\ldots,-\omega_c}_{\frac{n-1}{2}\text{ times}};\boldsymbol{\xi}_m), \tag{2.58}$$

can be considered as the mth-order nonlinear transfer function at frequency ω_c of a time-varying system, once the oscillator parameters A and ω_c have been fixed.

The above result is suited to analyze conversion gain/loss and intermodulation in mixers under a Volterra series perspective. For instance, if the baseband $z(t)$ is a single sinusoidal signal at frequency ω_0, equation (2.57) provides the desired mixer output at $\omega_c \pm \omega_0$ for the "linear" terms, so we can calculate the conversion gain. The response at $\omega_c \pm 3\omega_0$ returns the third-order intermodulation distortion. In general, we can calculate the higher-order intermodulation products at frequencies $\omega_c \pm m\omega_0$, for m odd.

2.6.1 The Double Volterra Series in the Analysis of Mixers

To exemplify the double Volterra series perspective in the nonlinear analysis of mixers, consider the two-port circuit shown in Figure 2.9. It consists of a linear network and a controlled nonlinear current source $i(v, u)$ dependent on the voltages of the input and output nodes v and u, respectively. Observe that the equivalent circuit structure of a FET fits this circuit, for instance, the simple circuit of Figure 2.6(a) analyzed in Section 2.3.2.

We particularize in a FET mixer to show how the double Volterra series approach can be used to analyze a two-input circuit, considering that the local oscillator $v(t)$ is applied to the gate port and the low-frequency $u(t)$ (or the RF signal, in down conversion) drives the drain port.

Realizing that the current source is a transfer nonlinearity depending on two voltages, as expressed by equation (1.17), and also that the relevant group is the third sum of terms with the cross products $v^n u^m$ in equation (1.18), the procedure to derive the nonlinear transfer functions for the gate $H_{g;nm}$ and the drain H_{nm} nodes can be straightforwardly applied.

The standard procedure, as explained in Section 2.3.1, consists first of the identification of the nonlinear currents contributed by the transfer nonlinearity, and then supposing input signals as sum of exponentials. Finally, the desired terms of the linear subnetwork outputs in the frequency-domain are defined as the nonlinear transfer functions.

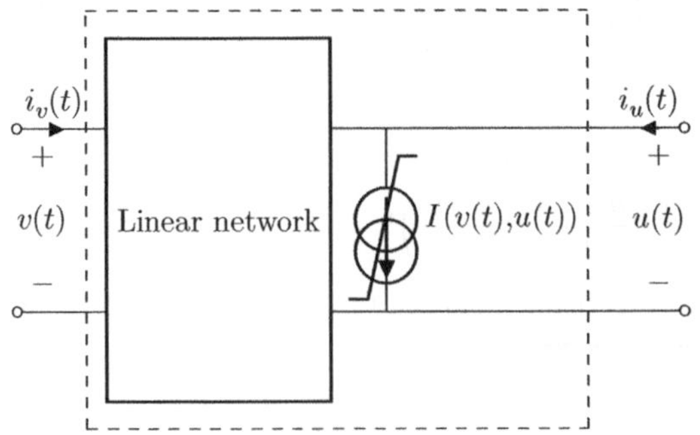

Figure 2.9 Nonlinear circuit with two input ports.

Fortunately, the similarity of the circuit allows to reuse the results obtained for the amplifier with the simple circuit of Figure 2.6, now with the condition that the nonlinear current source is dependent on two voltages.

Focusing on the relevant cross-terms, it implies that the strictly nonlinear current attached to the drain port of the linear subnetwork can be written as:

$$i_{\mathrm{NL}}(t) = \sum_{n=1}^{\infty} \sum_{m=1}^{\infty} g_{nm} v^n u^m. \tag{2.59}$$

The lth-order voltages are calculated by short-circuiting the input local oscillator and low-frequency sources and connecting successively the nonlinear current $i_l(t)$, which in this case is given by

$$i_l(t) = \sum_{n=1}^{l-1} \sum_{m=1}^{l-n} g_{nm} \Phi_{nm,l}, \tag{2.60}$$

where

$$\Phi_{nm,l} = \sum_{j=1}^{l-1} v_{j,n} u_{l-j,n}, \qquad l \geq m + n, \tag{2.61}$$

and $v_{j,n}$ is computed with the recursion (2.24). For the second-order nonlinear current, given by the term with $n = m = 1$, we have

$$i_2 = g_{11} v_1(t) u_1(t). \tag{2.62}$$

Combining these results with the harmonic input method and denoting F_{nm} the coefficient of the appropriate component in the generated current, the nonlinear transfer functions can be determined in terms of products of lower order ones. Following with the example, the second-order nonlinear current yields the corresponding coefficient

$$F_1 = g_{11} H_{g;10}(\omega_1) H_{01}(\xi_1), \tag{2.63}$$

where $H_{g;10}$ and H_{01} represent the first-order nonlinear transfer functions of nodes v and u, respectively.

It is pertinent to note that $H_{g;10}$ is obtained by computing the response of the linear subnetwork when it is driven only by the local oscillator signal, and for this particular example, the expression (2.32) applies. Moreover, given that the nonlinear source is attached to the drain node and the circuit of Figure 2.6 is unilateral, the input branch is linear and all node-v transfer functions of higher order are zero.

The procedure can be repeated for higher-order currents, in particular, the third order nonlinear current is

$$i_3 = g_{12} v_1(t) u_1^2(t) + g_{11} v_1(t) u_2(t), \tag{2.64}$$

which is used to compute the appropriate Fourier coefficient given by

$$F_3 = g_{12} H_{g;10}(\omega_1) H_{01}(\xi_1) H_{01}(\xi_1) + g_{11} H_{g;10}(\omega_1) \overline{H_{02}(\xi_1, \xi_2)}. \tag{2.65}$$

Compared with the coefficients of Table 2.3, it is clear that this term takes into account the distortion of the mixer response, allowing to evaluate the compression point and the third-order intermodulation products.

2.6.1.1 FET Resistive Mixer

The resistive mixer is a particular type of mixers, which exploits the weak non-linear channel resistance of a FET to deliver high output power at moderate local oscillator (LO) levels with very low intermodulation distortions.

A summary of the design guidelines of a microwave FET resistive mixer is encountered in the pioneering paper (Maas, 1987). Basically, the rules require an IF-RF filter designed to short-circuit the drain at the local oscillator frequency in order to avoid coupling the input voltage to the drain terminal. Assuming up conversion, this IF-RF filter should be also a short-circuit at all mixing frequencies except the IF band and the desired RF band.

Moreover, to avoid coupling of the low-frequency (IF) input signal to the gate, the input (local oscillator) filter should be designed to short-circuit the gate at RF and IF frequencies.

The circuit of Figure 2.6 that we are analyzing allows a drastic reduction in the number of terms to be considered. Firstly, the gate subcircuit is linear and does not generate nth-order terms with $n \geq 2$. In addition, the circuit is unilateral and this feature permits to conclude that $v_n(t) = 0$ for all $n \geq 2$, and $H_{g;10}(\omega)$ is the only non-zero transfer function.

A perspicacious reader can notice that a unilateral FET mixer in the real world is questionable. Thankfully, in a general resistive mixer, the condition $V_{DS} = 0$ and the use of the local oscillator input-port and IF-RF output-port filters produce likewise a reduction in the number of terms, providing a rationale for the assumption of approximately the same conditions.

Substituting these results in the recursion (2.61), the following simple equation is obtained for the nonlinear current source with $m = 1$

$$i_1(t) = \sum_{n=0}^{\infty} g_{n1} v_1^n(t) u_1(t). \tag{2.66}$$

We observe that this approach is equivalent to the established large-signal/small-signal analysis, and in the same form, the output can be derived following a standard procedure.

To illustrate the procedure, similar to Crespo-Cadenas and Reina-Tosina (2002), the double Volterra series approach has been applied to the analysis of a hypothetical down-converter with a FET resistive mixer. The FET under consideration corresponds to a packaged AT10650-5 device which has been characterized with an adequate nonlinear device model. We show some prediction results, in particular, the input/output characteristic in Figure 2.10(a), and the conversion loss in

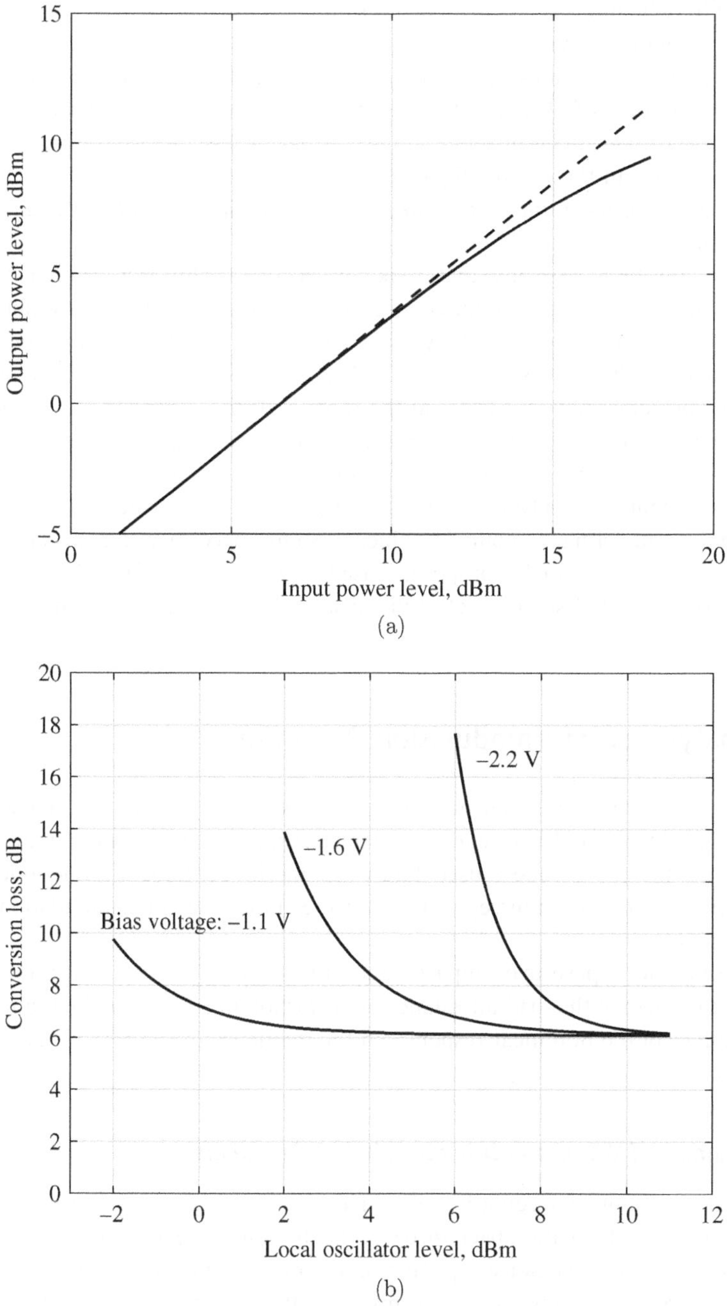

Figure 2.10 (a) Predicted input/output nonlinear response and (b) conversion loss of a FET resistive mixer.

Figure 2.10(b), which has been calculated for different oscillator levels and bias voltages as the figure shows.

According to the previous discussion, the formal double Volterra series approach can be successfully applied to the analysis of two-input ports networks, and in particular to evaluate mixers nonlinear behavior. Recalling the words of Bedrosian and Rice in their seminal paper, we can add that the double Volterra series approach "*has the virtue that nonlinear analysis in mixers can be treated in an orderly way.*"

But there are more reasons that make especially useful this approach. Modulators are important two-input circuits in the architecture of a transmitter, and demodulators in receivers, so they are suitable for a double Volterra series analysis. Recalling the more sophisticated FET models with electrothermal and charge-trapping subcircuits, we can add the more realistic case of an amplifier with an internal mechanism that generates a second variable that affects the overall response, and thus the double Volterra series is, again, a very suitable practical tool. Finally, the advantages of an analysis of the baseband equivalent systems makes modeling of complex-valued systems necessary, for which a formal Volterra model can be proposed based as well on a double Volterra series approach. We will see this set of practical and theoretical issues in later chapters.

2.7 Analysis of Intermodulation Distortion

From a practical perspective, an evaluation method to predict nonlinear distortions is invaluable to RF designers. Characterization of nonlinear circuits is focused on the most widely used figures of merit identified by one-tone and two-tone signals, the intercept point and the third-order intermodulation distortion, for instance.

In this section, we expose an example of procedure based on Volterra series, able to forecast the effect of the load impedance on distortion in a FET amplifier, and the dependence of the nonlinear response on the frequency or the FET physical parameters.

2.7.1 Example of Volterra IMD Analysis in FET Amplifiers

The equivalent circuit representation of a FET shown in Figure 1.9 consists of three nonlinearities: the gate capacitance C_{gs}, the drain nonlinear current source I_{ds}, and the feedback capacitance C_{gd}. This comprehensive model was simplified in Minasian (1980) neglecting the cross terms of the drain current source and moving the effects of the feedback capacitance to the gate and drain circuits. It makes

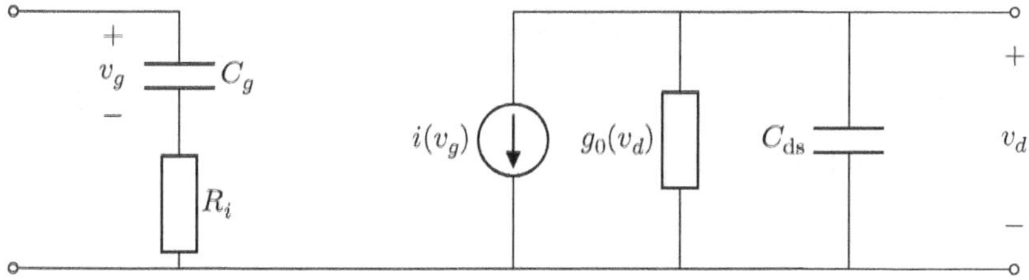

Figure 2.11 FET model used in the analysis.

possible a Volterra treatment leading to the obtention of closed-form expressions for the distortions.

The model to be analyzed, shown in Figure 2.11, is unilateral yet adequately representing FET behavior up to frequencies at which the feedback becomes significant. In this approximation, the associated linear circuit again has two nodes and is very similar to the one displayed in Figure 2.6, so there is no complication in deriving the linear transfer functions by repeating the procedure outlined in Section 2.3.2.

There are two additional nonlinear sources in this example. First, the nonlinear gate capacitance is represented as a current generator using the following expansion, equivalent to (1.15),

$$i_c = \frac{d}{dt}\sum_{n=1}^{\infty} c_{gn} v_g^n. \tag{2.67}$$

The second source is the result of splitting the drain source into a nonlinear transconductance (a transfer nonlinearity) and an output nonlinear conductance, which is written as

$$i_0 = \sum_{n=1}^{\infty} g_{0n} v_d^n. \tag{2.68}$$

Recall that, in this approximation, the third source with the cross-terms is neglected.

As we know, the first-order capacitance and conductance, c_{g1} and g_{01}, are part of the associated linear circuit, from whose analysis the following linear transfer functions are derived

$$H_{g1}(\omega) = \frac{1}{1 + j\omega c_{g1}\overline{R}_g}, \tag{2.69}$$

$$H_1(\omega) = \frac{g_{10}\overline{R}_g^{-1}}{Y_0(\omega)Y_i(\omega)}, \tag{2.70}$$

where $\overline{R}_g$ includes here the FET intrinsic resistance R_i,

$$Y_i(\omega) = j\omega c_{g1} + \overline{R}_g^{-1}, \tag{2.71}$$

is the input admittance, and

$$Y_0(\omega) = g_{01} + j\omega C_{ds} + \frac{1}{Z_L(\omega)}, \tag{2.72}$$

is the output admittance, where $Z_L(\omega)$ is the load impedance at the amplifier output.

Notwithstanding that the circuit is unilateral, the existence of the nonlinear capacitance C_g generates nonlinear responses in the gate node. First, we assume the second-order nonlinear currents attached to the corresponding nodes of the associated linear network, with the input signal source short-circuited, and then the second-order nonlinear transfer functions are calculated assuming an input signal of two exponentials. The second-order nonlinear transfer functions are

$$H_{g2}(\boldsymbol{\omega}_2) = -\frac{j(\omega_1 + \omega_2)c_{g2}H_{g1}(\omega_1)H_{g1}(\omega_2)}{Y_i(\omega_1 + \omega_2)}, \tag{2.73}$$

$$H_2(\boldsymbol{\omega}_2) = -\frac{H_{g1}(\omega_1)H_{g1}(\omega_2)}{Y_0(\omega_1 + \omega_2)}\left[g_{20} - \frac{j(\omega_1 + \omega_2)g_{10}c_{g2}}{Y_i(\omega_1 + \omega_2)} + \frac{g_{02}g_{10}^2}{Y_0(\omega_1)Y_0(\omega_2)}\right]. \tag{2.74}$$

The third-order responses are computed by iterating the procedure with the same associated linear circuit and assuming an input signal of three exponentials. Focusing on the output, the third-order nonlinear transfer function is given by

$$\begin{aligned} H_3(\boldsymbol{\omega}_3) = &-\frac{1}{Y_0(\omega')}\left[H_{g1}(\omega_1)H_{g1}(\omega_2)H_{g1}(\omega_3)\right.\\ &\times\left(g_{30} - \frac{j\omega' g_{10}c_{g3}}{Y_i(\omega')} + \frac{g_{03}g_{10}^3}{Y_0(\omega_1)Y_0(\omega_2)Y_0(\omega_3)}\right)\\ &+ \overline{H_{g1}(\omega_1)H_{g2}(\boldsymbol{\omega}_2)}\left(2g_{20} - \frac{2j\omega' g_{10}c_{g2}}{Y_i(\omega')}\right)\\ &\left.+ 2g_{02}\overline{H_1(\omega_1)H_2(\boldsymbol{\omega}_2)}\right], \end{aligned} \tag{2.75}$$

where $\omega' = \omega_1 + \omega_2 + \omega_3$. Observe that these expressions reduce to the nonlinear transfer functions of the simple amplifier analyzed in Section 2.3.2 when we neglect the nonlinearity in the gate subcircuit.

Repeating successively the same procedure, the response of any order can be found with the associated linear circuit to deduce the nth-order nonlinear transfer functions.

The contribution to the nonlinear distortions of components close to the fundamental band can be predicted if we consider the following two-tone signal

$$v_s(t) = A\cos(\omega_1 t) + A\cos(\omega_2 t), \tag{2.76}$$

applied to the amplifier input. Using the notation $H_1(\omega_1) = |H_1(\omega_1)|e^{j\theta_1(\omega_1)}$, the first-order output at ω_1 is

$$v_{d1}(t) = A|H_1(\omega_1)|\cos\left[\omega_1 t + \theta_1(\omega_1)\right], \tag{2.77}$$

and the third-order intermodulation distortion at frequency $2\omega_1 - \omega_2$ is expressed as:

$$v_{d3}(t) = \frac{3}{4}A^3|H_3(\omega_1,\omega_1,-\omega_2)|\cos\left[(2\omega_1-\omega_2)t + \theta_3(\omega_1,\omega_1,-\omega_2)\right], \tag{2.78}$$

where $H_3(\omega_1,\omega_1,-\omega_2) = |H_3(\omega_1,\omega_1,-\omega_2)|e^{j\theta_3(\omega_1,\omega_1,-\omega_2)}$.

With these expressions, the power of the amplifier response at each spectral component can be readily computed. Assuming that the frequency difference between the two tones is very small, $|\omega_1 - \omega_2| \ll (\omega_1+\omega_2)/2$, the third-order intermodulation distortion IM3 can be enunciated as

$$\mathrm{IM3} = 20\log\left[\frac{3}{4}A^2\frac{|H_3(\omega_1,\omega_1,-\omega_2)|}{|H_1(\omega_1)|}\right]. \tag{2.79}$$

The effect of device parameters on IM3 is disclosed if we recall the transfer functions (2.69), (2.70), and (2.75). Dependence of IM3 on frequency and circuit admittances is also explicitly obtained, allowing the comparison of estimated and experimental third-order intermodulation distortion (Minasian, 1980).

Table 2.4 tries to illustrate the different asymmetry situations that can arise for third-order intermodulation distortion components in two-tone tests depending on the baseband load impedance. Recall that an asymmetry refers to a difference between the upper and lower intermodulation distortion components. We will denote the upper intermodulation distortion component at $2\omega_2 - \omega_1$ as v_{d3}^u, and the lower intermodulation distortion component at $2\omega_1 - \omega_2$ as v_{d3}^l. According to (2.78), the upper and lower third-order intermodulation distortion components can be represented by phasors accounting for their magnitude and phase. We can also observe that they depend on the third-order nonlinear transfer function, that must be particularized for the circuit under analysis and generally takes the form of a sum of complex-valued components or phasors.

Let us focus on the FET model circuit of Figure 2.11 where the load impedance connected to the output port is denoted Z_L for baseband frequencies. In the present example, only two components are considered for simplicity, the first of them denoted as c_0. In Crespo-Cadenas et al. (2006), it is shown that the second component phasor depends on a constant c_1 given by the equivalent circuit parameters and the baseband load impedance for v_{d3}^u, that is, $c_1 Z_L$. On the other hand, for v_{d3}^l,

Table 2.4 Different asymmetry situations in the context of two-tone tests with respect to the baseband load impedance Z_L.

	$Z_L \in \mathbb{R}$	$Z_L \in \mathbb{C}$
$c_1 \in \mathbb{R}$	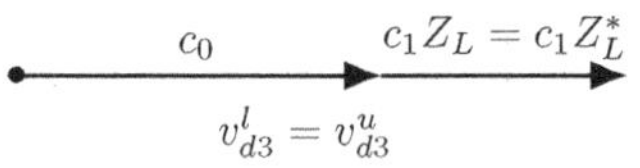 Symmetry	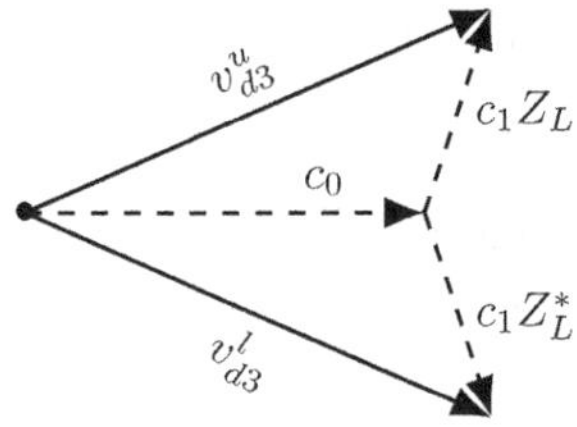 Symmetry in magnitude Asymmetry in phase
$c_1 \in \mathbb{C}$	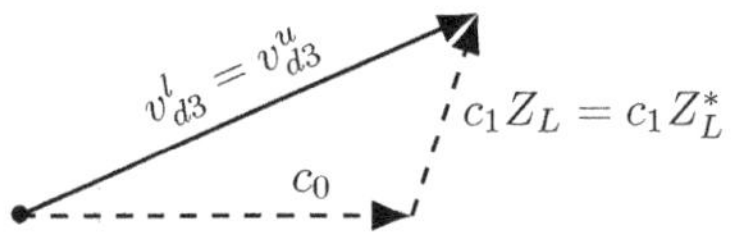 Symmetry	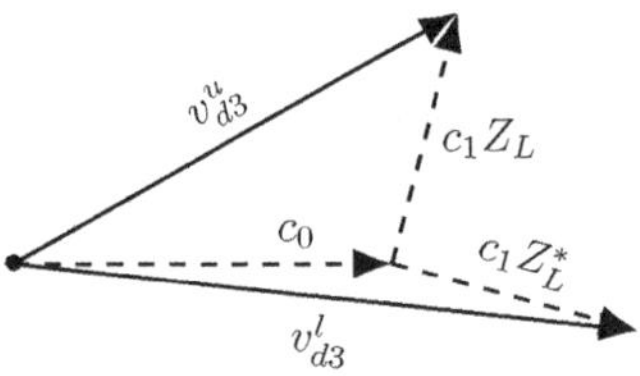 Asymmetry in magnitude and phase

the second component phasor depends on c_1 and the complex conjugate of the baseband load impedance, that is, $c_1Z_L^*$. As it can be observed in Table 2.4, when both c_1 and Z_L are complex-valued an asymmetry occurs, since v_{d3}^u and v_{d3}^l differ both in magnitude and in phase. However, if c_1 is real-valued, it is possible that the magnitude of v_{d3}^u equals the magnitude of v_{d3}^l (symmetry in magnitude), while their phases differ (asymmetry in phase).

2.8 Baseband Volterra Model

To complete the overview of the classical continuous-time Volterra models, we consider now the baseband equivalent representation and its convenience to analyze wireless communication nonlinear systems functioning over channels with restricted bandwidth. The discrete-time baseband equivalent models are left to be contemplated in the following chapters.

Most RF power amplifiers in the transmitter are operated at or near saturation and many types of adverse distortion effects arise from this nonlinear channel.

Perhaps, the most evident consequence is an out-of-band distortion produced by harmonic generation, which can be minimized by a RF bandpass filter centered at the fundamental frequency.

Another undesired spectral products generated by nonlinear distortions can be near or within the desired RF signal passband. These particular nonlinear impairments, denoted as in-band distortion, cannot be filtered out in the same way and affect considerably the communications system performance in two aspects.

The spectral distortion falling just over the band of the original RF signal is denoted as *co-channel distortion* and disturbs the quality of the transmitted signal. On the other hand, the spectral components appearing in the neighboring frequencies represent spectral regrowth that interferes with the communication of other users. These effects are commonly identified as *adjacent-channel distortion.*

In previous sections, we have discussed several useful results in the analysis of RF power amplifier circuits based on a Volterra series approach. Now we must realize that the effort has to be directed toward computing the amplifier output in the case of communication signals following the same perspective.

Referring to the power amplifier of the wireless system of Figure 2.12, if the RF input signal is $\tilde{x}(t)$, then we can express the transmitted signal $\tilde{y}(t)$ as (see equations (2.1) and (2.2))

$$\tilde{y}(t) = \sum_{n=1}^{\infty} \int_{-\infty}^{\infty} \tilde{h}_n(\boldsymbol{\tau}_n) \prod_{r=1}^{n} \tilde{x}(t - \tau_r) d\boldsymbol{\tau}_n. \tag{2.80}$$

where $\tilde{h}_n(\boldsymbol{\tau}_n)$ is the nth-order kernel of the amplifier. We use here the tilde to differentiate the real-valued RF waveforms and the respective complex envelopes. For example, we write the RF input signal as

$$\tilde{x}(t) = \Re\{x(t)e^{j\omega_c t}\} = \frac{1}{2}\left[x(t)e^{j\omega_c t} + x^*(t)e^{-j\omega_c t}\right], \tag{2.81}$$

where the signal complex envelope $x(t)$ is an equivalent lowpass signal, which contains all the amplitude and phase information of the modulated RF signal.

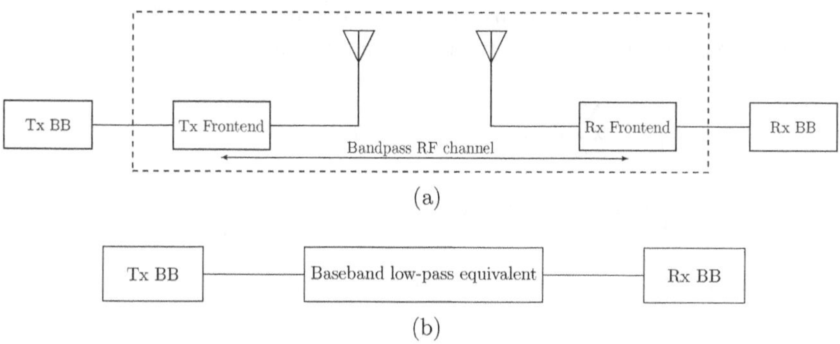

Figure 2.12 (a) Wireless communication channel and (b) baseband equivalent channel.

Ideally, the modulating waveform $x(t)$ can be processed in the baseband section of the transmitter to cope with spectral band restrictions, equalization and intersymbol interference. Indeed, that is the widely adopted solution, because low-frequency techniques are more tractable than those at radio frequency. In other words, the analysis of a lowpass equivalent nonlinear system with the input given by the complex envelope $x(t)$ to describe the output complex envelope $y(t)$ is formally a legitimate procedure (Benedetto et al., 1979; Raich and Zhou, 2002). Unfortunately, equation (2.80) cannot be directly assumed as a valid nonlinear model in that case, making necessary the deduction of a baseband Volterra model equivalent to the RF bandpass model (2.80).

For that purpose, if we substitute (2.81) in (2.80), the product of delayed signals inside the integral of the nth-order term becomes

$$\prod_{r=1}^{n}\tilde{x}(t-\tau_r)=\frac{1}{2^n}\prod_{r=1}^{n}[x(t-\tau_r)e^{j\omega_c(t-\tau_r)}+x^*(t-\tau_r)e^{-j\omega_c(t-\tau_r)}], \tag{2.82}$$

which results in a sum of 2^n terms with factors like

$$e^{j(n-2p)\omega_c t}\underbrace{x(t-\tau_1)\cdots x(t-\tau_{n-p})}_{n-p\text{ delayed signals}}\cdots\underbrace{x^*(t-\tau_{n-p+1})\cdots x^*(t-\tau_n)}_{p\text{ delayed conjugate signals}}, \tag{2.83}$$

where $n-p$ and p are, respectively, the number of delayed replicas of the input complex envelope $x(t)$ and the conjugate complex envelope $x^*(t)$.

Making use of equation (2.83), which stands for nonlinear responses around harmonic frequencies ranging from $-n\omega_c$ for $p=0$, to $n\omega_c$ for $p=n$, we spotlight only on the spectral components within the zone around the fundamental frequency ω_c, assuming those lying far from this fundamental band are filtered out. To preserve the terms that exhibit the exponential factor $e^{j\omega_c t}$, we will collect those for which the number of delayed replicas of $x(t)$ is $p+1$ and the number of delayed replicas of $x^*(t)$ is p, which means that the baseband model is restricted to having only terms whose order $n=2p+1$ is odd. A similar reasoning is valid for the complex conjugated terms with the exponential factor $e^{-j\omega_c t}$. Substituting in equation (2.80) and summing up those terms, we can write

$$\begin{aligned}\tilde{y}(t)=\Re\Bigg\{&e^{j\omega_c t}\sum_{p=0}^{\infty}\frac{1}{2^{2p}}\binom{2p+1}{p}\int_{-\infty}^{\infty}\tilde{h}_{2p+1}(\boldsymbol{\tau}_{2p+1})\\&\times\prod_{i=r}^{p+1}x(t-\tau_r)e^{-j\omega_c\tau_r}\prod_{r=p+2}^{2p+1}x^*(t-\tau_r)e^{j\omega_c\tau_r}d\boldsymbol{\tau}_{2p+1}\Bigg\}.\end{aligned} \tag{2.84}$$

If we define the lowpass equivalent Volterra kernels $h_{2p+1}(\boldsymbol{\tau}_{2p+1})$ as

$$h_{2p+1}(\boldsymbol{\tau}_{2p+1})=\frac{1}{2^{2p}}\binom{2p+1}{p}\tilde{h}_{2p+1}(\boldsymbol{\tau}_{2p+1})\prod_{r=1}^{p+1}e^{-j\omega_c\tau_r}\prod_{r=p+2}^{2p+1}e^{j\omega_c\tau_r}, \tag{2.85}$$

the response of the amplifier at ω_c is thoroughly described by the following equivalent baseband model, expressed as

$$y(t) = \sum_{p=0}^{\infty} \int_{-\infty}^{\infty} h_{2p+1}(\boldsymbol{\tau}_{2p+1}) \prod_{r=1}^{p+1} x(t-\tau_r) \prod_{r=p+2}^{2p+1} x^*(t-\tau_r)\, d\boldsymbol{\tau}_{2p+1}. \tag{2.86}$$

This rather remarkable input–output relationship is the universal Volterra model required to describe the behavior of an amplifier in terms of the input and output complex envelopes.

Unlike the original Volterra series (2.80) applied to the analysis of real-valued RF amplifiers, the baseband model (2.86) is made up exclusively of odd-order terms. It seems that there is a certain inconsistence to classify as Volterra models other approaches that include, not only odd-order terms, but also terms to which we cannot assign an odd order, for instance, the widely accepted generalized memory polynomial (GMP) model. Since many publications have validated the benefits of these *not-odd-order* terms in modeling and linearization, it sems a paradoxical situation. We will revisit this topic in Chapter 4 to demonstrate how these terms of some particular behavioral models can be considered proper under a Volterra series perspective.

Recalling that the Fourier transform of $x^*(t)$ is $X^*(-\omega)$, the output complex envelope can be also expressed by means of products of the input spectrum $X(\omega)$ as

$$y(t) = \sum_{p=0}^{\infty} \frac{1}{(2\pi)^n} \int_{-\infty}^{\infty} H_n(\boldsymbol{\omega}_n) \prod_{r=1}^{p+1} X(\omega_r) e^{j\omega_r t} \prod_{r=p+2}^{2p+1} X^*(-\omega_r) e^{j\omega_r t}\, d\boldsymbol{\omega}_{2p+1}, \tag{2.87}$$

where $H_n(\boldsymbol{\omega}_n)$ is the nth-order nonlinear transfer function of the baseband Volterra model.

To summarize, a basic overview of the Volterra series and its effectiveness in the analysis of nonlinear circuits in the continuous-time domain has been presented in this chapter. A noteworthy case is the evaluation of the response and intermodulation distortion (IMD) of nonlinear amplifiers making use of Volterra series. Other examples are the study of the equivalent baseband Volterra model in a RF bandpass scenario and the assessment of nonlinear behavior in mixers, modulators and demodulators with the aid of double Volterra series. In the following chapters, the analysis is focused in discrete-time Volterra models, and the demonstration of a complex-valued Volterra generic model is examined. Frequency-domain Volterra models and their benefits for OFDM communications are shown.

Bibliography

M. Allegue-Martínez, M.J. Madero-Ayora, J.G. Doblado, C. Crespo-Cadenas, J. Reina-Tosina, and V. Baena. Digital predistortion technique with in-band interference optimisation applied to DVB-T2. *Electronics Letters*, 48(10):566–568, 2012. doi: 10.1049/el.2012.0279.

E. Bedrosian and S.O. Rice. The output properties of Volterra systems (nonlinear systems with memory) driven by harmonic and gaussian inputs. *Proceedings of the IEEE*, 59(12):1688–1707, 1971. doi: 10.1109/PROC.1971.8525.

S. Benedetto, E. Biglieri, and R. Daffara. Modeling and performance evaluation of nonlinear satellite links — A Volterra series approach. *IEEE Transactions on Aerospace and Electronic Systems*, AES-15(4):494–507, 1979. doi: 10.1109/TAES.1979.308734.

S. Boyd and L. Chua. Fading memory and the problem of approximating nonlinear operators with Volterra series. *IEEE Transactions on Circuits and Systems*, 32(11):1150–1161, 1985. doi: 10.1109/TCS.1985.1085649.

A. Brihuega, L. Anttila, and M. Valkama. Frequency-domain digital predistortion for OFDM. *IEEE Microwave and Wireless Components Letters*, 31(6): 816–818, 2021. doi: 10.1109/LMWC.2021.3062982.

J.J. Bussgang, L. Ehrman, and J.W. Graham. Analysis of nonlinear systems with multiple inputs. *Proceedings of the IEEE*, 62(8):1088–1119, 1974. doi: 10.1109/PROC.1974.9572.

C. Crespo-Cadenas and J. Reina-Tosina. Analysis of FET resistive mixers with a double Volterra series approach. In *2002 32nd European Microwave Conference*, pages 1–4, 2002. doi: 10.1109/EUMA.2002.339275.

C. Crespo-Cadenas, J. Reina-Tosina, and M.J. Madero-Ayora. IM3 and IM5 phase characterization and analysis based on a simplified Newton approach. *IEEE Transactions on Microwave Theory and Techniques*, 54(1):321–328, 2006. doi: 10.1109/TMTT.2005.861659.

S.C. Cripps. *RF Power Amplifiers for Wireless Communications*. Artech House Microwave Library. Artech House, Boston, 1999. ISBN 9780890069899.

F.M. Ghannouchi, O. Hammi, and M. Helaoui. *Behavioral Modeling and Predistortion of Wideband Wireless Transmitters*. Wiley, Chichester, UK, 2015. ISBN 9781119004431.

K. Kundert, J. K. White, and A. Sangiovanni-Vincentelli. *Steady-State Methods for Simulating Analog and Microwave Circuits*. Kluwer Academic Publishers, Boston, 1990. ISBN 0792390695.

Z.-Q. Lang and S.A. Billings. Output frequency characteristics of nonlinear systems. *International Journal of Control*, 64(6):1049–1067, 1996. doi: 10.1080/00207179608921674.

Z.-Q. Lang and S.A. Billings. Output frequencies of nonlinear systems. *International Journal of Control*, 67(5):713–730, 1997. doi: 10.1080/002071797223965.

S.A. Maas. A GaAs MESFET mixer with very low intermodulation. *IEEE Transactions on Microwave Theory and Techniques*, 35(4):425–429, 1987. doi: 10.1109/TMTT.1987.1133665.

S.A. Maas. *Nonlinear Microwave and RF Circuits*. Artech House, Boston, 2nd edition, 2003. ISBN 9781580536110.

R.A. Minasian. Intermodulation distortion analysis of MESFET amplifiers using the Volterra series representation. *IEEE Transactions on Microwave Theory and Techniques*, 28(1):1–8, 1980. doi: 10.1109/TMTT.1980.1129998.

D.R. Morgan, Z. Ma, J. Kim, M.G. Zierdt, and J. Pastalan. A generalized memory polynomial model for digital predistortion of RF power amplifiers. *IEEE Transactions on Signal Processing*, 54(10):3852–3860, 2006. doi: 10.1109/TSP.2006.879264.

S. Narayanan. Transistor distortion analysis using Volterra series representation. *Bell System Technical Journal*, 46:991–1024, 1967. doi: 10.1002/j.1538-7305.1967.tb01723.x.

S. Narayanan. Application of Volterra series to intermodulation distortion analysis of transistor feedback amplifiers. *IEEE Transactions on Circuit Theory*, 17(4):518–527, 1970. doi: 10.1109/TCT.1970.1083157.

J.C. Pedro and N.B. Carvalho. *Intermodulation Distortion in Microwave and Wireless Circuits*. Artech House microwave library. Artech House, Boston, 2003. ISBN 9781580533560.

J.C. Pedro, N.B. Carvalho, and P.M. Lavrador. Modeling nonlinear behavior of band-pass memoryless and dynamic systems. In *IEEE MTT-S International Microwave Symposium Digest, 2003*, volume 3, pages 2133–2136, 2003. doi: 10.1109/MWSYM.2003.1210584.

J.C. Pedro, D.E. Root, J. Xu, and L.C. Nunes. *Nonlinear Circuit Simulation and Modeling: Fundamentals for Microwave Design*. The Cambridge RF and Microwave Engineering Series. Cambridge University Press, Cambridge, UK, 2018. ISBN 9781107140592.

R. Raich and G.T. Zhou. On the modeling of memory nonlinear effects of power amplifiers for communication applications. In *Proceedings of 2002 IEEE 10th Digital Signal Processing Workshop, 2002 and the 2nd Signal Processing Education Workshop*, pages 7–10, 2002. doi: 10.1109/DSPWS.2002.1231065.

S.O. Rice. Volterra systems with more than one input port — distortion in a frequency converter. *The Bell System Technical Journal*, 52(8):1255–1270, 1973. doi: 10.1002/j.1538-7305.1973.tb02019.x.

M. Schetzen. Nonlinear system modeling based on the Wiener theory. *Proceedings of the IEEE*, 69(12):1557–1573, 1981. doi: 10.1109/PROC.1981.12201.

D. Schreurs, M. O'Droma, A.A. Goacher, and M. Gadringer. *RF Power Amplifier Behavioral Modeling*. Cambridge University Press, Cambridge, UK, 2008. ISBN 9780511619960.

N. Wiener. Response of a nonlinear device to noise. Technical report v-16s, M.I.T. Radiation Lab, Apr. 1942.

J. Wood. *Behavioral Modeling and Linearization of RF Power Amplifiers*. Artech House Microwave Library. Artech House, Boston, 2014. ISBN 9781608071203.

3

Discrete-time Volterra Models

3.1 Introduction

In this chapter, the reader can learn specific aspects concerning the modeling of phenomena responsible for the generation of nonlinear distortions in communication systems. Several published books are focused on nonlinear modeling descriptions at device or circuit level, so that one purpose of this text is to exploit this knowledge to present a general procedure based on the useful Volterra series theoretical tool to derive behavioral models at system-level.

A principal objective is, therefore, the nonlinear modeling of the various blocks which constitute a wireless communications system in general, and a transmitter in particular. Recalling that the generation and amplification of radiofrequency signals is possible by exploiting the benefits of the electronic devices nonlinear characteristics, the emphasis will be put on those blocks most significantly contributing to the pernicious nonlinear effects: the power amplifier and the I/Q modulator.

The adverse effects generated by the nonlinear impairments have a twofold consequence in the design of a wireless communications transmitter. On the one hand, the engineers have the predominant requirement of generating a radiofrequency signal which must incorporate the baseband information with a high accuracy and, on the other hand, to comply with the restriction of transmitting a spectrum limited to the assigned radio channel. The disadvantageous consequence of the system nonlinear behavior in the transmitter has a serious impact on both requirements, perturbing the in-band signal fidelity as well as generating out-of-band spectral content causing interference in the adjacent channels. The present-day trend of highly efficient power amplifiers designs, with a more accentuated nonlinear behavior, contributes even more to make things worse.

To anticipate a solution to this awkward situation and alleviate the designer task, an insightful understanding of the transmitter blocks and their mathematical models seems to be an invaluable resource. Beginning with the typical power

A Volterra Approach to Digital Predistortion: Sparse Identification and Estimation, First Edition.
Carlos Crespo-Cadenas, María José Madero-Ayora, and Juan A. Becerra.

amplifier, for which the input and the output are real-valued radiofrequency signals, a classical Volterra series in the continuous-time domain can be a genuine option for modeling. However, in wireless systems, the signal to be transmitted is generated by incorporating first the discrete-time complex-valued baseband signal $x(k)$ to the radiofrequency continuous-time signal $\tilde{x}(t)$ in an I/Q modulator, and then this signal is applied to the power amplifier that delivers the radiofrequency continuous-time output signal $\tilde{y}(t)$. An idealized schematic is shown in Figure 3.1(a) to illustrate this fact.

Ideally, the sampled complex envelope of the power amplifier input is a scaled version of the baseband signal $x(k)$ and, therefore, it is more useful from a computational viewpoint to adopt a discrete-time Volterra model that connects $x(k)$ with the output complex envelope $y(k)$ in the discrete-time domain. This reasoning leads to the baseband equivalent representation depicted in Figure 3.1(b) for power amplifiers modeling and to the equivalent baseband Volterra model with odd-order terms discussed in Section 2.8, referred to in this text as *univariate* baseband Volterra model.

Even though a modern I/Q modulator is nearly ideal owing to the high optimization reached during its design, the implemented hardware introduces unavoidable linear and nonlinear impairments. To understand how these effects can be minimized, the focus of this text is addressed to the explanation of the I/Q modulator model, a matter extensible to any other type of nonlinear modules in transmitters or receivers, such as demodulators and mixers. In fact, all of them share the characteristic of being nonlinear subsystems in which the presence of one signal at the first input controls a second input signal. In the case of a modulator,

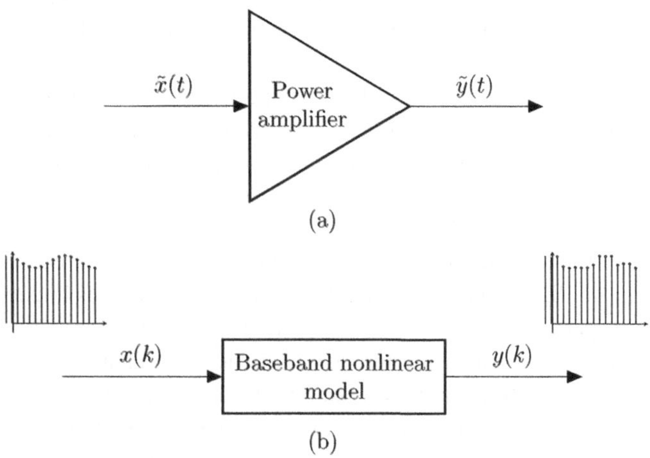

Figure 3.1 Real-valued PA model (a) and complex-valued baseband behavioral model (b).

the information of the discrete-time baseband signal, applied to the first input, is incorporated to the local oscillator continuous-wave carrier, applied to the second input, thus generating an output radiofrequency signal with amplitude and phase varying according the baseband signal.

Just as the univariate Volterra series is perhaps the best analytical approach to modeling weakly nonlinear phenomena in a power amplifier, the double Volterra series is possibly the most opportune method to predict nonlinear distortions in a bivariate system like the I/Q modulator. This is not the only reason why we have included the discrete-time version of the double Volterra series in the discussion of this chapter, but also because it will function in a further step as a theoretical foundation to formally deduce a comprehensive Volterra series for complex-valued signals. This complex Volterra series representation allows the modeling of an I/Q modulator with a complex-valued baseband signal $x(k)$ applied to the input and delivering the complex envelope of the radiofrequency output signal $y(k)$. Beyond this general widely nonlinear representation, we will later discuss in Chapter 4 the link between the I/Q impairments and the parameters of the specific model derived on the basis of the modulator architecture.

We complete our discussion by including in this chapter various topics, such as a summary of model complexity reduction approaches, the nontraditional Volterra–Parafac model, the discrete Volterra model in the frequency domain and an exposition of the figures of merit adopted in the case of evaluating communication signals distortions.

3.2 Discrete-time Volterra Models for Power Amplifiers

3.2.1 Volterra Models for Real-valued Systems

Developing the viewpoint of the Volterra series as one of the most useful theoretical tools to model modern power amplifiers for wireless communication systems and recalling that $\tilde{x}(k)$ and $x(k)$ denote the real-valued radiofrequency signal and its complex envelope, respectively, the response $\tilde{y}(k)$ at the fundamental frequency zone of a power amplifier driven by the input signal $\tilde{x}(k) = \Re\{x(k)e^{j\omega_c t_k}\}$, possibly corrupted with nonlinear distortions, is a function of the input samples $\tilde{x}(k)$.

If the system is assumed causal, the dependence is on the present and past samples, $\tilde{x}(k), \tilde{x}(k-1), \tilde{x}(k-2), \ldots$, and in the case of a system with finite memory, the response can be written as

$$\tilde{y}(k) = f\,(\tilde{x}(k), \tilde{x}(k-1), \ldots, \tilde{x}(k-Q)), \tag{3.1}$$

where Q is the memory length and $f(\cdot)$ is a nonlinear function with $Q+1$ variables. Supposing that $f(\cdot)$ is a differentiable function, it can be expressed as a

$(Q+1)$-dimensional Taylor-series expansion and the output can be written as

$$\tilde{y}(k) = \tilde{h}_0 + \sum_{q_1=0}^{Q} \tilde{h}_1(q_1)\tilde{x}(k-q_1) + \sum_{\mathbf{q}_2=\mathbf{0}}^{\mathbf{Q}_2} \tilde{h}_2(\mathbf{q}_2)\tilde{x}(k-q_1)\tilde{x}(k-q_2)$$
$$+ \sum_{\mathbf{q}_3=\mathbf{0}}^{\mathbf{Q}_3} \tilde{h}_3(\mathbf{q}_3)\tilde{x}(k-q_1)\tilde{x}(k-q_2)\tilde{x}(k-q_3) + \cdots. \tag{3.2}$$

This result enables to formally describe the output using a discrete-time Volterra series as (Mathews and Sicuranza, 2000)

$$\tilde{y}(k) = \tilde{h}_0 + \sum_{n=1}^{\infty} \sum_{\mathbf{q}_n=\mathbf{0}}^{\mathbf{Q}_n} \tilde{h}_n(\mathbf{q}_n)\tilde{x}(k-q_1)\tilde{x}(k-q_2)\cdots\tilde{x}(k-q_n), \tag{3.3}$$

where $\tilde{h}_n(\mathbf{q}_n)$ is the nth-order Volterra kernel of the power amplifier, $\mathbf{q}_n = [q_1, q_2, \dots, q_n]^T$ is the delays vector of the nth-order term, with $q_r = 0, 1, \dots, Q_n$, for all r, and $\mathbf{Q}_n$ is the vector of maximum delays. For the output of an amplifier with a blocking capacitor, $\tilde{h}_0 = 0$.

Each term of the Volterra model contains a product of the variables $\tilde{x}(k-q_r)$, and possibly powers of these variables with nonnegative exponents. These terms, referred to as monomials, have an assigned degree defined as the sum of the exponents.[1] The Volterra representation (3.3) is a linear regression model with basis functions, or *Volterra regressors*, given by

$$\xi_{\mathbf{q}_n}(k) = \prod_{r=1}^{n} \tilde{x}(k-q_r) = \tilde{x}(k-q_1)\tilde{x}(k-q_2)\cdots\tilde{x}(k-q_n), \tag{3.4}$$

where the product $\prod_{r=1}^{n}$ is understood to have the value 1 when $n = 0$.

All the products with order n in (3.3) can be grouped to form the nth-order *homogeneous* Volterra term given by

$$\tilde{y}_n(k) = \sum_{\mathbf{q}_n=\mathbf{0}}^{\mathbf{Q}_n} \tilde{h}_n(\mathbf{q}_n)\xi_{\mathbf{q}_n}(k) = \sum_{\mathbf{q}_n=\mathbf{0}}^{\mathbf{Q}_n} \tilde{h}_n(\mathbf{q}_n)\tilde{x}(k-q_1)\tilde{x}(k-q_2)\cdots\tilde{x}(k-q_n). \tag{3.5}$$

To conclude this section, let us summarize some relevant properties of the Volterra models. Focusing on the Volterra model (3.3), it can be written in compact form as

$$\tilde{y}(k) = \tilde{H}\left[\tilde{x}(k)\right], \tag{3.6}$$

1 Although in systems dependent only on the signal $\tilde{x}(k)$, the term *order* is often used as equivalent to *degree*, owing to the fact that the natural nonlinear ordering is given by the degree, this equivalence cannot be directly extended to systems with more than one input signal.

where $\hat{H}[\cdot]$ operates over the input signal. It presents some distinctive properties that we briefly enumerate below:

1. The system $\hat{H}[\cdot]$ is completely defined by the kernels $\tilde{h}_n(\mathbf{q}_n)$.
2. The power amplifier is a *system invariant to an integer shift*, i.e., it satisfies

$$\tilde{y}(k+m) = \hat{H}\left[\tilde{x}(k+m)\right]. \tag{3.7}$$

3. Given that $0 \leq q_r \leq Q_r$ for all $r = 1, \dots, n$, the system is *causal with finite memory*.
4. The Volterra kernels $\tilde{h}_n(\mathbf{q}_n)$ are *multidimensional and symmetrical* discrete functions that completely describe the power amplifier. They can be considered as a generalized nth-order impulse response of the system.
5. The number of coefficients of the generic kernel $\tilde{h}_n(\mathbf{q}_n)$ is $n_c = (Q_n + 1)^n$. However, because of the symmetry property, several regressors are redundant allowing the total number of coefficients to be reduced to

$$n_c = \frac{(Q_n + n)!}{n!Q_n!}. \tag{3.8}$$

6. The Volterra model (3.3) is *linear with respect to the kernels.*
7. According to the linear property, if the kernels are expressed as $\tilde{h}_n(\mathbf{q}_n) = \tilde{h}_n^{(a)}(\mathbf{q}_n) + \tilde{h}_n^{(b)}(\mathbf{q}_n)$, any Volterra model can be conceived as the superposition of two models $\hat{H}\left[\tilde{x}(k)\right] = \hat{H}^{(a)}\left[\tilde{x}(k)\right] + \hat{H}^{(b)}\left[\tilde{x}(k)\right]$.
8. If the input changes to $a\tilde{x}(k)$, with a a real scalar, the nth-order homogeneous Volterra term is scaled by a^n.
9. The nth-order term of the Volterra model can be computed as the n-dimensional convolution (3.5).
10. For a given power amplifier, a Volterra model *exists* if every bounded-input signal produces a bounded-output signal, i.e., if it is stable in the bounded-input bounded-output (BIBO) sense. A sufficient condition for the BIBO stability of a nth-order Volterra term is

$$\sum_{\mathbf{q}_n=\mathbf{0}}^{\mathbf{Q}_n} |\tilde{h}_n(\mathbf{q}_n)| < \infty. \tag{3.9}$$

    In that case, the mathematical model is analytical.

3.2.2 The Equivalent Baseband Volterra Model

Although the Volterra series approach is a natural choice when nonlinear effects of real-valued systems need to be modeled, in communication systems, a nonlinear baseband model is more convenient to express the relationship between the complex envelopes of the input $x(k)$ and the fundamental zone output $y(k)$.

Starting from the Volterra series RF model of a power amplifier, Benedetto, Biglieri, and Daffara derived in Benedetto et al. (1979) the equivalent baseband model (2.86). Written in discrete-time form, the baseband Volterra input–output relationship truncated to the odd nonlinear order N is

$$y(k) = \sum_{p=0}^{(N-1)/2} \sum_{\mathbf{q}_{2p+1}=\mathbf{0}}^{\mathbf{Q}_{2p+1}} h_{2p+1}(\mathbf{q}_{2p+1}) \prod_{r=1}^{p+1} x(k-q_r) \prod_{r=1}^{p} x^*(k-q_{p+1+r}), \tag{3.10}$$

where only odd-order terms with indices $n = 2p + 1$ contribute to the sum. The relationship between the baseband Volterra kernel $h_{2p+1}(\mathbf{q}_{2p+1})$ and the RF kernel $\tilde{h}_{2p+1}(\mathbf{q}_{2p+1})$ is

$$h_{2p+1}(\mathbf{q}_{2p+1}) = \frac{1}{2^{2p}} \binom{2p+1}{p} \tilde{h}_{2p+1}(\mathbf{q}_{2p+1}) \left(\prod_{r=1}^{p+1} e^{-j\Omega_c q_r} \right) \left(\prod_{r=1}^{p} e^{j\Omega_c q_{p+r+1}} \right), \tag{3.11}$$

where $\Omega_c = 2\pi f_c t_s$ if f_c is the carrier frequency and t_s is the sampling time (Raich and Zhou, 2002). Relationship (3.10), referred to as *full Volterra* (FV) model, has been truncated to a maximum nonlinear order for practical reasons.

The nth-order kernel $h_n(\mathbf{q}_n)$ can be visualized as a discrete grid forming an n-dimensional (n–D) hypercube with separable symmetry, i.e., it is symmetric under any permutation of its first $p + 1$ indices, and it is also separately symmetric under any permutation of its last p indices.

If the kernels $h_n(\mathbf{q}_n)$ are viewed as n-way arrays, $h_1(\mathbf{q}_1)$ is a vector, the third-order kernel $h_3(\mathbf{q}_3)$ is a 3-way array, the fifth-order kernel $h_5(\mathbf{q}_5)$ is a 5-way array, and so on. The third-order kernel grid for $q_r = \{0, 1, 2\}$, is represented in Figure 3.2 with the hope that the readers' imagination can extend the vision to higher dimensions.

Observe that redundancy of regressors associated to some coefficients allows a reduction of the discrete grid points. For instance, the q_2–q_3 facet vertex is associated to the same regressor $|x(k-2)|^2 x(k)$ of the q_1–q_3 facet vertex and can be discarded. The rejected coefficients are drawn with empty dots showing a reduction from 27 initial coefficients to 18 active coefficients in this example.

3.2.3 Multidimensional Signal Processing

To develop the study of the discrete-time Volterra models, we need to discuss the necessary fundamental concepts for multidimensional signal processing. Bearing in mind that some readers may be unfamiliar with those notions, this section is dedicated to extending the basics of one-dimensional discrete-time systems to multidimensional ones in a simple way.

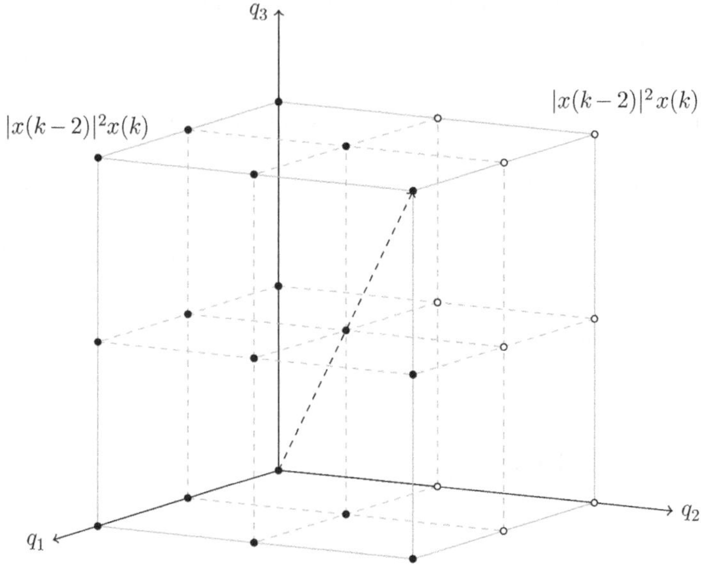

Figure 3.2 Grid of the third-order kernel for the baseband Volterra model. Symmetry allows to discard coefficients associated to redundant regressors (empty dots) to reduce the model complexity.

3.2.3.1 Frequency-domain Characterization of Discrete Signals and Systems

Consider a multidimensional periodic sampling of a n-dimensional continuous-time signal $x_a(\mathbf{t}_n) \equiv x_a(t_1, t_2, \dots, t_n)$ with a regular sampling period t_s. Then, the discrete signal $x(\mathbf{k}_n)$ is obtained by rectangular sampling

$$x(\mathbf{k}_n) \equiv x(k_1, k_2, \dots, k_n) = x_a(k_1 t_s, k_2 t_s, \dots, k_n t_s), \tag{3.12}$$

where $k_r = 0, 1, \dots, K_r$, for all $r = 1, \dots, n$. Notice that we can define the vector $\mathbf{K}_n = \left[K_1, K_2, \dots, K_n\right]$.

The definitions of the multidimensional Fourier transform relations for continuous signals are given by

$$X_a(\boldsymbol{\omega}_n) = \int_{-\infty}^{\infty} x_a(\mathbf{t}_n) e^{-j\boldsymbol{\omega}_n^T \mathbf{t}_n} d\mathbf{t}_n, \tag{3.13}$$

and

$$x_a(\mathbf{t}_n) = \frac{1}{(2\pi)^n} \int_{-\infty}^{\infty} X_a(\boldsymbol{\omega}_n) e^{j\boldsymbol{\omega}_n^T \mathbf{t}_n} d\boldsymbol{\omega}_n, \tag{3.14}$$

where $\boldsymbol{\omega}_n = [\omega_1, \dots, \omega_n]^T$. In the case of bandlimited signals, the Fourier transform $X_a(\boldsymbol{\omega}_n)$ is zero outside some region of finite extent in the $\boldsymbol{\omega}_n$ plane, defined by $|\omega_i| \leq \pi / t_s = \pi B$, and we can conclude for the Fourier transform of $x(\mathbf{k}_n)$ that

$$X(\boldsymbol{\omega}_n t_s) = B^n X_a(\boldsymbol{\omega}_n). \tag{3.15}$$

Therefore, the discrete-time signal $x(\mathbf{k}_n)$ in (3.12) can be written as

$$x(\mathbf{k}_n) = \frac{1}{(2\pi B)^n} \int_{-\pi B}^{\pi B} X(\boldsymbol{\omega}_n t_s) e^{j\mathbf{k}_n^T \boldsymbol{\omega}_n t_s} d\boldsymbol{\omega}_n, \tag{3.16}$$

and the n-dimensional Fourier transform can be evaluated using

$$X(\boldsymbol{\omega}_n t_s) = \sum_{\mathbf{k}_n=\mathbf{0}}^{\mathbf{K}_n} x(\mathbf{k}_n) e^{-j\mathbf{k}_n^T \boldsymbol{\omega}_n t_s}. \tag{3.17}$$

Note that $X(\boldsymbol{\omega}_n t_s)$ is periodic and its values are not limited to the original signal bandwidth. Inverting (3.15), we can recover $X_a(\boldsymbol{\omega}_n)$

$$X_a(\boldsymbol{\omega}_n) = \begin{cases} t_s^n X(\boldsymbol{\omega}_n t_s), & |\omega_i| \leq \pi B \\ 0, & \text{otherwise,} \end{cases} \tag{3.18}$$

or

$$X_a(\boldsymbol{\omega}_n) = t_s^n X(\boldsymbol{\omega}_n t_s) \prod_{i=1}^{n} \text{rect}\left(\frac{\omega_i}{2\pi B}\right), \tag{3.19}$$

where

$$\text{rect}(\alpha) = \begin{cases} 1, & |\alpha| \leq \frac{1}{2} \\ 0, & |\alpha| > \frac{1}{2}. \end{cases} \tag{3.20}$$

It is also possible to recover the continuous-time signal $x_a(\mathbf{t}_n)$ from the discrete-time signal:

$$x_a(\mathbf{t}_n) = \sum_{\mathbf{k}_n=\mathbf{0}}^{\mathbf{K}_n} x(\mathbf{k}_n) \prod_{r=1}^{n} \text{sinc}[B(t_r - k_r t_s)] = \sum_{\mathbf{k}_n=\mathbf{0}}^{\mathbf{K}_n} x(\mathbf{k}_n) f(\mathbf{t}_n - \mathbf{k}_n t_s). \tag{3.21}$$

We have used the definition $\text{sinc}(a) = \sin(\pi a)/(\pi a)$. The interpolating function $f(\mathbf{t}_n)$ allows to construct the values of $x_a(\mathbf{t}_n)$ at points between the sample locations given by $\mathbf{t}_n = \mathbf{k}_n t_s$.

For a linear shift-invariant (LSI) discrete multidimensional system, we can write the output sequence as

$$y(\mathbf{k}_n) = \sum_{\mathbf{q}_n=\mathbf{0}}^{\mathbf{Q}_n} h(\mathbf{q}_n) x(\mathbf{k}_n - \mathbf{q}_n), \tag{3.22}$$

where $h(\mathbf{q}_n)$ is the response of the LSI system to a unit impulse. If the system frequency response is defined as

$$H(\boldsymbol{\omega}_n t_s) = \sum_{\mathbf{q}_n=\mathbf{0}}^{\mathbf{Q}_n} h(\mathbf{q}_n) e^{-j\mathbf{q}_n^T \boldsymbol{\omega}_n t_s}, \tag{3.23}$$

the impulse response derived from this frequency response is expressed as

$$h(\mathbf{q}_n) = \frac{1}{(2\pi B)^n} \int_{-\pi B}^{\pi B} H(\boldsymbol{\omega}_n t_s) e^{j\mathbf{q}_n^T \boldsymbol{\omega}_n t_s} d\boldsymbol{\omega}_n. \tag{3.24}$$

Although (3.24) is very useful to explain LSI systems, to compute and implement these systems, it is more practical to adopt an expression that avoids the laborious evaluation of the multidimensional integrals and employs, instead, finite summations. The discrete Fourier transform is an effective instrument to compute and implement these truncated models.

If the n-dimensional frequency response is periodically sampled with a regular frequency separation of $\Delta f = B/N$, then equations (3.23) and (3.24) adopt the form of the discrete N points Fourier transform

$$H(\mathbf{m}_n) = \sum_{\mathbf{q}_n=\mathbf{0}}^{\mathbf{Q}_n} h(\mathbf{q}_n)e^{-j(2\pi/N)\mathbf{q}_n^T\mathbf{m}_n}, \tag{3.25}$$

for $m_r = 0, 1, \ldots, N-1$, and the inverse discrete Fourier transform

$$h(\mathbf{q}_n) = \frac{1}{N^n}\sum_{\mathbf{m}_n=\mathbf{0}}^{\mathbf{M}_n} H(\mathbf{m}_n)e^{j(2\pi/N)\mathbf{m}_n^T\mathbf{q}_n}, \tag{3.26}$$

for $q_r = 0, 1, \ldots, N-1$. We have two alternative ways of expressing the output of an LSI system: in the time domain, using the impulse response (3.26), or in the frequency domain, using the discrete Fourier transform (3.25).

The intrinsic nature of the Volterra models is the n-dimensional character of the kernels $h_n(\mathbf{q}_n)$ or the transfer functions $H_n(\omega_n t_s)$, regardless of the time-domain or frequency-domain representation we are using. In Section 3.4, a new mathematical formulation is employed to derive specific models under a multidimensional Volterra–Parafac perspective.

3.3 Reducing the Volterra Model Complexity

3.3.1 Need for Model Pruning

The FV model is a computationally expensive multidimensional filter requiring a huge number of coefficients that grow exponentially with the filter order n and the maximum delay Q. The large number of coefficients that (3.10) demands, even if it is truncated to a maximum nonlinear order, is the main origin of this model complexity and a formidable limitation to its application to power amplifiers behavioral modeling. In other words, while the output of a given power amplifier driven by a narrowband signal can be described by a low-complexity memoryless model, the output of a wide bandwidth input signal would require a considerable memory, which may lead to an unacceptable complexity.

This great number of parameters poses the challenge of model order reduction. To reduce the model complexity, several approaches have been proposed (Zhu and Brazil, 2004). Bearing in mind that the primary cause of the computational cost is the large number of components in the n-way array assembling the kernel $h(\mathbf{q}_n)$, an immediate proposal is to neglect some components of these kernels following an aprioristic criterion. A different approach is to reduce the model according to the previous knowledge of the device equivalent circuit. Other proposals have been based on the assumption of a sparse delay tap structure (Ku and Kenney, 2003). In all cases, a cut and try procedure is necessary to obtain an optimized model, often based on the brainwave of a clever researcher.

Finally, a more rigorous procedure is to start with the richest set of regressors available, for example, the FV model, and carrying on a pursuit of the active regressors, a search that is stopped according to a given criterion. This efficient method is suitable in sparse systems with a reduced number of active regressors, as it is the case in power amplifiers models. The final chapters of this book are dedicated to a detailed discussion of the search and identification of the active regressors, and coefficient estimation procedures for an optimized model.

3.3.2 Heuristic Reduction of the Volterra Model Complexity

The simplest solution to the high computational complexity problem seems to lie in reducing the number of model coefficients, supposing an *a priori* manageable kernel structure (Zhu and Brazil, 2004). Moving that way, we reach model formats for which no previous information is necessary about the physical or equivalent circuit models. Some simple models are described below, deduced if some simplifications of the kernel structure of the univariate Volterra relationship (3.10) are assumed *a priori*.

3.3.2.1 The Univariate Zero-memory Volterra Model

The baseband output of a power amplifier with zero memory can be explained if all $\mathbf{Q}_n$ are set to zero in (3.10), yielding the following complex polynomial relationship

$$y(k) = \sum_{\substack{n=1 \\ (n\,\text{odd})}}^{N} h_n(0)|x(k)|^{n-1}x(k). \tag{3.27}$$

Figure 3.3(a) illustrates this *univariate* zero-memory simple structure composed only of terms with n odd, or equivalently, only terms with even power exponent $n-1$ of the envelope $|x(k)|$. Observe, however, that if terms with n even are also included in (3.27) (i.e., odd exponents $n-1$), the structure becomes the widespread *memoryless* model. The structures with odd and even envelope orders, referred to in this text as *bivariate* memoryless Volterra models, are discussed in Chapter 4.

To be precise, the model is strictly-memoryless only if the kernels are real-valued and the power amplifier has no AM–PM conversion. Nevertheless, it is widely

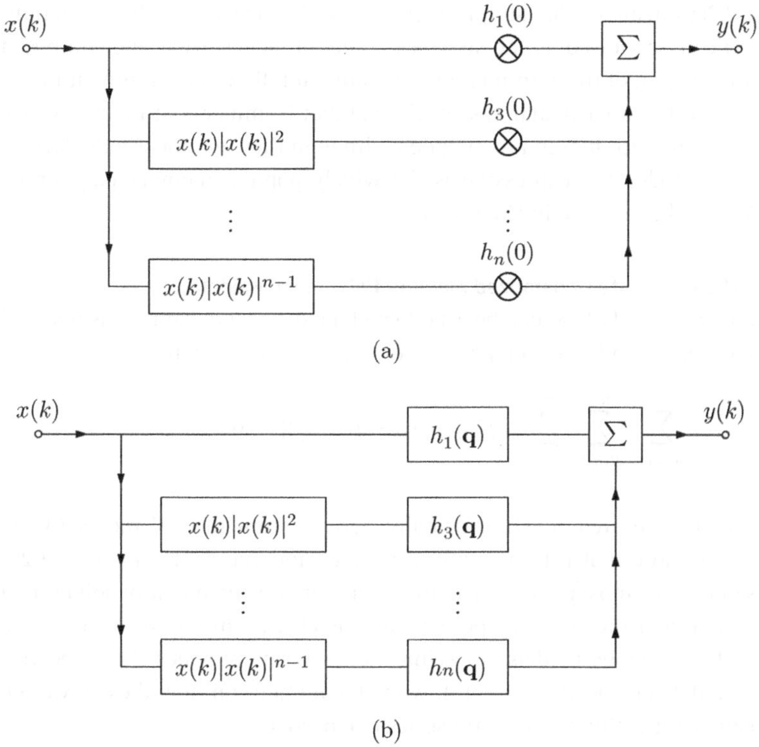

Figure 3.3 Volterra models with aprioristic reduced complexity. (a) The univariate zero-memory model and (b) the univariate memory polynomial model.

accepted that the model is designed as memoryless if the condition $\mathbf{Q}_n = \mathbf{0}$ is met, even in the case that the kernels are complex-valued.

3.3.2.2 The Univariate Memory Polynomial Model

The simplest form to include memory in the model of a power amplifier is considering that $q_1 = q_2 = \cdots = q_n = q$ for all n in (3.10), resulting in the relation

$$y(k) = \sum_{\substack{n=1 \\ (n\,\mathrm{odd})}}^{N} \sum_{q=0}^{Q} h_n(q)|x(k-q)|^{n-1}x(k-q). \tag{3.28}$$

It is straightforward to see that the composite structure of the kernel has been reduced to $[h_n(0), h_n(1), \ldots, h_n(Q)]^T$, a vector given by the diagonal elements of the n-way array $h_n(\mathbf{q}_n)$. For the sake of illustration, recall the structure of $h_3(\mathbf{q}_3)$ shown in Figure 3.2.

Figure 3.3(b) displays this simple memory model structure, whose relationship (3.28) can be viewed as the sum of parallel branches, each one composed of a nonlinear term of the zero-memory polynomial followed by a finite impulse response filter, i.e., a particular case of the parallel Hammerstein model. Notice also that if terms with n even are included, although they are not strictly derived from the FV model, the expression is the widely popular memory polynomial model (MP), to be revisited in Chapter 4.

3.3.2.3 The Univariate Generalized Memory Polynomial Model

A generalized form of (3.28) can be established if a delay of q_2 samples is inserted between the signal and its envelope, providing the model given by

$$y(k) = \sum_{\substack{n=1 \\ (n\,\text{odd})}}^{N} \sum_{q_1=0}^{Q_1} \sum_{q_2=0}^{Q_2} h_n(q_1, q_2)|x(k - q_1 - q_2)|^{n-1} x(k - q_1). \tag{3.29}$$

This is an orthodox relationship formally derived from the FV model (3.10). Observe once more that if terms with n even are incorporated, equation (3.29) becomes one of the most popular structures, the generalized memory polynomial (GMP) model, and the accuracy performance is clearly improved. In the same way as in the previous models, allowing n-even terms provides the respective bivariate model version. In Chapter 4, we will comment on how these bivariate models can be set within the Volterra series perspective.

3.4 Discrete-time Double Volterra Model

An indispensable block in a wireless communications transmitter is the I/Q modulator, a system with a topology composed of two branches, the in-phase branch and the quadrature branch, exemplified in Figure 3.4(a). Focusing on the in-phase branch, it can be considered as a bivariate nonlinear (Bi-NL) system with two real-valued inputs, the in-phase signal $x_I(t)$ and the continuous-wave carrier signal of the local oscillator. Similar reflections can be made for the quadrature branch, whose contribution is added to the in-phase response to produce the RF modulated signal. This output RF signal is generated by the modulator with possibly linear and nonlinear distortions and, just as in the continuous-time domain, a discrete-time double Volterra model is necessary to thoroughly describe the modulator behavior. This is true for any system driven by two-input signals and, hence, a formal method to the study of these systems is of high concern.

Consider the nonlinear system with memory, which has two-input ports, schematically shown in Figure 3.4(b). Under the assumption of a Volterra series

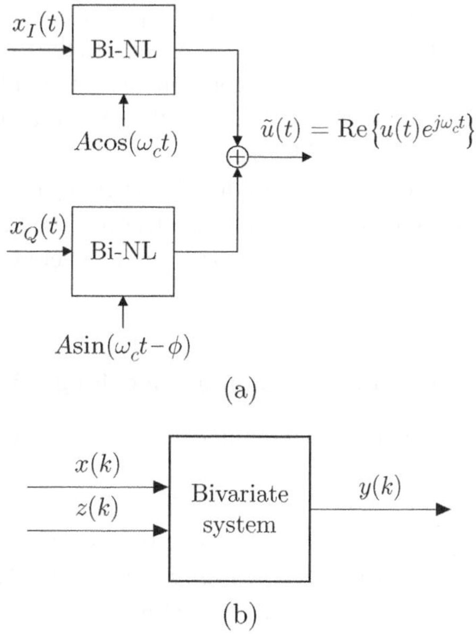

Figure 3.4 Schematic representations of (a) the I/Q modulator modeled with bivariate nonlinear (Bi-NL) blocks and (b) the bivariate nonlinear system.

approach (Rice, 1973), the search of a model for this system suggests the extension of the conventional univariate problem toward a Bi-NL representation.

According to this methodology, the output of a shift-invariant Bi-NL system with real-valued inputs $x(k)$ and $z(k)$ can be expressed as a double Volterra series, written as:

$$
\begin{aligned}
y(k) = & \sum_{n=1}^{\infty} \sum_{\mathbf{q}_n=\mathbf{0}}^{\mathbf{Q}_n} h_{n0}(\mathbf{q}_n) \prod_{r=1}^{n} x(k-q_r) \\
& + \sum_{n=1}^{\infty} \sum_{m=1}^{\infty} \sum_{\mathbf{q}_n=\mathbf{0}}^{\mathbf{Q}_n} \sum_{\mathbf{p}_m=\mathbf{0}}^{\mathbf{P}_m} h_{nm}(\mathbf{q}_n, \mathbf{p}_m) \prod_{r=1}^{n} x(k-q_r) \prod_{s=1}^{m} z(k-p_s) \\
& + \sum_{m=1}^{\infty} \sum_{\mathbf{p}_m=\mathbf{0}}^{\mathbf{P}_m} h_{0m}(\mathbf{p}_m) \prod_{s=1}^{m} z(k-p_s),
\end{aligned}
\tag{3.30}
$$

where $\mathbf{q}_n = [q_1, q_2, \dots, q_n]^T$ and $\mathbf{p}_m = [p_1, p_2, \dots, p_m]^T$ are vectors of delays defined as in the univariate Volterra model (3.3). Likewise, $\mathbf{Q}_n$ and $\mathbf{P}_m$ denote the vectors of maximum delays.

The functions $h_{n0}(\mathbf{q}_n)$ and $h_{0m}(\mathbf{p}_m)$ are standard Volterra kernels of order n and m, respectively. The second group of sums contains bivariate terms with cross products of $x(k-q_r)$ by $z(k-p_s)$ and the multidimensional functions $h_{nm}(\mathbf{q}_n, \mathbf{p}_m)$ are designated bivariate Volterra kernels.

This model can be directly applied to describe the contributions of the modulator in-phase or quadrature branches. In the case of the in-phase branch, the inputs are the baseband signal component $x_I(k)$ and the continuous-wave signal of the local oscillator $\tilde{z}(k)$. The sum of the in-phase and quadrature contributions generates the modulated RF signal.

The detailed procedure to develop particular models for other blocks of a wireless communications system, as it is the case of an I/Q modulator, a demodulator, a nonlinear post-equalizer, or an envelope tracking power amplifier, is discussed in Chapter 4.

3.4.1 Double Volterra Model Properties

The double Volterra model (3.30) can be represented as

$$y(k) = \hat{H}\left[x(k), z(k)\right], \tag{3.31}$$

where $\hat{H}[\cdot,\cdot]$ operates over the two-input signals. As the univariate model, this bivariate model is *invariant to an integer shift*, *causal with finite memory* and *linear with respect to the kernels*. Other properties are:

1. The system $\hat{H}[\cdot,\cdot]$ is completely defined by the kernels $h_{n0}(\mathbf{q}_n)$, $h_{0m}(\mathbf{p}_m)$ and $h_{nm}(\mathbf{q}_n, \mathbf{p}_m)$.
2. The kernel $h_{nm}(\mathbf{q}_n, \mathbf{p}_m)$ is related to the terms with $n + m$ degree.
3. The Volterra kernels $h_{n0}(\mathbf{q}_n)$ and $h_{0m}(\mathbf{q}_m)$ are equivalent to the *multidimensional and symmetrical* kernels of the conventional model. The kernels $h_{nm}(\mathbf{q}_n, \mathbf{p}_m)$ are symmetric with respect to the first n indices on the one hand, and its last m indices on the other hand, thus they are characterized by two structures with *separate symmetry*.
4. The *homogeneous* bivariate Volterra term of degree $n + m$ is given by

$$y_{n+m}(k) = \sum_{\mathbf{q}_n=\mathbf{0}}^{\mathbf{Q}_n} \sum_{\mathbf{p}_m=\mathbf{0}}^{\mathbf{P}_m} h_{nm}(\mathbf{q}_n, \mathbf{p}_m) \prod_{r=1}^{n} x(k - q_r) \prod_{s=1}^{m} z(k - p_s). \tag{3.32}$$

5. If the inputs change to $a_1 x(k)$ and $a_2 z(k)$, with a_1 and a_2 real scalars, the homogeneous bivariate Volterra term of degree $n + m$ is scaled by $a_1^n a_2^m$.

3.5 Volterra–Parafac Model

The baseband model (3.10) was further developed by exploiting the multidimensional essence of the Volterra kernels viewed as tensors and introduced in Favier et al. (2012), and Bouilloc and Favier (2012) as the Volterra–Parafac model. To deal with the Volterra–Parafac model in this section, let us first discuss the mathematical formulations given in Comon et al. (2008), and Cichocki et al. (2009) for tensors and n-way arrays.

3.5.1 Basis of Tensor Algebra

A tensor can be defined as a multidimensional matrix, the order of the tensor being the number of its dimensions. Therefore, a tensor of order N is an N-way array. In particular, a zero-order tensor is a scalar, a first-order tensor is a vector, a second-order tensor is a matrix, and so on.

3.5.1.1 Special Forms of Tensors

Rank-one Tensor: As special cases, the outer product of two vectors yields a rank-one matrix

$$\mathbf{A} = \mathbf{a} \circ \mathbf{b} = \mathbf{a}\mathbf{b}^T, \tag{3.33}$$

and the outer product of three vectors yields a third-order rank-one tensor

$$\underline{\mathbf{H}} = \mathbf{a} \circ \mathbf{b} \circ \mathbf{c}, \tag{3.34}$$

being the tensor generic element computed as

$$h(q_1, q_2, q_3) = a_{q_1} b_{q_2} c_{q_3}. \tag{3.35}$$

- In the Volterra model (3.10), the kernel $h_3(\mathbf{q}_n)$ is an example of third-order tensor. Another example is given if we define the vector $\mathbf{x}_k^{(3)} = [x(k), x(k-1), \dots, x(k-Q_3)]^T$ and construct a third-order tensor with the outer product

$$\underline{\mathbf{X}}_k^{(3)} = \mathbf{x}_k^{(3)} \circ \mathbf{x}_k^{(3)} \circ \left(\mathbf{x}_k^{(3)}\right)^*, \tag{3.36}$$

  for which the generic element is $\xi_{q_1,q_2,q_3}(k) \equiv x(k-q_1)x(k-q_2)x^*(k-q_3)$.

A tensor $\underline{\mathbf{H}}$ of order n has rank-one if it can be written as an outer product of n vectors, i.e.,

$$\underline{\mathbf{H}} = \mathbf{a}^{(1)} \circ \mathbf{a}^{(2)} \circ \cdots \circ \mathbf{a}^{(n)}. \tag{3.37}$$

The rank of a tensor is defined as the minimal number R of rank-one tensors $\underline{\mathbf{H}}^{(r)}$ such that $\underline{\mathbf{H}} = \sum_{r=1}^{R} \underline{\mathbf{H}}^{(r)}$ (Comon et al., 2008, Cichocki et al., 2009).

Symmetric Tensors: An n-way array is called symmetric if its entries do not change under any permutation of its indices. When all the n vectors $\mathbf{a}^{(j)}$ are equal to $\mathbf{g}$, the outer product results in a symmetric rank-one tensor

$$\underline{\mathbf{H}} = \underbrace{\mathbf{g} \circ \mathbf{g} \circ \cdots \circ \mathbf{g}}_{n \text{ times}}. \tag{3.38}$$

According to the cited work Cichocki et al. (2009), we can express a rank-R symmetric tensor as

$$\underline{\mathbf{H}} = \sum_{r=1}^{R} \underline{\mathbf{H}}^{(r)} = \sum_{r=1}^{R} \mathbf{g}^{(r)} \circ \mathbf{g}^{(r)} \circ \cdots \circ \mathbf{g}^{(r)}, \tag{3.39}$$

and recalling the notation introduced in equation (3.3), the generic element can be computed as

$$h(\mathbf{q}_n) \equiv h(q_1, \ldots, q_n) = \sum_{r=1}^{R} g_{q_1}^{(r)} g_{q_2}^{(r)} \cdots g_{q_n}^{(r)} = \sum_{r=1}^{R} \prod_{s=1}^{n} g_{q_s}^{(r)}. \tag{3.40}$$

This tensor representation by a linear combination of rank-one tensors is denoted Parafac (Parallel factor decomposition) model.

3.5.2 Baseband Volterra–Parafac Model

The discrete-time input and output complex envelopes of a communications power amplifier, $x(k)$ and $y(k)$, can be related by the baseband Volterra model (3.10). Exploiting the complexity reduction of the Volterra kernels viewed as tensors, the baseband Volterra–Parafac model for nonlinear communication channels was introduced by Favier et al. (2012), Bouilloc and Favier (2012). The output of the Volterra–Parafac model is written as[2]

$$y(k) = \sum_{\substack{n=1 \\ (n\,\text{odd})}}^{N} \sum_{r=1}^{R_n} \left(\mathbf{x}_k^{(n)T} \mathbf{a}_r^{(n)}\right)^{p+1} \left(\mathbf{x}_k^{(n)H} \mathbf{b}_r^{(n)}\right)^{p}, \tag{3.41}$$

where $p = (n-1)/2, \mathbf{x}_k^{(n)} = [x(k), x(k-1), \ldots, x(k-Q_n)]^T$ and R_n is the symmetric rank of the Volterra kernel ($R_1 = 1$). The vectors $\mathbf{a}_r^{(n)}$ and $\mathbf{b}_r^{(n)}$ are the rth column of the factor matrices with complex-valued entries related to the Volterra kernels by a double symmetric Parafac decomposition.

This Volterra–Parafac approach was further developed in Crespo-Cadenas et al. (2014) considering that the nth-order Volterra kernel in (3.3) can be seen as an n-way symmetrical array. Using (3.40) and defining a new symmetrical array with elements given by

$$\hat{h}_n(\mathbf{q}_n) = \frac{1}{2^{n-1}} \binom{n}{p} \tilde{h}_n(\mathbf{q}_n) = \sum_{r=1}^{R_n} \prod_{s=1}^{n} g_{q_s,r}^{(n)}, \tag{3.42}$$

with $g_{q_s,r}^{(n)}$ real-valued, we can substitute in (3.11) to obtain

$$h_n(\mathbf{q}_n) = \sum_{r=1}^{R_n} \prod_{s=1}^{p+1} \left(g_{q_s,r}^{(n)} e^{-j\Omega_c q_s}\right) \prod_{s=1}^{p} \left(g_{q_{p+1+s},r}^{(n)} e^{j\Omega_c q_{p+1+s}}\right). \tag{3.43}$$

2 The notation here uses $p+1$ copies of the signal and p copies of the conjugate signal.

If we denote $a^{(n)}_{q_s,r} = g^{(n)}_{q_s,r} e^{-j\Omega_c q_s}$, then

$$h_n(\mathbf{q}_n) = \sum_{r=1}^{R_n} \prod_{s=1}^{p+1} \left(a^{(n)}_{q_s,r}\right) \prod_{s=1}^{p} \left(a^{(n)}_{q_{p+1+s},r}\right)^*. \tag{3.44}$$

Therefore, the input–output relationship given by (3.10) for the truncated Volterra–Parafac model becomes

$$y(k) = \sum_{\substack{n=1 \\ (n\,\text{odd})}}^{N} \sum_{\mathbf{q}_n=\mathbf{0}}^{\mathbf{Q}_n} \sum_{r=1}^{R_n} \prod_{s=1}^{p+1} \left(a^{(n)}_{q_s,r} x(k-q_s)\right) \prod_{s=1}^{p} \left(a^{(n)}_{q_{p+1+s},r} x(k-q_{p+1+s})\right)^*. \tag{3.45}$$

After re-arranging the order of the summations, the output is

$$y(k) = \sum_{\substack{n=1 \\ (n\,\text{odd})}}^{N} \sum_{r=1}^{R_n} \prod_{s=1}^{p+1} \left(\sum_{q_s=0}^{Q_s} a^{(n)}_{q_s,r} x(k-q_s)\right) \prod_{s=1}^{p} \left(\sum_{q_{p+1+s}=0}^{Q_{p+1+s}} a^{(n)}_{q_{p+1+s},r} x(k-q_{p+1+s})\right)^*. \tag{3.46}$$

Assuming $Q_1 = Q_2 = \cdots = Q_n = Q$ and recalling that

$$\mathbf{x}_k^T \mathbf{a}_r^{(n)} = a^{(n)}_{0,r} x(k) + \cdots + a^{(n)}_{Q,r} x(k-Q), \tag{3.47}$$

represents the response of a linear filter with coefficients given by the vector $\mathbf{a}_r^{(n)} = [a^{(n)}_{0,r}, \ldots, a^{(n)}_{Q,r}]$ to an input $x(k)$, the modified Volterra–Parafac model can be simplified using vectorial notation as

$$y(k) = \sum_{\substack{n=1 \\ (n\,\text{odd})}}^{N} \sum_{r=1}^{R_n} \left(\mathbf{x}_k^T \mathbf{a}_r^{(n)}\right)^{p+1} \left(\left(\mathbf{x}_k^T \mathbf{a}_r^{(n)}\right)^*\right)^p, \tag{3.48}$$

or written in a more compact form as

$$y(k) = \sum_{\substack{n=1 \\ (n\,\text{odd})}}^{N} \sum_{r=1}^{R_n} \left|\mathbf{x}_k^T \mathbf{a}_r^{(n)}\right|^{n-1} \mathbf{x}_k^T \mathbf{a}_r^{(n)}. \tag{3.49}$$

Figure 3.5 shows the schematic representation of the baseband Volterra–Parafac model (3.49). Recalling the zero-memory model (3.27), the structure can be viewed as an arrangement of parallel Wiener models for whom the complex envelope $x(k)$ is the input of linear filters with coefficients given by the vector $\mathbf{a}_r^{(n)}$, and the corresponding outputs are then applied to memoryless nth order Volterra nonlinearities NLn given by $|\mathbf{x}_k^T \mathbf{a}_r^{(n)}|^{n-1} \mathbf{x}_k^T \mathbf{a}_r^{(n)}$. Two immediate examples are the linear branch output, given by $\mathbf{x}_k^T \mathbf{a}_1^{(1)}$ and the third-order and rank r branch output

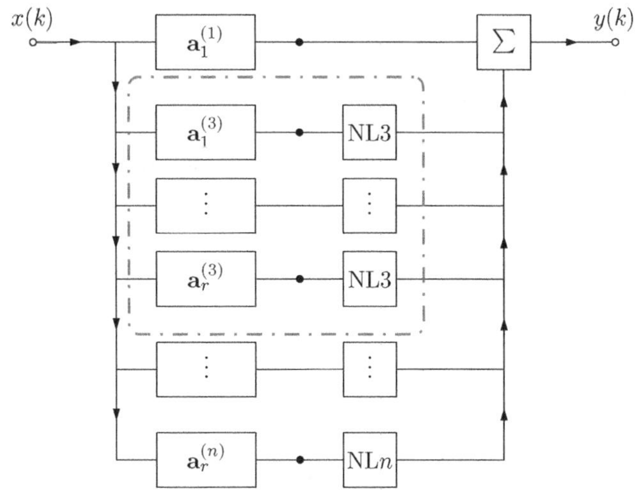

Figure 3.5 Block diagram of the proposed Volterra–Parafac model.

(framed in the figure), computed by the filtered response $\mathbf{x}_k^T\mathbf{a}_r^{(3)}$ applied to the nonlinearity NL3.

Comparing the Volterra–Parafac (3.41) and the modified Volterra–Parafac (3.49) models, we observe that $b_{q_s,r}^{(n)} = (a_{q_s,r}^{(n)})^*$, a result that has been experimentally verified with acquired data of a power amplifier. A 7.6 MHz bandwidth signal with a carrier frequency of 850 MHz was applied to a comercial device and the parameters $\mathbf{a}_r^{(n)}$ and $\mathbf{b}_r^{(n)}$ of equation (3.41) were empirically evaluated. The more significant parameters of the model (3.49) were also estimated and represented in the polar plot of Figure 3.6. In compliance with this graph, the parameters $\mathbf{a}_r^{(n)}$ are equivalent for both models and the relation between parameters $\mathbf{b}_r^{(n)} = \mathbf{a}_r^{(n)*}$ is also demonstrated.

Accordingly, the model complexity can be evaluated if we observe that the number of parameters of the Volterra–Parafac model (3.41),

$$N_{VP} = Q_1 + 2 \sum_{\substack{n=3 \\ (n\,\text{odd})}}^{N} Q_n r_n, \tag{3.50}$$

is reduced to

$$N_{MVP} = Q_1 + \sum_{\substack{n=3 \\ (n\,\text{odd})}}^{N} Q_n r_n, \tag{3.51}$$

parameters of the modified Volterra–Parafac model (3.49).

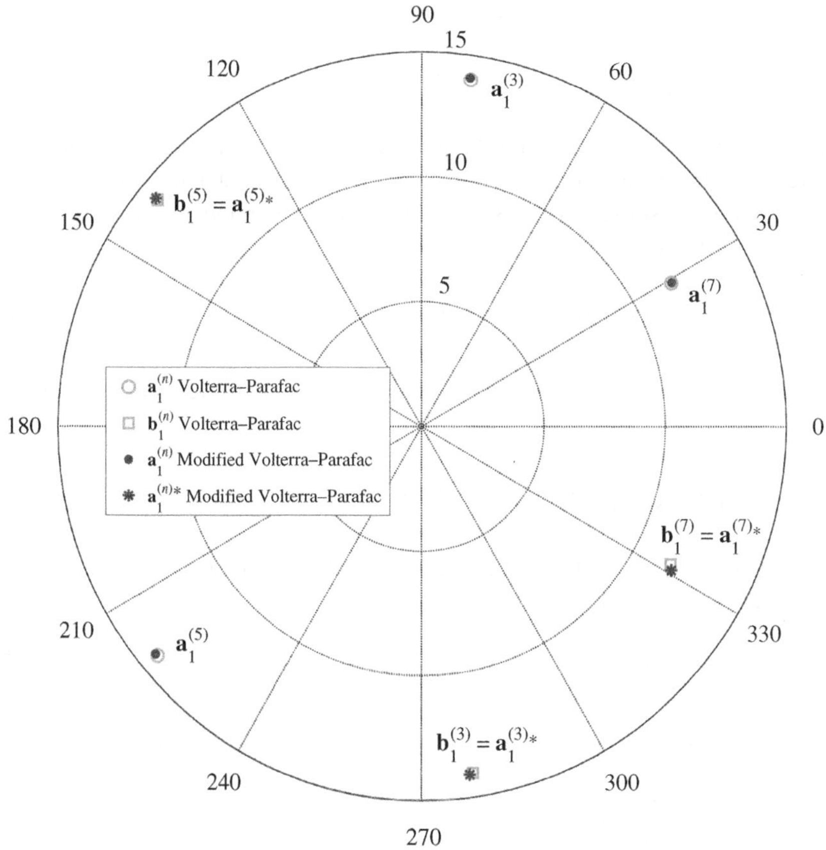

Figure 3.6 Experimental evaluation of the Volterra–Parafac parameters.

3.6 Volterra Models in the Frequency Domain

It is possible to make use of the Fourier transform of discrete-time Volterra representations to generate alternative models in the frequency domain to take advantage of the additional capabilities they can provide with respect to models in the time domain.

The familiar discrete Fourier transformation (DFT) for a unidimensional real-valued input signal $\tilde{x}(k)$, given by

$$\tilde{x}(k) = \frac{1}{N}\sum_{m=0}^{N-1} \tilde{X}(m)e^{j2\pi mk/N}, \qquad k = 0, 1, \ldots, N-1, \tag{3.52}$$

and the inverse discrete Fourier transform

$$\tilde{X}(m) = \sum_{k=0}^{N-1} \tilde{x}(k) e^{-j2\pi mk/N}, \qquad m = 0, 1, \ldots, N-1, \tag{3.53}$$

can be extended to the multidimensional case, as it was discussed in Section 3.2.3.

Referring to the model (3.3), and recalling the n-dimensional discrete Fourier transform of the kernel $\tilde{h}_n(\mathbf{q}_n)$ (3.25), denoted as the nonlinear transfer function of order n, we can write

$$\tilde{H}_n(\mathbf{m}_n) = \sum_{\mathbf{q}_n=\mathbf{0}}^{N-1} \tilde{h}_n(\mathbf{q}_n) e^{-j(2\pi/N)\mathbf{q}_n^T \mathbf{m}_n}, \tag{3.54}$$

and the inverse discrete Fourier transformation yields the Volterra kernel of order n, i.e.,

$$\tilde{h}_n(\mathbf{q}_n) = \frac{1}{N^n} \sum_{\mathbf{m}_n=\mathbf{0}}^{N-1} \tilde{H}_n(\mathbf{m}_n) e^{j(2\pi/N)\mathbf{m}_n^T \mathbf{q}_n}. \tag{3.55}$$

To get a reliable insight into the Volterra model in the frequency domain, let us write the system output (3.3) as

$$\tilde{y}(k) = \sum_{n=1}^{\infty} \tilde{y}_n(k), \tag{3.56}$$

where we have used the concept of the nth-order component defined as

$$\tilde{y}_n(k) = \sum_{\mathbf{q}_n=\mathbf{0}}^{\mathbf{Q}_n} \tilde{h}_n(\mathbf{q}_n) \tilde{x}(k-q_1)\tilde{x}(k-q_2)\cdots\tilde{x}(k-q_n). \tag{3.57}$$

Expressing the delayed copies of the input signal as a function of the input spectrum $\tilde{X}(m)$,

$$\tilde{x}(k-q) = \frac{1}{N} \sum_{m=0}^{N-1} \tilde{X}(m) e^{-j2\pi mq/N} e^{j2\pi mk/N}, \tag{3.58}$$

equation (3.57) can be written as

$$\tilde{y}_n(k) = \frac{1}{N^n} \sum_{\mathbf{m}_n=\mathbf{0}}^{N-1} \left[\sum_{\mathbf{q}_n=\mathbf{0}}^{\mathbf{Q}_n} \tilde{h}_n(\mathbf{q}_n) e^{-j(2\pi/N)\mathbf{q}_n^T \mathbf{m}_n} \right] \prod_{s=1}^{n} \tilde{X}(m_s) e^{j(2\pi/N)m_s k}. \tag{3.59}$$

Recalling that the term in brackets is the nonlinear transfer function of order n given by (3.54), we get

$$\tilde{y}_n(k) = \frac{1}{N^n} \sum_{\mathbf{m}_n=\mathbf{0}}^{N-1} \tilde{H}_n(\mathbf{m}_n) \prod_{s=1}^{n} \tilde{X}(m_s) e^{j(2\pi/N)m_s k}. \tag{3.60}$$

Without loss of generality, the periodicity of the DFT when N is assumed even, and the summation limits can be changed for convenience as

$$\tilde{y}_n(k) = \frac{1}{N^n} \sum_{\mathbf{m}_n=-N/2+1}^{N/2} \tilde{H}_n(\mathbf{m}_n) \prod_{s=1}^{n} \tilde{X}(m_s) e^{j(2\pi/N)m_s k}, \tag{3.61}$$

which describes the nth-order term of the system output as a function of the input spectrum $\tilde{X}(m)$. A complete representation in the frequency domain is achieved after Fourier transforming (3.61), relating the input and the nth-order output spectra

$$\tilde{Y}_n(m) = \sum_{\substack{\mathbf{m}_n=-N/2+1 \\ (m=m_1+\cdots+m_n)}}^{N/2} \tilde{H}_n(\mathbf{m}_n) \prod_{s=1}^{n} \tilde{X}(m_s), \tag{3.62}$$

where the sums over the indices m_s are restricted to the condition $m = m_1 + \cdots + m_n$. Therefore, the output spectrum $\tilde{Y}(m)$ is

$$\tilde{Y}(m) = \sum_{n=1}^{\infty} \tilde{Y}_n(m). \tag{3.63}$$

It is opportune to observe that (3.61) can be generalized if we define the auxiliary multidimensional discrete-time signal

$$\tilde{y}_n(\mathbf{k}_n) = \frac{1}{N^n} \sum_{\mathbf{m}_n=-N/2+1}^{N/2} \tilde{H}_n(\mathbf{m}_n) \prod_{s=1}^{n} \tilde{X}(m_s) e^{j(2\pi/N)m_s k_s}, \tag{3.64}$$

which can be also written in relation with its n-dimensional Fourier transform, $\tilde{Y}_n(\mathbf{m}_n)$,

$$\tilde{y}_n(\mathbf{k}_n) = \frac{1}{N^n} \sum_{\mathbf{m}_n=-N/2+1}^{N/2} \tilde{Y}_n(\mathbf{m}_n) e^{j(2\pi/N)\mathbf{m}_n^T \mathbf{k}_n}. \tag{3.65}$$

Comparing the two expressions, it follows that

$$\tilde{Y}_n(\mathbf{m}_n) = \tilde{H}_n(\mathbf{m}_n) \prod_{s=1}^{n} \tilde{X}(m_s). \tag{3.66}$$

This result states that the nth-order output spectrum (3.62) can be calculated by an element-wise product of the n-way arrays with elements $\tilde{H}_n(\mathbf{m}_n)$ and $\prod_{s=1}^{n} \tilde{X}(m_s)$, subject to the constraint $m = m_1 + m_2 + \cdots + m_n$.

3.6.1 The Baseband Volterra Model in the Frequency Domain

After having addressed the theoretical features of the real-valued Volterra model in the discrete-frequency domain, let us now deal with the equivalent baseband

Volterra model (3.10), a case of most practical interest to microwave and wireless systems designers.

For the sake of conciseness, we base on the results demonstrated above for the real-valued case. Correspondingly, we consider that the system output in the fundamental zone can be written as the sum of nonlinear terms

$$y(k) = \sum_{\substack{n=1 \\ (n\,\text{odd})}}^{\infty} y_n(k), \tag{3.67}$$

where the nth-order term of the output discrete-time signal can be also described as a function of the input baseband spectrum $X(m)$, with the expression

$$y_n(k) = \frac{1}{N^n} \sum_{\mathbf{m}_n=-N/2+1}^{N/2} H_n(\mathbf{m}_n) e^{j(2\pi/N)\overline{m}k} \prod_{s=1}^{p+1} X(m_s) \prod_{s=1}^{p} X^*(-m_{p+1+s}), \tag{3.68}$$

and $\overline{m}_n = m_1 + m_2 + \cdots + m_n$. Recall that the spectrum has a period of N (even) and the discrete Fourier transform of $x^*(k)$ is given by $X^*(-m)$.

The corresponding multidimensional discrete-time signal can be defined as

$$y_n(\mathbf{k}_n) = \frac{1}{N^n} \sum_{\mathbf{m}_n=-N/2+1}^{N/2} H_n(\mathbf{m}_n) e^{j(2\pi/N)\mathbf{m}_n^T\mathbf{k}_n} \prod_{s=1}^{p+1} X(m_s) \prod_{s=1}^{p} X^*(-m_{p+1+s}), \tag{3.69}$$

and after a n-fold Fourier transform, the discrete baseband multispectral density is written as

$$Y_n(\mathbf{m}_n) = H_n(\mathbf{m}_n) \prod_{s=1}^{p+1} X(m_s) \prod_{s=1}^{p} X^*(-m_{p+1+s}). \tag{3.70}$$

Therefore, the input–output relation for the nth-order spectra is

$$Y_n(m) = \sum_{\substack{\mathbf{m}_n=-N/2+1 \\ (m_1+\cdots+m_n=m)}}^{N/2} H_n(\mathbf{m}_n) \prod_{s=1}^{p+1} X(m_s) \prod_{s=1}^{p} X^*(-m_{p+1+s}). \tag{3.71}$$

This operation can be divided, first, in an element-wise product of the partially symmetric n-way array with elements $H_n(\mathbf{m}_n)$ and the rank-one n-way array $\mathbb{X}_n$ with elements

$$\prod_{s=1}^{p+1} X(m_s) \prod_{s=1}^{p} X^*(-m_{p+1+s}), \tag{3.72}$$

and then, summing up all the elements that satisfy $m_1 + m_2 + \cdots + m_n = m$. Equivalently, the elements of each array satisfying this condition can be first selected and then multiplied.

3.6.2 Volterra–Parafac Models in the Frequency Domain

The system representation in the frequency domain is clearly conceived with mathematical operations involving multidimensional arrays, as it has been demonstrated above. Although there are other forms of modeling in the frequency domain, an explanation based on the Volterra–Parafac approach is very attractive.

Under this scenary, let us consider a particular simple baseband system for which the Volterra kernels are diagonal, and therefore, the elements $h_n(\mathbf{q}_n) \neq 0$ only if $q_1 = \cdots = q_n = q$. Substituting this condition in (3.44), the elements of the corresponding n-way array can be written as

$$h_n(\mathbf{q}_n) = \sum_{r=1}^{R_n} |a_n^{(r)}(q)|^{n-1} a_n^{(r)}(q), \tag{3.73}$$

and its n-dimensional discrete Fourier transform is (see [3.25])

$$H_n(\mathbf{m}_n) = \sum_{r=1}^{R_n} \sum_{q=0}^{Q} |a_n^{(r)}(q)|^{n-1} a_n^{(r)}(q) e^{-j(2\pi/N)q\overline{m}_n}. \tag{3.74}$$

Stated in this way, the discrete-frequency transfer function for a rank-one system is actually the one-dimensional discrete Fourier transform of the sequence $|a_n^{(r)}(q)|^{n-1} a_n^{(r)}(q)$, i.e.,

$$H_n(\overline{m}_n) = \sum_{q=0}^{Q} |a_n^{(r)}(q)|^{n-1} a_n^{(r)}(q) e^{-j(2\pi/N)q\overline{m}_n}, \tag{3.75}$$

and we immediately can substitute this useful result in (3.70) to demonstrate that the nth-order discrete-frequency multispectral density is given by

$$Y_n(\mathbf{m}_n) = H_n(\overline{m}_n) \prod_{s=1}^{p+1} X(m_s) \prod_{s=1}^{p} X^*(N - m_{p+1+s}). \tag{3.76}$$

The nth-order output spectrum is then evaluated with the expression

$$Y_n(m) = H_n(\overline{m}_n) \underbrace{\sum_{\mathbf{m}_n=\mathbf{0}}^{N-1}}_{\overline{m}_n=m} \prod_{s=1}^{p+1} X(m_s) \prod_{s=1}^{p} X^*(N - m_{p+1+s}), \tag{3.77}$$

a very convenient way to calculate the output in the frequency domain, given by the element-wise product of the "nth-order input spectrum"

$$X_n(m) = \sum_{\substack{\mathbf{m}_n=\mathbf{0} \\ (\overline{m}_n=m)}}^{N-1} \prod_{s=1}^{p+1} X(m_s) \prod_{s=1}^{p} X^*(N - m_{p+1+s}), \tag{3.78}$$

and the vector of coefficients $H_n(m)$, for all frequency points m. Then,

$$Y_n(m) = H_n(m)X_n(m), \tag{3.79}$$

and the complete output spectrum is computed as a sum of the filtered nth-order input spectra

$$Y(m) = \sum_{n=1}^{\infty} H_n(m)X_n(m). \tag{3.80}$$

Before we conclude this section, it is appropriate to briefly discuss some fundamental properties derived from the frequency domain representation.

Let us consider first the particular case of the static memoryless model, for which the nth-order output spectrum (3.79) is reduced to

$$Y_n^{\mathrm{ML}}(m) = H_n(0)X_n(m), \tag{3.81}$$

and define the nth-order *dynamic factor* as the ratio between the output spectrum of the system to the output spectrum of the memoryless model, for the nth-order terms

$$F_n(m) \triangleq \frac{Y_n(m)}{Y_n^{ML}(m)}. \tag{3.82}$$

According to this definition, the nth-order output spectrum can be regarded as the nth-order output spectrum of the memoryless system, filtered by $F_n(m)$. We can conclude that the nonlinear system can be represented by parallel branches, each one formed by an homogeneous nth-order memoryless nonlinearity and a linear filter.

The nth-order dynamic factor $F_n(m)$ is a theoretical concept, but it also provides some important advantages in the practical analysis of nonlinear systems. We can enumerate the following properties:

1. $F_n(m) = 1$ for a memoryless system.
2. In a general system, $F_n(m)$ only depends on the input signal dynamics, but not on its average amplitude. It means that, generally, the nth-order dynamic factor is different for a continuous wave signal or a wideband modulated signal.
3. It contains all the information about the system memory and, for example, the spectrum asymmetries. Specifically, $|F_n(m)|^2$ gives information about the spectrum asymmetry.
4. $F_1(m) = 1$ for a wideband system.

We can also describe the output spectrum of the system as the filtered version of the output spectrum of the memoryless system, i.e.,

$$Y(m) = F(m)Y^{\mathrm{ML}}(m), \tag{3.83}$$

where $F(m)$ is referred to as the *dynamic factor* for the entire system. Recalling (3.82), the output spectrum is given by

$$Y(m) = \sum_{n}^{\infty} F_n(m) Y_n^{\mathrm{ML}}(m), \tag{3.84}$$

and it is possible to express the dynamic factor as

$$F(m) = \sum_{n}^{\infty} F_n(m) \frac{Y_n^{\mathrm{ML}}(m)}{Y^{\mathrm{ML}}(m)}. \tag{3.85}$$

Although it is reasonable to interpret the effect of $F_n(m)$ in terms of a conventional linear filtering of the memoryless spectrum $Y_n^{\mathrm{ML}}(m)$, it should be emphasized that in a general system this factor depends on both the frequency characteristics of the system and the input spectrum. Only in the case of systems with a diagonal structure of the Volterra kernels, $F_n(m) = H_n(m)/H_n(0)$ is a filter nondependent on the input signal, as it can be deduced from (3.80).

3.6.3 Application of a Frequency Domain MP Model to Linearization in OFDM Transmissions

A simple frequency-domain approach to compensate for the power amplifier distortions for the orthogonal frequency division multiplexing (OFDM) transmission scheme shown in Figure 3.7 was proposed in Allegue-Martínez et al. (2012). According to (3.83), the power amplifier can be regarded a memoryless nonlinear block, denoted as G, cascaded with a linear filter $F(m)$. Therefore, linearization can be performed with a two-block predistorter, a linear filter $F_{\mathrm{PD}}(m)$, and a memoryless block G_{PD}. Ideally, the composed response of the two cascaded nonlinear blocks, $G_{\mathrm{PD}}G$, obeys a linear characteristic, and therefore, the coefficients of $F_{\mathrm{PD}}(m)$ are adjusted to equalize the overall response.

This frequency-domain technique was directly applied to the linearization of a power amplifier driven with a complex envelope of a digital video broadcasting – terrestrial 2 (DVB-T2) 32K-mode signal with OFDM symbols. The input level was chosen to operate the amplifier in a significant nonlinear region and one symbol was used to identify the coefficients of the model. Once the model has

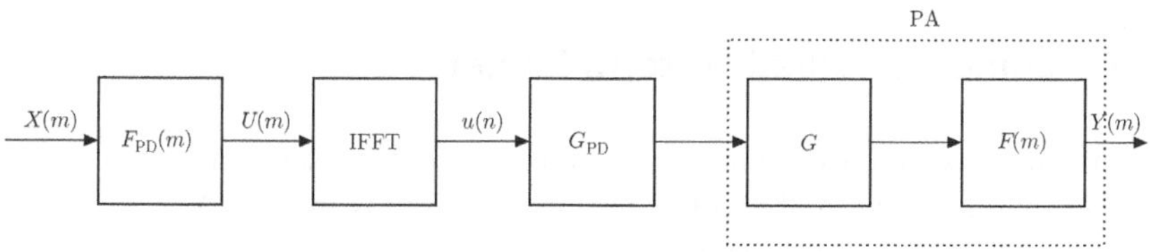

Figure 3.7 Frequency domain linearization in OFDM transmission.

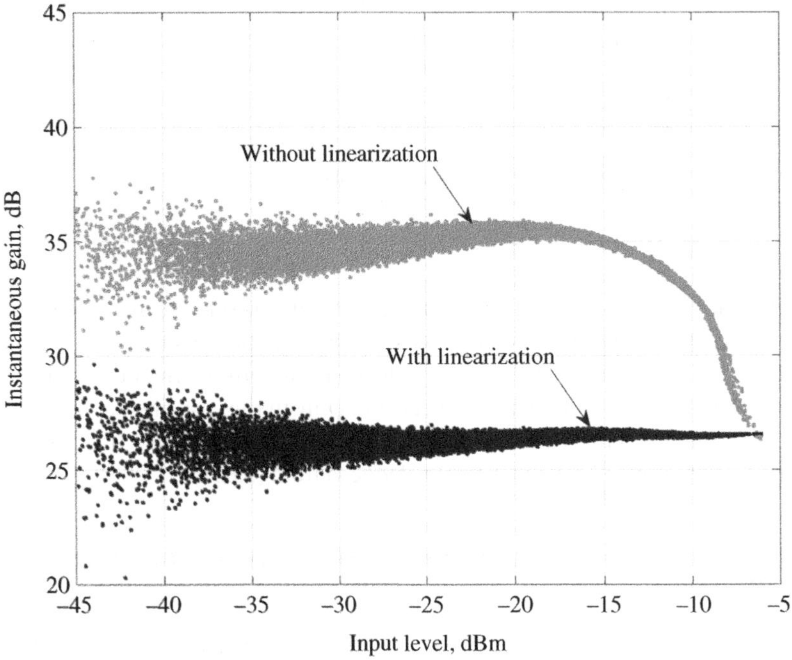

Figure 3.8 Frequency-domain linearization of an OFDM signal in the DVB-T2 standard.

been estimated, the OFDM signal is digitally processed with the corresponding predistorter and sent to be transmitted. The resulting AM–AM characteristic of Figure 3.8 shows a complete linearization of the power amplifier output. A noticeable feature of this procedure is the direct digital processing of the OFDM symbols before inverse Fourier transforming to the time domain.

Another frequency-domain digital predistortion solution for OFDM is proposed in Brihuega et al. (2021). The approach allows to emphasize the in-band signal quality within the allocated channel bandwidth, as well as weaken this requirement at the adjacent channels in order to reduce the complexity of the algorithms, a feature of special interest in the millimeter-wave bands.

3.7 Complex-valued Volterra Model

In the previous sections, we have considered the Volterra series model (3.3) as a significant tool for the analysis of real-valued RF systems and the FV model (3.10) as its equivalent baseband representation. Observe, however, that the FV model cannot adequately describe the behavior of a general complex-valued system.

Consider, for example, the I/Q modulator of Figure 3.4(a), perhaps with quadrature imbalance, driven by the in-phase and quadrature components $x_I(k)$ and $x_Q(k)$. The modulator can be considered as a nonlinear system with the complex-valued input $x(k) = x_I(k) + jx_Q(k)$ and the corresponding FV prediction of the output contains a linear term given by the convolution

$$y(k) = h_{10}(k) * x(k). \tag{3.86}$$

As a consequence of impairments, the output is actually a linear combination of the input and its conjugate, i.e.,

$$y(k) = h_{10}(k) * x(k) + h_{01}(k) * x^*(k). \tag{3.87}$$

This *widely linear transformation* (Schreier and Scharf, 2010) includes the explicit dependence on the conjugate signal $x^*(k)$, not present in conventional Volterra models. Moreover, the internal nonlinearities of the modulator generate more terms inexistent in the FV model (3.10).

In addition to the baseband Volterra model of a power amplifier, some researchers have proposed specific approaches to model other systems in different fields, such as the third-order Volterra model reported for array processing and beamforming (Chevalier et al., 1991, 2011), and the nonlinear extension of the widely linear transformation to a widely nonlinear representation in I/Q modulators modeling (Crespo-Cadenas et al., 2015).

Each of the above approaches is adequate to represent the system in its respective field but is unsuitable to show a good performance when applied to other situations. In this section, we demonstrate a Volterra model capable of solving general problems where signals and systems are complex-valued. The deduction is based on the Wirtinger calculus used to evaluate the output of a nonlinear function with a complex argument (Wirtinger, 1927). This approach performs the computations and derivations in the complex domain, dealing with expressions that are very similar to the real-valued case, in a more elegant and efficient manner (Adali et al., 2011).

Under the Wirtinger calculus perspective, the output $y(k)$ of a system with a complex-valued input $x(k)$ depends also on the complex conjugate $x^*(k)$ and both signals are treated as independent variables. The schematic of a complex-valued nonlinear system is shown in Figure 3.9. Accordingly, the system can be described

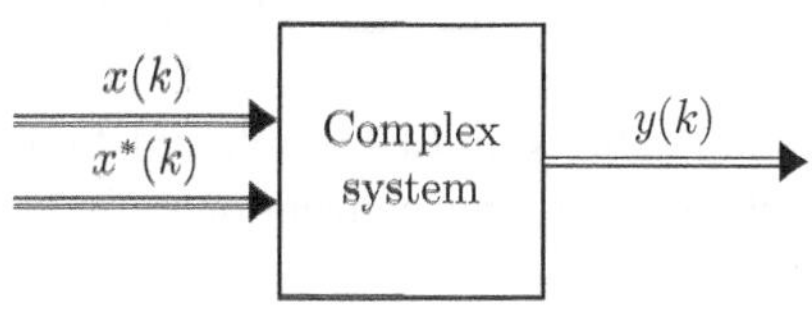

Figure 3.9 Schematic of a complex-valued nonlinear system.

by the two input representation (3.30) replacing $z(k)$ with $x^*(k)$. Therefore, the output can be evaluated as

$$y(k) = \sum_{n=1}^{\infty} \sum_{\mathbf{q}_n=\mathbf{0}}^{\mathbf{Q}_n} h_{n0}(\mathbf{q}_n) \prod_{r=1}^{n} x(k-q_r)$$

$$+ \sum_{n=1}^{\infty} \sum_{m=1}^{\infty} \sum_{\mathbf{q}_n=\mathbf{0}}^{\mathbf{Q}_n} \sum_{\mathbf{p}_m=\mathbf{0}}^{\mathbf{P}_m} h_{nm}(\mathbf{q}_n, \mathbf{p}_m) \prod_{r=1}^{n} x(k-q_r) \prod_{s=1}^{m} x^*(k-p_s)$$

$$+ \sum_{m=1}^{\infty} \sum_{\mathbf{q}_m=\mathbf{0}}^{\mathbf{Q}_m} h_{0m}(\mathbf{q}_m) \prod_{s=1}^{m} x^*(k-q_s). \tag{3.88}$$

Although the Wirtinger calculus formally regards the model as dependent on two signals, $x(k)$ and $x^*(k)$, in reality, the only input signal is $x(k)$ and we can refer to the degree $\nu = n$ in the first sum, $\nu = m$ in the last sum, and $\nu = n + m$ in the sum with bivariate terms, as the nonlinear order of the respective terms.

The input–output relationship can be reorganized after grouping the terms with the same nonlinear order ν, summing up those for which the index $m = 1, 2, \dots, \nu - 1$, and finally summing all nonlinear orders, yielding

$$y(k) = \sum_{\nu=1}^{\infty} \sum_{\mathbf{q}_\nu=\mathbf{0}}^{\mathbf{Q}_\nu} h_{\nu 0}(\mathbf{q}_\nu) \prod_{r=1}^{\nu} x(k-q_r)$$

$$+ \sum_{\nu=2}^{\infty} \sum_{m=1}^{\nu-1} \sum_{\mathbf{q}_{\nu-m}=\mathbf{0}}^{\mathbf{Q}_{\nu-m}} \sum_{\mathbf{p}_m=\mathbf{0}}^{\mathbf{P}_m} h_{\nu-m,m}(\mathbf{q}_{\nu-m}, \mathbf{p}_m) \prod_{r=1}^{\nu-m} x(k-q_r) \prod_{s=1}^{m} x^*(k-p_s)$$

$$+ \sum_{\nu=1}^{\infty} \sum_{\mathbf{q}_\nu=\mathbf{0}}^{\mathbf{Q}_\nu} h_{0\nu}(\mathbf{q}_\nu) \prod_{s=1}^{\nu} x^*(k-q_s). \tag{3.89}$$

To conclude, after recovering the index notation $\nu \to n$ for the nonlinear order, the Volterra model for complex systems is

$$y(k) = \sum_{n=1}^{\infty} \sum_{\mathbf{q}_n=\mathbf{0}}^{\mathbf{Q}_n} h_{n0}(\mathbf{q}_n) \prod_{r=1}^{n} x(k-q_r)$$

$$+ \sum_{n=2}^{\infty} \sum_{m=1}^{n-1} \sum_{\mathbf{q}_{n-m}=\mathbf{0}}^{\mathbf{Q}_{n-m}} \sum_{\mathbf{p}_m=\mathbf{0}}^{\mathbf{P}_m} h_{n-m,m}(\mathbf{q}_{n-m}, \mathbf{p}_m) \prod_{r=1}^{n-m} x(k-q_r) \prod_{s=1}^{m} x^*(k-p_s)$$

$$+ \sum_{n=1}^{\infty} \sum_{\mathbf{q}_n=\mathbf{0}}^{\mathbf{Q}_n} h_{0n}(\mathbf{q}_n) \prod_{s=1}^{n} x^*(k-q_s). \tag{3.90}$$

The relationship (3.90), referred to as **complex Volterra series** model, is the extension of the conventional real-valued Volterra series to complex-valued systems.

Some features can be highlighted:

- If the nonlinear terms of the system can be neglected, the complex Volterra series model reduces to the widely linear transformation (3.87). The entire complex Volterra series model (3.90) can be considered, therefore, a *widely nonlinear transformation.*
- The expression includes terms with n odd and even.
- The FV (3.10) is a particular case of the complex Volterra model when only terms associated to the fundamental frequency zone are taken into account. Accordingly, the FV expression includes the linear terms, $n = 1$, of the first and third blocks of the model (3.90), and those terms with $n = 2p + 1$ and $m = p$ of the cross products in the second block.
- In the memoryless version of (3.90), the even-order terms have the form $x^{n-m}(k)(x^*(k))^m$ with $n - m$ and m integers. For instance, the 2nd-order terms are $x^2(k)$, $x(k)x^*(k) = |x(k)|^2$, and $(x^*(k))^2$.
- This complex Volterra series model is a formal tool to describe not only power amplifiers but also other particular systems. One example is the model proposed for nonlinear detection and estimation in narrowband array processing (Chevalier et al., 1991).
- It can be applied, as well, to the analysis of I/Q demodulators and modulators, or even a complete wireless communications transmitter.
- In particular, this is a useful model for the design of a digital predistorter for the joint compensation of the modulator and the power amplifier impairments.

It is evident that there is an increase in the number of regressors in the complex Volterra model (3.90) with respect to the FV model (3.10), aggravating the computational complexity. On the positive side, this is a very beneficial model with a richer initial set of regressors to be used as the starting point with sparse identification techniques (Reina-Tosina et al., 2013, 2015), an approach we will discuss in Chapters 5 and 6.

3.8 Figures of Merit for Experimental Methods in Modeling and Linearization

To address the description of nonlinear distortions caused by the blocks of a wireless communications system on the basis of empirical results, the definition of new widely accepted figures of merit, suited to modulated communication signals, is clearly beneficial.

Classical figures of merit defined for the evaluation of the most important nonlinear distortion effects have been proposed for one-tone and two-tone tests. For example, the 1-dB compression point or the third-order intercept point,

IP3, are widely used in the nonlinear evaluation of microwave circuits with continuous-wave probing signals. However, when our objective is the analysis of the nonlinear distortions generated in a wireless system driven by a digitally modulated signal, the modern laboratory tests should also include parameters to characterize the power spectral density of the communication signals.

While the output spectrum of a linear system is a scaled copy of the input spectrum, a nonlinear system generates a spectrum with many other components, a phenomenon usually referred to as spectral regrowth. Since the wireless service provider is allowed to occupy just the spectral band inside the communications channel, any out-of-band spectral regrowth that may produce interference to other transmissions must be kept below a given security threshold, especially in the adjacent channels.

A laboratory setup similar to that depicted in Figure 3.10 is widely employed to experimentally characterize a nonlinear block of a wireless transmitter, like a power amplifier, for example. The complex-valued baseband input data originated by software in the computer is downloaded to the vector signal generator, which produces the radiofrequency test signal in its internal I/Q modulator. This test signal is then applied to the device under test (DUT) and the output it produces is down-converted and acquired by the vector signal analyzer. The first direct observation that we can make is the comparison of the output samples to the scaled input samples. In a perfectly linear system the difference is negligible, an indication of an errorless signal transmission. On the contrary, a nonlinear system

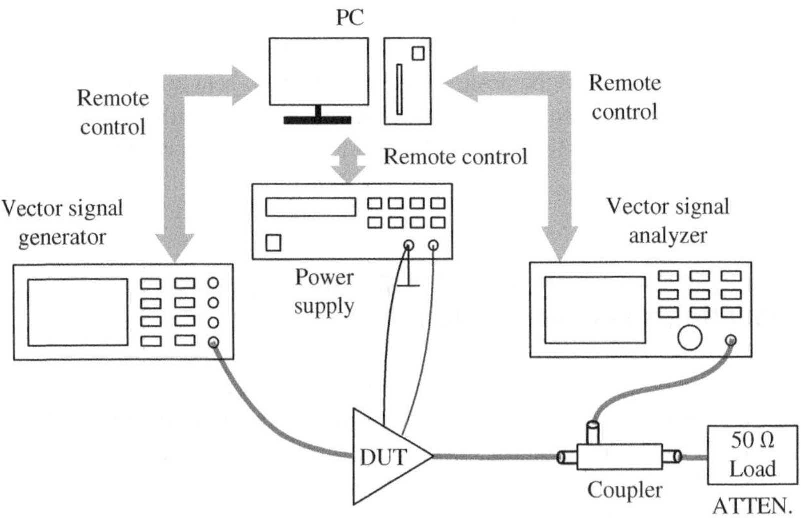

Figure 3.10 Framework of experimental equipment for the characterization of nonlinear systems and the estimation of their distortions.

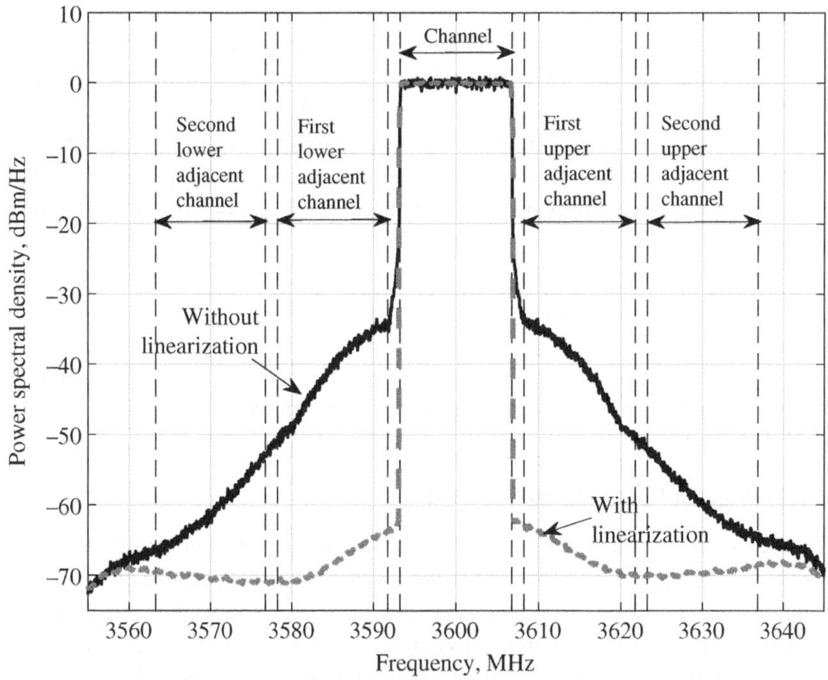

Figure 3.11 Power spectral density of an output signal exhibiting spectral regrowth and producing adjacent channel interference, together with the linearized output.

introduces an error that can be used as a figure of merit of the signal fidelity. If this is the case, the output of the signal analyzer will also show spectral regrowth, and other figures of merit associated to the spectrum of the acquired signal, like the adjacent channel power ratio (ACPR) or the noise power ratio, can be defined to qualify the nonlinear distortions of the evaluated system. Obviously, the ACPR is not only a parameter that evaluates the nonlinear distortions of the system, but also an objective figure to be reduced in the case of linearization techniques like digital predistortion.

Figure 3.11 illustrates the impact of a nonlinear system on the signal spectrum, displaying the spectral regrowth in the adjacent channels (solid trace) and how this regrowth can be reduced by linearization through the use of a digital predistorter.

3.8.1 Normalized Mean Squared Error

As it was already commented, the output of a Volterra model is linear with respect to its coefficients, which makes it feasible to identify a given model employing a straightforward procedure based on the use of arbitrary modulated sampled input and output signals. To directly assess the predictive accuracy of the model, a very useful time-domain waveform metric is the normalized mean squared error (NMSE) (Muha et al., 1999, Ghannouchi et al., 2015).

If $y(k)$ represents the sequence of output samples acquired by the experimental setup and $\hat{y}(k)$ is the corresponding sequence predicted by the model, then the NMSE is a verification metric to evaluate the model's fidelity in the time domain

$$\mathrm{NMSE[dB]} = 10\log\left\{\frac{\sum_{k=0}^{N-1} |y(k)-\hat{y}(k)|^2}{\sum_{k=0}^{N-1} |y(k)|^2}\right\}. \tag{3.91}$$

This metric is the total power of the error vector between the measured and modeled waveforms, normalized to the measured signal power.

3.8.2 Adjacent Channel Power Ratio

Another metric used to evaluate the transmitter output signal quality is the measure of the distortion produced in the adjacent channel because of all spectral components falling on this band. The ACPR is the figure of merit proposed to characterize the out-of-band emission level caused by the nonlinearity. It is specified in all wireless standards involving digital modulation, where typically the signal power is integrated over the occupied bandwidth. In order to share the spectrum with other users, wireless standards also specify a certain frequency offset from the center of the considered channel to locate the center frequency of other adjacent channels with higher or lower frequencies. The integration bandwidth is usually smaller than the frequency offset between the channels, since a certain guard band is considered to minimize interference among signals in adjacent channels (McCune, 2010).

The ACPR is defined as the ratio of the signal power integrated over the occupied bandwidth of the considered adjacent channel to the signal power integrated over the occupied bandwidth of the nominal channel.

$$\mathrm{ACPR[dBc]} = 10\log\frac{P_{\text{Adj. channel}}}{P_{\text{In-channel}}} = 10\log\frac{\int_{\text{Adj. channel}} S_Y(f)df}{\int_{\text{In-channel}} S_Y(f)df}, \tag{3.92}$$

where $S_Y(f)$ denotes the power spectral density function of the output signal.

According to this definition, we can define two ACPR figures, one for the *lower* adjacent channel and another for the *upper* adjacent channel. Moreover, we are implicitly considering the nearest adjacent channels, but the definition can be also applied to the second adjacent channels (lower and upper), also referred to as alternate channels. Figure 3.11 illustrates these adjacent channels and how spectral regrowth can introduce interference in them, whose level is assessed with the help of ACPR.

3.8.3 Noise Power Ratio

A measure of a transmitter dynamic range without spurious nonlinear responses is the noise power ratio (NPR). To evaluate the NPR, the digitally modulated signal is no longer applied with the full signal bandwidth, but with a narrow band in the center of the channel set to zero, as illustrated in Figure 3.12. Due to the nonlinear distortions, the output spectrum will have frequency components trespassing the notched band. In other words, the power spectral density function in this notched band experiments a spectral regrowth that is thus a measure of the co-channel distortion (Pedro and Carvalho, 2003).

The NPR is defined as the ratio of the power spectral density measured in the notched channel to the power spectral density measured in an equivalent filled band in the proximity of the notch.

$$\mathrm{NPR[dB]} = 10\log\frac{P_{\text{Notch band}}}{P_{\text{Filled band}}}. \tag{3.93}$$

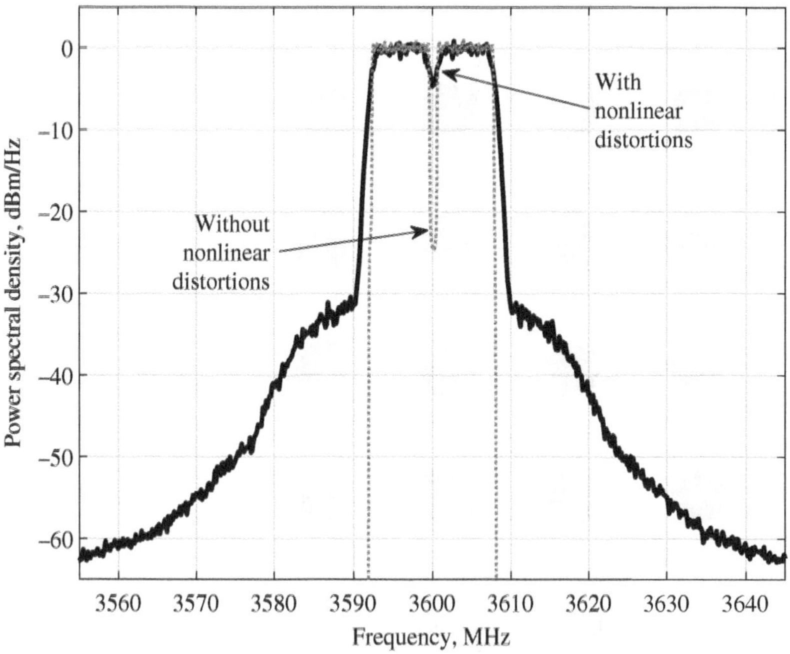

Figure 3.12 Power spectral density for a signal with a notch in the center of the channel to evaluate noise power ratio, with and without nonlinear distortions. The output signal exhibiting nonlinear distortions presents both out-of-band and in-band spectral regrowth.

3.8.4 Adjacent Channel Error Power Ratio

There is an alternative figure of merit to NMSE in order to quantify the predictive accuracy of the model, referred to as adjacent channel error power ratio (ACEPR) and related to the spectrum of the error (Isaksson et al., 2006). To calculate it, the discrete-time error signal is obtained between the sequence of output samples acquired by the experimental setup $y(k)$ and the corresponding sequence predicted by the model $\hat{y}(k)$, that is, $e(k) = y(k) - \hat{y}(k)$. The ACEPR is defined as the ratio of the error signal power located in the band of the adjacent channel to the signal power in the band of the nominal channel.

$$\text{ACEPR[dB]} = 10\log\frac{P_{\text{Error adj. channel}}}{P_{\text{Signal in-channel}}} = 10\log\frac{\int_{\text{Adj. channel}} S_E(f)df}{\int_{\text{In-channel}} S_Y(f)df}, \tag{3.94}$$

where $S_E(f)$ and $S_Y(f)$ denote the power spectral density function of the error signal and the measured output signal, respectively.

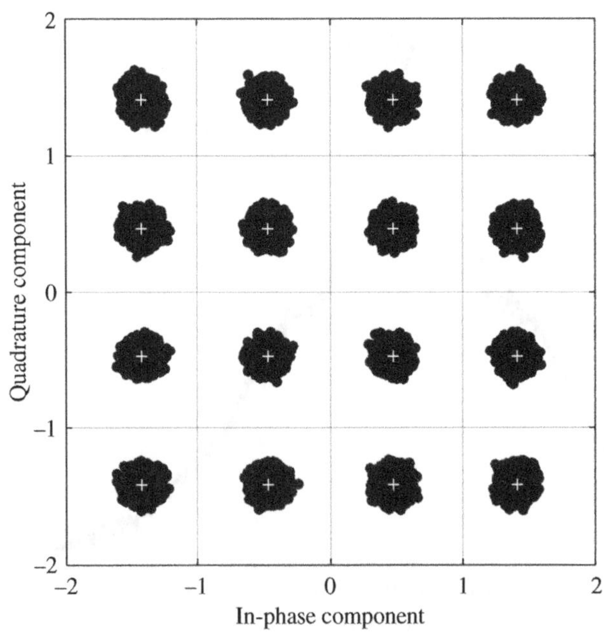

Figure 3.13 Constellation of an output signal where nonlinear distortions produce spreading of the received symbols around their reference positions.

The main difference between the two figures of merit employed for assessing modeling accuracy is that the ACEPR only accounts for the out-of-band error, while NMSE considers both in-band and out-of-band error.

3.8.5 Error Vector Magnitude

To quantify nonlinear distortions that affect inside the signal channel, referred to as *in-band nonlinear distortions*, the error vector magnitude (EVM) is an additional specification in the case of modulated wireless signals. This significant figure of merit measures the alteration experienced by the signal constellation because of the system nonlinearities, an example of which can be observed in Figure 3.13. The error vector is the difference between the distorted constellation point of the output signal and the corresponding ideal point of the input signal. Therefore, the root mean squared value of the EVM is defined as

$$\mathrm{EVM}[\%] = 100 \cdot \sqrt{\frac{\sum_{k=0}^{N-1} |y(k) - y_{\mathrm{Ideal}}(k)|^2}{\sum_{k=0}^{N-1} |y_{\mathrm{Ideal}}(k)|^2}}. \tag{3.95}$$

Bibliography

T. Adali, P.J. Schreier, and L.L. Scharf. Complex-valued signal processing: The proper way to deal with impropriety. *IEEE Transactions on Signal Processing*, 59(11):5101–5125, 2011. doi: 10.1109/TSP.2011.2162954.

M. Allegue-Martínez, M.J. Madero-Ayora, J.G. Doblado, C. Crespo-Cadenas, J. Reina-Tosina, and V. Baena. Digital predistortion technique with in-band interference optimisation applied to DVB-T2. *Electronics Letters*, 48(10):566–568, 2012. doi: 10.1049/el.2012.0279.

S. Benedetto, E. Biglieri, and R. Daffara. Modeling and performance evaluation of nonlinear satellite links — A Volterra series approach. *IEEE Transactions on Aerospace and Electronic Systems*, AES-15(4):494–507, 1979. doi: 10.1109/TAES.1979.308734.

T. Bouilloc and G. Favier. Nonlinear channel modeling and identification using baseband Volterra–Parafac models. *Signal Processing*, 92(6):1492–1498, 2012. doi: 10.1016/j.sigpro.2011.12.007.

A. Brihuega, L. Anttila, and M. Valkama. Frequency-domain digital predistortion for OFDM. *IEEE Microwave and Wireless Components Letters*, 31(6): 816–818, 2021. doi: 10.1109/LMWC.2021.3062982.

P. Chevalier, P. Duvaut, and B. Picinbono. Complex transversal Volterra filters optimal for detection and estimation. In *[Proceedings] ICASSP 91: 1991*

International Conference on Acoustics, Speech, and Signal Processing, pages 3537–3540 vol.5, 1991. doi: 10.1109/ICASSP.1991.150234.

P. Chevalier, A. Oukaci, and J.-P. Delmas. Third order widely non linear Volterra MVDR beamforming. In *2011 IEEE International Conference on Acoustics, Speech and Signal Processing (ICASSP)*, pages 2648–2651, 2011. doi: 10.1109/ICASSP.2011.5947029.

A. Cichocki, R. Zdunek, A.H. Phan, and S. Amari. *Nonnegative Matrix and Tensor Factorizations: Applications to Exploratory Multi-way Data Analysis and Blind Source Separation*. Wiley, Hoboken, NJ, 2009. ISBN 9780470747285.

P. Comon, G. Golub, L.-H. Lim, and B. Mourrain. Symmetric tensors and symmetric tensor rank. *SIAM Journal on Matrix Analysis and Applications*, 30(3):1254–1279, 2008. doi: 10.1137/060661569.

C. Crespo-Cadenas, P. Aguilera-Bonet, J.A. Becerra-González, and S. Cruces. On nonlinear amplifier modeling and identification using baseband Volterra-Parafac models. *Signal Processing*, 96: 401–405, 2014. doi: 10.1016/j.sigpro.2013.09.028.

C. Crespo-Cadenas, M.J. Madero-Ayora, J. Reina-Tosina, and J.A. Becerra-González. A widely nonlinear approach to compensate impairments in I/Q modulators. In *2015 European Microwave Conference (EuMC)*, pages 506–509, 2015. doi: 10.1109/EuMC.2015.7345811.

G. Favier, A.Y. Kibangou, and T. Bouilloc. Nonlinear system modeling and identification using Volterra-PARAFAC models. *International Journal of Adaptive Control and Signal Processing*, 26(1):30–53, 2012. doi: 10.1002/acs.1272.

F.M. Ghannouchi, O. Hammi, and M. Helaoui. *Behavioral Modeling and Predistortion of Wideband Wireless Transmitters*. Wiley, Chichester, UK, 2015. ISBN 9781119004431.

M. Isaksson, D. Wisell, and D. Ronnow. A comparative analysis of behavioral models for RF power amplifiers. *IEEE Transactions on Microwave Theory and Techniques*, 54(1):348–359, 2006. doi: 10.1109/TMTT.2005.860500.

H. Ku and J. S. Kenney. Behavioral modeling of nonlinear RF power amplifiers considering memory effects. *IEEE Transactions on Microwave Theory and Techniques*, 51(12):2495–2504, 2003. doi: 10.1109/TMTT.2003.820155.

V.J. Mathews and G.L. Sicuranza. *Polynomial Signal Processing*. Wiley Series in Telecommunications and Signal Processing. Wiley, New York, 2000. ISBN 9780471034148.

E. McCune. *Practical Digital Wireless Signals*. The Cambridge RF and Microwave Engineering Series. Cambridge University Press, Cambridge, UK, 2010. ISBN 9781139484732.

M.S. Muha, C.J. Clark, A.A. Moulthrop, and C.P. Silva. Validation of power amplifier nonlinear block models. In *1999 IEEE MTT-S International Microwave Symposium Digest*, volume 2, pages 759–762, 1999. doi: 10.1109/MWSYM.1999.779870.

J.C. Pedro and N.B. Carvalho. *Intermodulation Distortion in Microwave and Wireless Circuits*. Artech House Microwave Library. Artech House, Boston, 2003. ISBN 9781580533560.

R. Raich and G.T. Zhou. On the modeling of memory nonlinear effects of power amplifiers for communication applications. In *Proceedings of 2002 IEEE 10th Digital Signal Processing Workshop, 2002 and the 2nd Signal Processing Education Workshop*, pages 7–10, 2002. doi: 10.1109/DSPWS.2002.1231065.

J. Reina-Tosina, M. Allegue-Martínez, M.J. Madero-Ayora, C. Crespo-Cadenas, and S. Cruces. Digital predistortion based on a compressed-sensing approach. In *2013 European Microwave Conference*, pages 408–411, 2013. doi: 10.23919/EuMC.2013.6686678.

J. Reina-Tosina, M. Allegue-Martínez, C. Crespo-Cadenas, C. Yu, and S. Cruces. Behavioral modeling and predistortion of power amplifiers under sparsity hypothesis. *IEEE Transactions on Microwave Theory and Techniques*, 63(2): 745–753, 2015. doi: 10.1109/TMTT.2014.2387852.

S.O. Rice. Volterra systems with more than one input port — distortion in a frequency converter. *The Bell System Technical Journal*, 52(8):1255–1270, 1973. doi: 10.1002/j.1538-7305.1973.tb02019.x.

P.J. Schreier and L.L. Scharf. *Statistical Signal Processing of Complex-Valued Data: The Theory of Improper and Noncircular Signals*. Cambridge University Press, Cambridge, UK, 2010. ISBN 9781139487627.

W. Wirtinger. Zur formalen theorie der funktionen von mehr komplexen veranderlichen. *Mathematische Annalen*, 97:357–375, 1927.

A. Zhu and T.J. Brazil. Behavioral modeling of RF power amplifiers based on pruned Volterra series. *IEEE Microwave and Wireless Components Letters*, 14(12):563–565, 2004. doi: 10.1109/LMWC.2004.837380.

4

Volterra Models Pruning Based on Circuit Knowledge

4.1 Introduction

The convenience of the Volterra series approach to model nonlinear distortions in wireless communication systems has been discussed before, and a discrete-time full Volterra model was specifically deduced to describe the nonlinear behavior of a power amplifier. Having in mind the nonlinear characterization of other communication circuits beyond the power amplifiers, mixers and modulators, for instance, the univariate baseband representation was augmented in Chapter 3 with a comprehensive Volterra model for complex-valued systems and signals.

Being aware of the great importance the previously discussed Volterra approach possesses as a mathematical tool, we are also conscious of the huge number of coefficients that these models contain and the need to reduce their complexity. That being so, the main purpose of this chapter is precisely a further development of the Volterra models based on knowledge of the device equivalent circuit that allows to achieve a rational simplification of the model structure.

Under this perspective, we first address a discussion on well-known heuristic model pruning approaches and the consequent coefficients reduction they provide. Next, model reduction techniques that can be additionally accomplished by deepening into the nexus between the model parameters and the corresponding equivalent circuit of the power amplifier are discussed. Based on this viewpoint, we advance some improved models that include supplementary circuit knowledge.

In a first proposal, we demonstrate a Volterra behavioral model for wideband amplifiers, and then we consider possible equivalent circuits with additional subnetworks which reflect internal mechanisms in power amplifiers that generate additional variables. Examples of internal mechanisms are the equivalent electrothermal or the charge-trapping subnetworks in transistors, and the envelope generation in envelope tracking power amplifiers, which gives rise to a structure that we have denoted bivariate circuit knowledge Volterra (CKV) model.

A Volterra Approach to Digital Predistortion: Sparse Identification and Estimation, First Edition.
Carlos Crespo-Cadenas, María José Madero-Ayora, and Juan A. Becerra.

A relevant feature of this result is substantiated by observing that the popular GMP model can be considered a particular case of this bivariate-CKV model.

Apart from the case of pruned models for power amplifiers, the complex Volterra model is used here as a widely nonlinear approach to examine the impairments in I/Q modulators and to derive the model parameters on the basis of the system architecture.

4.2 Heuristic Pruning of Volterra Models

The full Volterra model (3.10) discussed in Chapter 3 is suitable to characterize power amplifiers but with the inherent drawback of a large number of coefficients, even in the case that the structure is truncated to a given nonlinear order. However, the number of coefficients of the model can be dramatically reduced if some redundant kernels are pruned.

It is possible to make a distinction between pruning procedures based on previous information, available according to the knowledge of the device properties or its equivalent circuit, that will be discussed in the next section, and the procedures based on the mathematical advantages provided by the pruned model, referred to as heuristic pruning. First, we describe some of the more significant heuristic approaches.

A common pruning approach has been the input–output relationships advanced in Chapter 2 for the Wiener and Hammerstein two-block models. In addition, different schemes can be proposed by restricting the complete structure of the kernels in the full Volterra model (3.10). The simplest example is the baseband model (3.27) for a quasi-memoryless power amplifier, deduced in Chapter 3 after selecting $q_1 = q_2 = \cdots = q_n = 0$ for all n in the full Volterra kernels.

4.2.1 Memory Polynomial (MP) Model

A direct way to include memory in the model is by selecting $q_1 = q_2 = \cdots = q_n = q$ for all n to derive the univariate memory polynomial (MP) model (3.28). Both models include only odd-order terms, i.e., terms with even power exponents of the envelope $|x(k)|$. However, by heuristically including terms with odd-power exponents of the envelope as well, the baseband model is described by the following complex polynomial

$$y(k) = \sum_{n=1}^{N} \sum_{q=0}^{Q} h_n(q)|x(k-q)|^{n-1}x(k-q). \tag{4.1}$$

The above expression is the widely accepted MP model proposed in Kim and Konstantinou (2001) to take into account memory effects and improve performance of digital predistorters.

Since all off-diagonal terms of the kernel structure are zero, this pruned Volterra model offers a substantial reduction in the number of parameters at the cost of a possibly significant decrease in model fidelity because the off-diagonal terms are more important than the diagonal terms in some amplifiers.

4.2.2 Generalized Memory Polynomial (GMP) Model

A straight way to introduce off-diagonal terms is by means of the so-called cross terms involving products with different time shifts, like $|x(k-l-m)|^p x(k-l)$ where a delay of m samples is inserted between the signal and its exponentiated envelope. Including these cross terms with leading and lagging envelope in the MP structure of equation (4.1), yields the popular *generalized memory polynomial* (GMP) model (Morgan et al., 2006), expressed as

$$
\begin{aligned}
y(k) = & \sum_{p\in\mathcal{P}_a}\sum_{l\in\mathcal{L}_a} a_p(l)x(n-l)|x(n-l)|^p \\
& + \sum_{p\in\mathcal{P}_b}\sum_{l\in\mathcal{L}_b}\sum_{m\in\mathcal{M}_b} b_p(l,m)x(n-l)|x(n-l-m)|^p \\
& + \sum_{p\in\mathcal{P}_c}\sum_{l\in\mathcal{L}_c}\sum_{m\in\mathcal{M}_c} c_p(l,m)x(n-l)|x(n-l+m)|^p. \qquad (4.2)
\end{aligned}
$$

The set of parameters $\mathcal{P}_a, \mathcal{L}_a, \mathcal{P}_b, \mathcal{L}_b, \mathcal{M}_b, \mathcal{P}_c, \mathcal{L}_c$, and $\mathcal{M}_c$ of the model are defined in the abovementioned reference. This structure is one of the most commonly used models owing to its superior performance, basically due to a richer basis set with the incorporation of odd-order envelope power terms.

We can recall that the baseband structure of the univariate generalized polynomial model (3.29), formally derived from the Volterra bandpass model, does not contain odd-order envelope power terms, suggesting the idea that the GMP should not be classified as a Volterra model. Another debatable aspect is the presence of the so-called noncausal terms in the second group of sums of (4.2). From a mathematical point of view, the incorporation of these beneficial terms gives the GMP a notable significance in digital predistorter designs. Besides exploiting these favorable features, a curious reader could be interested in the physical origin and the equivalent circuit mechanisms that explains these terms, and how this model is in line with the Volterra series perspective, two questions related to the discussion of bivariate Volterra models in the next sections of this chapter.

4.2.3 Dynamic Deviation Reduction (DDR) Model

Another approach to prune the highly complex structure of a general Volterra model is based on the separation of static kernels from those with memory effects (Filicori and Vannini, 1991). The concept was extended to the discrete time domain in Zhu et al. (2006) where the authors proposed the *dynamic deviation reduction* (DDR) Volterra model.

In addition to the maximum nonlinear order and delay Q_n parameters, the order of the dynamics r is a singular parameter of the DDR proposal that controls the model reduction in a way that tends to the full Volterra model as r increases. Conversely, the GMP increments of the maximum delay parameters raises the number of terms, never reaching the structure of the full Volterra model.

4.2.4 Other Heuristic Pruning Proposals

The customary procedure with these heuristic models is to include output signal information of *a posteriori* measurements to estimate *all* the coefficients of the whole structure, pruned in a first instance by following a deductive reasoning.

Therefore, a relevant merit of the pruned model is the reduced number of coefficients it has *a priori*, compared to, lets say, the full Volterra model. For example, the MP is a truncation of the full Volterra model that consists of diagonal kernels with unit-delay taps, providing a significantly reduced set of parameters compared to the general full Volterra set.

A further reduction of the number of coefficients is possible by adopting the proposal of a sparse delay tap structure that reduces the parameter space required for model identification (Ku and Kenney, 2003).

Inevitably, the selection procedure of the optimum pruned model leads to a compromise between the *a priori* reduced complexity of the structure and the fidelity of the modeled output signal.

Alternatively, we can start the procedure with a candidate model composed of the highest number of coefficients, and therefore, with a richer basis set including likely beneficial terms. Then, a search methodology is applied in order to identify the active basis functions and discard the irrelevant ones. The result is a sparse model with a structure made up of the truly significant terms. This efficient methodology will be examined in Chapter 6.

4.3 Pruning Based on Equivalent Circuit Knowledge

As it was previously commented, it is likely that the coefficients reduction of the *a priori* pruned structures, like MP or GMP, affects the accuracy of the output signal modeling owing to the possible absence of significant terms of the full Volterra model.

An alternative approach, where the incorporation of available device knowledge, physical or equivalent circuit information, for instance, is utilized to deduce the pruned model. Examples of proposals following this approach have been published in Crespo-Cadenas et al. (2006, 2007), and Zhu et al. (2007).

In particular, the reasoning in Crespo-Cadenas et al. (2006, 2007) makes use of the Volterra transfer functions deduced by analyzing the equivalent circuit of an amplifier with the procedure demonstrated in Section 2.7.

4.3.1 Structure of the Kernels

The kernel structure of an amplifier is traceable to its particular equivalent circuit, represented here in Figure 4.1 for an amplifier with a single field-effect transistor (FET).

Assuming an amplifier having an operating band wider than the narrow bandwidth of the input signal, its output complex envelope samples $y(k)$ can be acquired with negligible aliasing if a suitable sampling rate is adopted. Owing to the spectral regrowth originated by the nonlinear nth-order terms, the output bandwidth is n times larger than the input bandwidth B, with $n = \infty$ in theory. However, these effects decay for higher orders and we can consider adequate a sampling rate of nB for a finite n. As a rule of thumb, the criterion to decide the value of n is given by the maximum order of the terms with spectral content above the experimental noise level. Considering that it is very common to find reported spectra with significant values in the second adjacent channels, i.e., inside a $5B$ band, it is opportune to select at least an order $n \geq 5$ and a sampling rate $f_s \geq 5B$.

Looking back on equation (3.68) derived in the frequency domain discussion of Section 3.6, the input–output relationship for the nth-order term can be written as

$$y_n(k) = \frac{1}{M^n} \sum_{\mathbf{m}_n=0}^{M-1} H_n(\mathbf{m}_n) e^{j(2\pi/M)\overline{m}k} \prod_{r=1}^{p+1} X(m_r) \prod_{r=p+2}^{2p+1} X^*(M - m_r), \tag{4.3}$$

where $n = 2p + 1$ and $\overline{m} = m_1 + m_2 + \cdots + m_n$. The input baseband M-periodic spectrum is $X(m)$ and $H_n(\mathbf{m}_n)$ is the discretized nonlinear transfer function of order n, related to the kernel $h_n(\mathbf{q}_n)$ by the discrete Fourier transform

$$H_n(\mathbf{m}_n) = \sum_{\mathbf{q}_n=0}^{M-1} h_n(\mathbf{q}_n) e^{-j(2\pi/M)\mathbf{q}_n^T \mathbf{m}_n}. \tag{4.4}$$

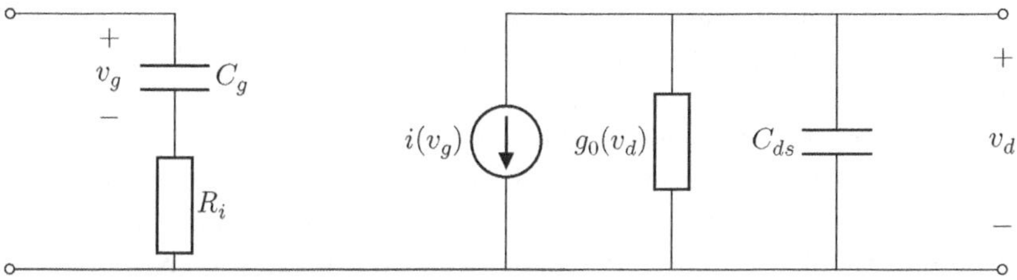

Figure 4.1 FET model used in the analysis.

The Volterra series perspective enables us to calculate the transfer functions of the circuit represented in Figure 4.1 and then compute the Volterra kernel of order n using the inverse discrete Fourier transform

$$h_n(\mathbf{q}_n) = \frac{1}{M^n} \sum_{\mathbf{m}_n=0}^{M-1} H_n(\mathbf{m}_n) e^{j(2\pi/M)\mathbf{m}_n^T \mathbf{q}_n}. \tag{4.5}$$

4.3.1.1 First-Order Kernel

For convenience, we change the RF frequency variables ω_i^{RF} (previously written without the superscript RF in Section 2.7), to baseband frequency variables ω_i, i.e., $\omega_i^{\mathrm{RF}} \rightarrow \pm\omega_c + \omega_i$ for components in the frequency zone around the carrier frequency ω_c. Since we are mainly interested in the response inside this fundamental frequency zone where ω_c is fixed, only the explicit dependence on frequencies ω_i is shown from now on.

Assuming that the gate capacitance of the model is linear, the coefficients of expansion (2.67) are $c_{g1} = C_g$ and $c_{gn} = 0$ for $n \geq 2$. Substituting in (2.70), the linear transfer function is given by

$$H_1(\omega) = \frac{g_{10} Z_0(\omega_c + \omega)}{1 + j(\omega_c + \omega) C_g \overline{R}_g}. \tag{4.6}$$

First, the linear transfer function is discretized $H_1(\omega) \rightarrow H_1(m)$ and, after its association to the kernel $h_1(q)$ recalling the discrete Fourier transformation (4.4) with $n = 1$, substituted in equation (4.3) to obtain the linear term written as

$$y_1(k) = \sum_{q=0}^{Q} h_1(q) \left[\frac{1}{M} \sum_{m=0}^{M-1} X(m) e^{j(2\pi/M)m(k-q)} \right] = \sum_{q=0}^{Q} h_1(q) x(k-q), \tag{4.7}$$

where Q is the length of the finite memory kernel.

In a first approximation, a narrow-band RF signal centered at the carrier frequency ω_c is assumed, so that $Z_0(\omega_c + \omega) \approx Z_0(\omega_c)$ and $j(\omega_c + \omega) C_g \overline{R}_g \approx j\omega_c C_g \overline{R}_g$. Accordingly, the linear output transfer function is constant inside the operating band and the linear term is quasi-memoryless.

4.3.1.2 Third-Order Kernel

Following the conventional procedure reviewed in the example of Section 2.7 and recalling that n is odd, the third-order output transfer function of the amplifier is obtained in terms of the second-order transfer function.

Remembering that the only nonlinearities in the circuit of Figure 4.1 are the voltage-controlled current source $i(v_g)$ and the drain conductance $g_0(v_d)$, with coefficients g_{n0} and g_{0n} in their respective Taylor series expansions, the second-order nonlinear transfer function calculated with (2.74) is given by

$$H_2(\omega_1, \omega_2) = -g'_{20} Z_0\left((\pm\omega_c + \omega_1) + (\pm\omega_c + \omega_2) \right) H_{g1}(\pm\omega_c) H_{g1}(\pm\omega_c), \tag{4.8}$$

where

$$g'_{20} \approx \begin{cases} g_{20} + g_{02}g_{10}^2|Z_0(\omega_c)|^2 & \text{in the DC zone,} \\ g_{20} + g_{02}g_{10}^2 Z_0^2(\omega_c) & \text{in the 2nd harmonic zone.} \end{cases} \tag{4.9}$$

This result indicates a dependence of the second-order transfer function given by

$$H_{20}(\omega_1 + \omega_2) = -g'_{20}Z_0\left(\omega_1 + \omega_2\right)|H_{g1}(\omega_c)|^2 \quad \text{in the DC zone,} \tag{4.10}$$

for an input signal content at frequencies $\omega_c + \omega_1$ and $-\omega_c + \omega_2$, and

$$H_{22}(\omega_c) = -g'_{20}Z_0\left(2\omega_c\right)H_{g1}(\omega_c)H_{g1}(\omega_c) \quad \text{in the 2nd harmonic zone,} \tag{4.11}$$

for an input signal content at frequencies $\omega_c + \omega_1$ and $\omega_c + \omega_2$. Observe that, if the condition $\omega_1 + \omega_2 \ll 2\omega_c$ is satisfied in the second harmonic zone, then $Z_0\left(2\omega_c + \omega_1 + \omega_2\right) \approx Z_0\left(2\omega_c\right)$ can be considered constant. However, we cannot state the same in the DC zone where a frequency dependence is feasible.

The second-order terms do not directly contribute to the fundamental zone output, but the presence of the nonlinear conductance in the drain subcircuit makes possible the translation of their generated low-frequency spectral content to the fundamental zone, giving rise to the third-order transfer function dependence. The third-order nonlinear transfer function for the output in the fundamental zone can be written as

$$\begin{aligned} H_3(\omega_1, \omega_2, \omega_3) = & -g'_{30}Z_0(\omega_c)|H_{g1}(\omega_c)|^2 H_{g1}(\omega_c) \\ & + 2g_{02}Z_0(\omega_c)H_1(\omega_c)H_{20}(\omega_1 + \omega_2). \end{aligned} \tag{4.12}$$

The first component of (4.12) is frequency independent and generates a memoryless term, whereas the second component depends on the sum of frequencies $\omega_1 + \omega_2$ and is instrumental in the memory of the model. Denoting this component with memory as $H_{31}(\omega_1 + \omega_2)$, the discretized version according to (4.4) can be written as

$$H_{31}(m) = \sum_{q=0}^{Q} h_3(q)e^{-j(2\pi/M)qm}. \tag{4.13}$$

where $m = m_1 + m_2$. Substituting in (4.3), the contribution to the third-order term is given by

$$\begin{aligned} y_{31}(k) = & \frac{1}{M}\sum_{m_3=0}^{M-1} X(m_3)e^{j(2\pi/M)m_3 k}\sum_{q=0}^{Q} h_3(q)e^{-j(2\pi/M)(m_1+m_2)q} \\ & \times \frac{1}{M^2}\sum_{m_1=0}^{M-1}\sum_{m_2=0}^{M-1} e^{j(2\pi/M)(m_1+m_2)k}X(m_1)X^*(M - m_2), \end{aligned} \tag{4.14}$$

or

$$y_{31}(k) = \frac{1}{M}\sum_{m_3=0}^{M-1} X(m_3)e^{j(2\pi/M)m_3 k}\sum_{q=0}^{Q} h_3(q)$$
$$\times \frac{1}{M}\sum_{m_1=0}^{M-1} X(m_1)e^{j(2\pi/M)m_1(k-q)}\frac{1}{M}\sum_{m_2=0}^{M-1} X^*(M-m_2)e^{j(2\pi/M)m_2(k-q)}. \quad (4.15)$$

For convenience, we have swapped the indices m_2 and m_3 to accommodate the notation. Using the inverse discrete Fourier transform, we finally obtain

$$y_{31}(k) = \sum_{q=0}^{Q} h_3(q)|x(k-q)|^2 x(k), \quad (4.16)$$

demonstrating an "out-of-diagonal" structure of the kernel, in contrast with the conventional diagonal structure of the MP model. Notice that this third-order kernel structure is a special case of the heuristically proposed GMP model.

4.3.1.3 Fifth-Order Kernel

Even though deduction of closed-form expressions for higher order transfer functions is a laborious task, their frequency dependence can be analyzed with reasonable facility (Crespo-Cadenas et al., 2007). In the case of the more relevant terms with respect to our exposition, the frequency dependence can be summarized noticing that the fourth-order transfer function in the DC zone is given by

$$H_{40}(\omega_1+\omega_2,\omega_3+\omega_4) = \alpha\left(\omega_1+\omega_2+\omega_3+\omega_4\right)\mathcal{H}_{40}(\omega_1+\omega_2)\mathcal{H}_{40}(\omega_3+\omega_4), \quad (4.17)$$

with ω_1, ω_3 and ω_2, ω_4 defined around ω_c and $-\omega_c$, respectively. In these expressions, $\alpha(\omega)$ is a generic function with dependence on the sum of four baseband frequencies and $\mathcal{H}_{40}(\omega)$ is a function dependent on the sum of two baseband frequencies. As in the third-order transfer function, the nonlinear conductance translates this low-frequency dependence to the fundamental zone, owing in this case to the spectral content generated by the fifth-order transfer function. Accordingly, the relevant component of the fifth-order nonlinear transfer function is denoted as $H_{51}(\omega_1+\omega_2,\omega_3+\omega_4)$, and the discretized version is

$$H_{51}(\overline{m}_1,\overline{m}_2) = \sum_{q_1=0}^{Q}\sum_{q_2=0}^{Q} h_3(q_1,q_2)e^{-j(2\pi/M)(q_1\overline{m}_1+q_2\overline{m}_2)}. \quad (4.18)$$

where $\overline{m}_1 = m_1+m_2$ and $\overline{m}_2 = m_3+m_4$. After substitution in (4.3), the contribution to the fifth-order term can be written as

$$y_{51}(k) = \frac{1}{M}\sum_{m_5=0}^{M-1} X(m_5)e^{j(2\pi/M)m_5 k}\sum_{q_1=0}^{Q}\sum_{q_2=0}^{Q} h_3(q_1,q_2)$$

$$\times \frac{1}{M}\sum_{m_1=0}^{M-1} X(m_1)e^{j(2\pi/M)m_1(k-q_1)} \frac{1}{M}\sum_{m_2=0}^{M-1} X^*(M-m_2)e^{j(2\pi/M)m_2(k-q_1)}$$

$$\times \frac{1}{M}\sum_{m_3=0}^{M-1} X(m_3)e^{j(2\pi/M)m_3(k-q_2)} \frac{1}{M}\sum_{m_4=0}^{M-1} X^*(M-m_4)e^{j(2\pi/M)m_4(k-q_2)}, \quad (4.19)$$

yielding a fifth-order response given by

$$y_{51}(k) = \sum_{q_1=0}^{Q}\sum_{q_2=0}^{Q} h_5(q)|x(k-q_1)|^2|x(k-q_2)|^2 x(k). \quad (4.20)$$

This result reiterates the "out-of-diagonal" structure of the model, in this case for the fifth-order kernel.

4.3.2 Volterra Behavioral Model for Wideband Amplifiers

The preceding discussion demonstrates the practicability of an approach to infer an input–output relationship formally derived from a conventional nonlinear analysis of the amplifier circuit. It is possible to propose the extension of the structure to higher order terms, resulting in the following Volterra behavioral model for wideband amplifiers, or VBW model, expressed as

$$y(k) = h_1x(k) + \sum_{p=1}^{\infty}\sum_{\mathbf{q}_p=0}^{\mathbf{Q}_p} h_{2p+1}(\mathbf{q}_p)\prod_{r=1}^{p}|x(k-q_r)|^2 x(k). \quad (4.21)$$

This VBW model is organized with only odd-order terms, and therefore, we can refer to it as a univariate baseband Volterra model (see Chapter 3).

In addition to its traceable physical motivations and the formal derivation of the order reduction with respect to the full Volterra model, the VBW model discloses some other original features. The most noticeable is the already discussed difference of the kernels with respect to the MP model, consisting in the revelation of the so-called "out-of-diagonal" terms. On the other hand, since we have assumed a basic circuit to facilitate the deduction of equation (4.21), the model provides a rationale for amplifiers with predominantly nonlinear memory and negligible linear memory.

A clear virtue of this approach, based on the amplifier circuit knowledge, is its capability to formally predict some terms advanced in an ad hoc manner for the GMP model. For example, the presence of odd-order causal terms with $l = 0$ in the GMP (4.2) is justified by the VBW model.

4.3.2.1 Extension of the VBW

If the amplifier does not fulfill the hypothesis of a flat linear response, the memory of the first-order terms can no longer be neglected. In that case, it is possible to extend the VBW model by following ad hoc procedures.

- One immediate solution to include memory was proposed with the incorporation of the MP diagonal kernels to the VBW structure in the extended behavioral model for wideband amplifiers (EVBW) Crespo-Cadenas et al. (2008). The EVBW output is described as

$$y(k) = \sum_{p=0}^{P} \sum_{q=0}^{Q} h_{2p+1} |x(k-q)|^{2p} x(k-q) + \sum_{p=1}^{P} \sum_{\mathbf{q}_p \neq 0}^{\mathbf{Q}_p} h_{2p+1}(\mathbf{q}_p) \prod_{r=1}^{p} |x(k-q_r)|^2 x(k). \tag{4.22}$$

  The new diagonal terms with memory contribute to the reported greater performance of the EVBW model with respect to the MP and the VBW models. By contrast with the order reduction of the MP model, the EVBW is a compromise to balance accuracy and complexity. For example, a fifth-order MP model is composed of $3(Q+1)$ coefficients compared to the $Q(Q+1)/2 + 3(Q+1) + 2Q$ coefficients of the EVBW model. While the EVBW number of coefficients is obviously higher than the number of coefficients of the MP model, the accuracy is significantly better, with performances of NMSE = −41 dB for the MP and NMSE = −44 dB for the EVBW. Furthermore, the EVBW model order reduction is better than that presented by other published models with equivalent performance (Wisell and Isaksson, 2007).
- A more sophisticated ad hoc approach to include memory in the structure is the radially pruned Volterra (RPV) model advanced in Crespo-Cadenas et al. (2010b). This procedure retains only the radial branches of the full Volterra model that emanate from the origin of the kernel hypercube structure, like that represented in Figure 3.2 for a third-order kernel. The reader will have noticed that the MP, VBW, and EVBW models have a radial structure, so they can be considered as particular cases (and source of motivation) of the RPV proposal.

  To exemplify the idea concerning this radial pruning, consider the simple case of the third-order kernel with unit-delay ($Q = 1$) represented in Figure 4.2. The

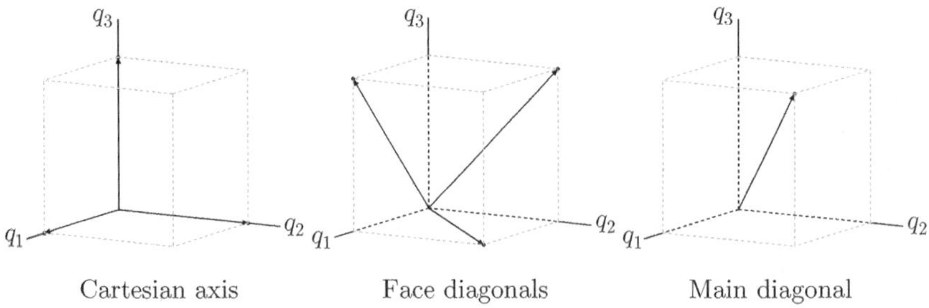

Figure 4.2 Grid of the third-order kernel domain.

radial regressors are associated to the Cartesian axis directions

$$x^2(k)x^*(k-q) \quad \text{and} \quad |x(k)|^2x(k-q),$$

the diagonals of the cube faces

$$x^2(k-q)x^*(k) \quad \text{and the VBW regressor} \quad |x(k-q)|^2x(k),$$

and the main diagonal

$$|x(k-q)|^2x(k-q), \quad \text{a MP regressor.}$$

Therefore, we have reduced the structure of the third-order term to five one-dimensional filters with delays q. The number of coefficients will grow linearly, rather than exponentially, as the maximum delay Q increases, being the effect of more effective pruning.

After selection of the radial directions for all kernels, the RPV model can be expressed as

$$y_{\mathrm{RPV}}(k) = \sum_{\substack{n=1\\(n\,\mathrm{odd})}}^{N} \sum_{r=1}^{n} \sum_{s=1}^{S_{n,r}} \sum_{q=0}^{Q} h_{n,r,s}(q)\Pi_{n,r,s}\{x(k),q\}, \tag{4.23}$$

where $\Pi_{n,r,s}\{x(k),q\}$ is an nth-order regressor of the branch s belonging to the radial direction r.

In some cases, it is desirable to further reduce the RPV model complexity at the cost of an acceptable performance worsening. This can be achieved with a procedure that limits the radial directions of the nth-order kernels, $n > 3$, to only the five radial directions of the third-order kernel. This simplified RPV (SRPV) model can be written as

$$y_{\mathrm{SRPV}}(k) = \sum_{n=1}^{N} \sum_{r=1}^{5} \sum_{q=0}^{Q} h_{n,r}(q)\Pi_{n,r}\{x(k),q\}. \tag{4.24}$$

The RPV model methodology has the advantage of a reduced number of coefficients compared to the full Volterra model because it is based on retaining only a few directions of the delay space. For a RPV model truncated to order $N = 2P + 1$, the number of coefficients is given by

$$M = (P+1)(Q+1) + (3+P)PQ. \tag{4.25}$$

showing a linear increment with respect to Q.
For the corresponding SRPV model, the number of coefficients is expressed as

$$M = (P+1)(Q+1) + 4PQ. \tag{4.26}$$

It is worth noting that the number of coefficients in the SRPV model shows also a linear dependence with respect to the model order, $N = 2P + 1$.

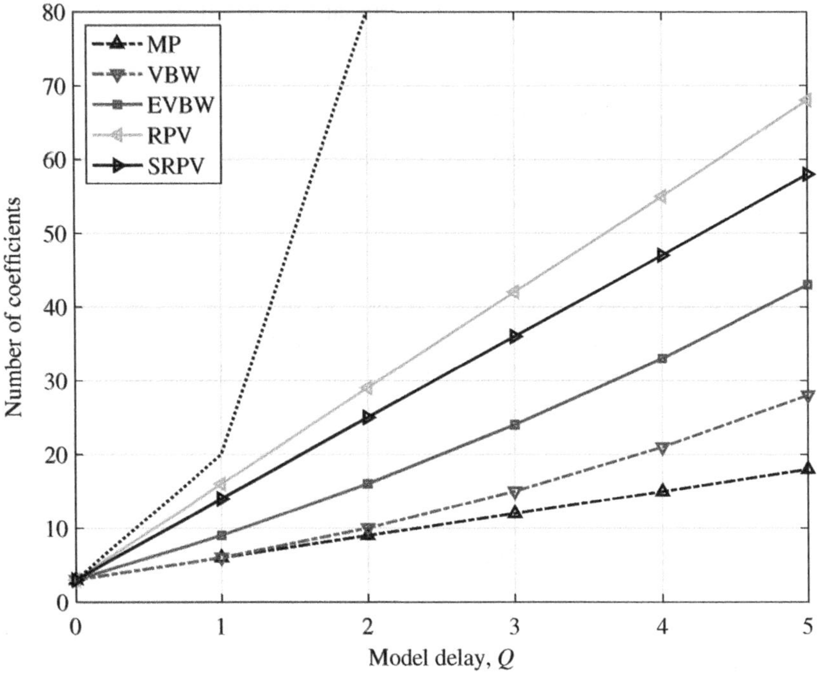

Figure 4.3 Number of coefficients for different fifth-order behavioral models. Only odd-order terms have been considered in all models.

Different proposals in terms of complexity are plotted in Figure 4.3, where the number of coefficients has been represented versus the maximum delay Q for different fifth-order models. The extreme cases are the FV model, with an exorbitant number of coefficients, and the univariate MP model (3.28), with the smallest number of coefficients linearly dependent on Q. Recall that in these models we are considering only odd-order terms. The VBW model exhibits a slightly higher complexity than the (univariate) MP. The RPV model is computationally more expensive and the EVBW model has a moderate complexity.

A comparison of the NMSE values associated with each of the fifth-order models is shown in Figure 4.4. The device under test was a commercial class-AB amplifier driven into a clearly nonlinear regime by an OFDM signal. As expected, the RPV model obtains the best overall performance in all cases. It is also notable that, even though EVBW has a lower complexity, its performance is almost comparable to SRPV.

Besides the proposals discussed above to enhance the model structure on an ad hoc basis, an alternative approach that further develops the incorporation of equivalent circuit information is also possible.

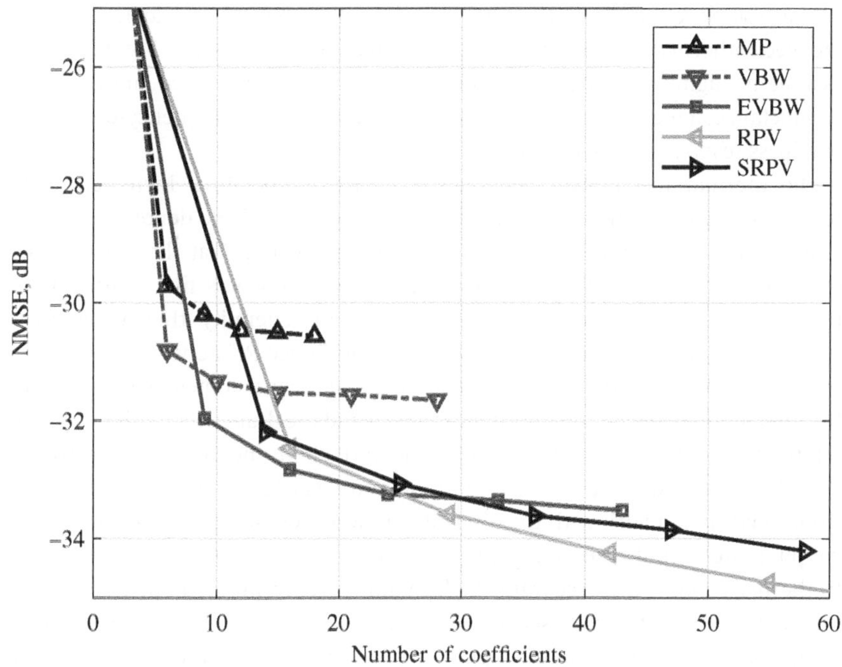

Figure 4.4 NMSE as a function of the number of coefficients.

An evident improvement is the inclusion, in the analysis, of the amplifier frequency dependence instead of a flat response, which explains in a very direct way the MP diagonal as well as improved VBW terms.

The complete analysis, including also the electrothermal effects of the amplifier, gives rise to the enriched model based on circuit knowledge discussed in detail below.

4.4 Circuit Knowledge Model with Electrothermal Effects

A Volterra-based behavioral model for power amplifiers that takes into account both short and long-term memory effects has been derived from a circuit-level representation (Crespo-Cadenas et al., 2010a, 2013). Although the level of detail of the deduction is involved due to the publication requirements in specialized scientific journals, for the sake of clarity here we summarize the most relevant points to simplify the discussion.

The procedure starts with the equivalent circuit of a typical FET device in which the electrothermal subcircuit has been inserted, as it is shown in Figure 4.5. We assume negligible nonlinear contribution of the gate capacitance and the drain conductance, a memoryless input matching network, and a frequency-dependent bandpass response at the output.

The thermal mechanism generates a device temperature increment with respect to the DC temperature, an additional internal variable denoted as θ. The temperature increment θ can be regarded as the "voltage" at the output node of the electrothermal subcircuit modeled by an *RC*-lowpass filter with several time constants (Dai et al., 2003). Without loss of generality, this subcircuit is schematically represented in Figure 4.5 as a single-stage *RC*-network.

The novelty now is that we are considering two unaccustomed aspects. Firstly, the interaction between the internal variable θ and the drain subcircuit by including the temperature increment as a second variable in the drain current source $i(v_g, \theta)$ and secondly, the association of the nonlinear "current" driving the electrothermal subnetwork with the dissipated power increment. In this way, we are coupling the output response to the "voltage" θ at the output node of the electrothermal subnetwork. According to the nonlinear currents method, the temperature increment can be expressed as $\theta = \theta_1 + \theta_2 + \theta_3 + \ldots$, where θ_n is the nth-order component of θ. Notice that the current source is a transfer nonlinearity depending on two variables, as described by equation (1.17), with a Taylor series that includes cross products of the voltage v_g and the temperature increment θ.

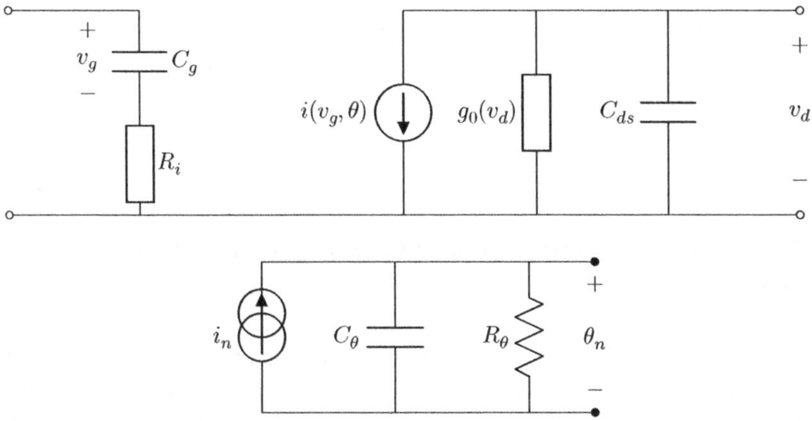

Figure 4.5 Nonlinear FET equivalent circuit with self-heating thermal subnetwork. The electrothermal coupling between subnetworks arises from the dependence of the drain current source $i(v_g, \theta)$ on the temperature increment $\theta = \theta_1 + \theta_2 + \theta_3 + \cdots$.

The linear transfer function of the output node derived in (2.70) can be reused directly in this circuit as

$$H_1(\omega_1) \approx g_{10} H_{g1}(\omega_c) Z_0(\omega). \tag{4.27}$$

Here, the impedance $Z_0(\omega)$ is dependent on frequency, straightforwardly explaining the memory of the first-order kernel. In the analysis of this circuit, it is also necessary to consider the thermal transfer functions $H_{\theta n}$ relating θ to the input signal. Since the temperature increment is caused by power dissipation, there is no linear relation between θ and the input voltage, so that $H_{\theta 1} = 0$ and $\theta_1 = 0$.

4.4.1 Third-Order Kernel

The second-order nonlinear transfer functions are calculated by applying the corresponding nonlinear currents to the output and electrothermal subnetworks. The output subnetwork analysis is standard, evaluating the drain nonlinear currents i_n as in Table 2.2. On the other hand, the electrothermal subnetwork is driven by the dissipated power, which is roughly equal to a scaled version of the nonlinear drain currents. For example, the second-order thermal "current source" approximated by $i_2 = g_{20} v_{g1}^2$, see equation (2.21), is applied to the "load impedance" of the *RC*-thermal subnetwork

$$Z_\theta(\omega) = R_\theta \left(1 + j\omega R_\theta C_\theta\right)^{-1} \tag{4.28}$$

to generate the second-order temperature increment θ_2.

Considering that the spectral content of the dissipated power is concentrated at DC and second harmonic zones, and that the thermal filter $Z_\theta(\omega)$ presents a low-pass characteristic, the spectrum of θ_2 is significant only at baseband frequencies, and the self-heating can be directly related to the low-frequency components of the nonlinear currents. Therefore, the second-order thermal transfer function is given by

$$H_{\theta 2}(\omega_1, \omega_2) = -g_{20} |H_{g1}|^2 Z_\theta(\omega_1 + \omega_2). \tag{4.29}$$

Observe the explicit frequency dependence on the sum $\omega_1 + \omega_2$, which produces long-term memory effects in the output response because of the electrothermal coupling. For the interested readers, a more detailed description of the procedure was published in Crespo-Cadenas et al. (2013).

The third-order transfer functions are obtained following the standard approach, calculating first the third-order nonlinear current

$$i_3 = g_{30} v_{g1}^2 + \gamma_{11} v_{g1} \theta_2, \tag{4.30}$$

where γ_{11} is a second-order coefficient of the cross products in the current source Taylor expansion. The corresponding expression for the output is given by

$$H_3(\omega_1, \omega_2, \omega_3) = -\left[g_{30}|H_{g1}|^2 H_{g1} + \gamma_{11} H_{g1} H_{\theta 2}(\omega_2 + \omega_3)\right] Z_0(\omega_1 + \omega_2 + \omega_3) \tag{4.31}$$

inside the fundamental zone and is zero elsewhere. Moreover, the thermal transfer function $H_{\theta 3}(\omega_1, \omega_2, \omega_3)$ is zero because of the lowpass characteristic of the *RC*-subnetwork.

Notice that, in the fundamental spectral zone, the drain transfer function $H_3(\omega_1, \omega_2, \omega_3)$ is composed of a short-term memory part depending on the output impedance $Z_0(\omega)$, and a long-term memory part, which additionally contains the frequency dependence of the electrothermal subnetwork produced by $H_{\theta 2}(\omega_2 + \omega_3)$. This result can be compared to equation (4.12) for the third-order nonlinear transfer function of the VBW, where the frequency dependence provided by $Z_0(\omega)$ is missing.

Denoting the first component with memory as $H_{31}(\omega_1 + \omega_2 + \omega_3)$ and recalling (4.4), the discretized version can be written as

$$H_{31}(m) = \sum_{q=0}^{Q} h_{31}(q) e^{-j(2\pi/M)qm}, \tag{4.32}$$

where $m = m_1 + m_2 + m_3$. Repeating the procedure advanced for the VBW model, after substitution in (4.3) and making use of the inverse discrete Fourier transform, we can deduce the contribution to the third-order term, given by

$$y_{31}(k) = \sum_{q=0}^{Q} h_{31}(q)|x(k-q)|^2 x(k-q). \tag{4.33}$$

This part of the third-order kernel demonstrates the diagonal structure of the MP model.

Executing again the steps of the methodology for the other component of (4.31), denoted as $H_{32}(\omega_1, \omega_2, \omega_3)$, we deduce a second part of the third-order kernel, given by

$$y_{32}(k) = \sum_{q_1=0}^{Q} \sum_{q_2=0}^{Q} h_{32}(q_1, q_2)|x(k - q_1 - q_2)|^2 x(k - q_1). \tag{4.34}$$

Observe that this third-order kernel structure coincides with the univariate GMP model (3.29), a GMP model restricted to causal odd-order terms.

4.4.2 Fifth-Order Kernel

Analyzing the more relevant terms, we notice that the fourth-order transfer function in the DC zone is given by the product of two functions dependent on the sum of two baseband frequencies, $\mathcal{H}_{40}(\omega_1+\omega_2)\mathcal{H}_{40}(\omega_3+\omega_4)$, as in (4.17).

The analysis of the fifth-order transfer function indicates that this baseband dependence is translated to the fundamental zone with a relevant component given by

$$H_{51}(\omega_1+\omega_2,\omega_3+\omega_4+\omega_5) = Z_0\left(\omega_1+\omega_2+\omega_3+\omega_4+\omega_5\right) \times \mathcal{H}_{50}(\omega_1+\omega_2)\mathcal{H}_{50}(\omega_3+\omega_4). \tag{4.35}$$

Repeating the procedure exposed above, the fifth-order term is expressed as

$$y_{51}(k) = \sum_{q_1=0}^{Q}\sum_{q_2=0}^{Q} h_{51}(q_1,q_2,q_3)|x(k-q_1-q_2)|^2|x(k-q_1-q_3)|^2x(k-q_1). \tag{4.36}$$

If we collect all the terms, the amplifier output we can be expressed as

$$\begin{aligned} y(k) = & \sum_{\substack{n=1\\(n\,\text{odd})}}^{N}\sum_{q=0}^{Q} h_1(q)x(k-q) \\ & + \sum_{\substack{n=1\\(n\,\text{odd})}}^{N}\sum_{q_1=0}^{Q}\sum_{q_2=0}^{Q} h_3(q_1,q_2)|x(k-q_1-q_2)|^2x(k-q_1) \\ & + \sum_{\substack{n=1\\(n\,\text{odd})}}^{N}\sum_{q_1=0}^{Q}\sum_{q_2=0}^{Q}\sum_{q_3=0}^{Q} h_5(q_1,q_2,q_3)|x(k-q_1-q_2)|^2 \\ & \times |x(k-q_1-q_3)|^2x(k-q_1)+\cdots. \end{aligned} \tag{4.37}$$

This representation is here referred to as *univariate* circuit knowledge Volterra (CKV) model.

Let us emphasize some relevant reflections. Firstly, the heuristic univariate GMP model (3.29) is a particular case of the univariate CKV model when $q_2 = q_3 = \cdots = q_n$, and therefore it can be also regarded as the result of a formal deduction grounded on the knowledge of the original circuit model of the amplifier.

It is also opportune to state that the popular GMP model (4.2) includes terms with odd powered envelopes, like $|x(k-q_1-q_2)|^{n-1}x(k-q_1)$ with n even, which cannot be classified as even-order terms. The reason is illustrated with the observation that the envelope $|x(k)| = \sqrt{|x(k)|^2}$ is not a single term with a definite order, but a function with a Taylor series expansion at any point $x(k) \neq 0$. Clearly, the

GMP model can be considered as a bivariate Volterra model dependent on the two variables $x(k)$ and $|x(k)|$.

At this point, an important question arises: can we deduce from a circuit knowledge basis a model that incorporates these beneficial terms? The procedure to derive this model is the topic discussed in the next section.

4.5 Circuit Knowledge in Bivariate Volterra Models

The electrothermal mechanism that generates a temperature increment as an internal state variable has been taken into account above to demonstrate the univariate CKV model.

As we already commented, the univariate CKV model was formally deduced based on the original circuit model of the amplifier yielding the causal odd-order terms of the widely accepted GMP model under a physical knowledge perspective. However, a physical justification for the noncausal and non-odd-order terms has not been discussed yet. We consider in this section the presence of an additional mechanism to give a physical justification to all the terms.

Two cases of mechanisms which are responsible of the signal envelope generation, a new internal variable, can be identified in power amplifiers. The first example, and perhaps the most evident, is the high-efficiency envelope tracking (ET) power amplifier, shown schematically in Figure 4.6.

Considered as a whole, the ET amplifier incorporates an internal mechanism that generates a variable similar to the RF signal envelope, denoted as $z(t)$ in the figure, which is used to provide a dynamically adjusted supply voltage to operate the basic power amplifier in a more efficient way. Notice that the basic power amplifier is driven by two signals, the RF input signal $\tilde{x}(t)$ and the supply voltage $z(t)$, thus the formal description should be based on a double Volterra series approach.

Another example of mechanism generating the envelope as an internal variable is the charge-trapping subnetwork, which is the cause of drain-lag effects in FET

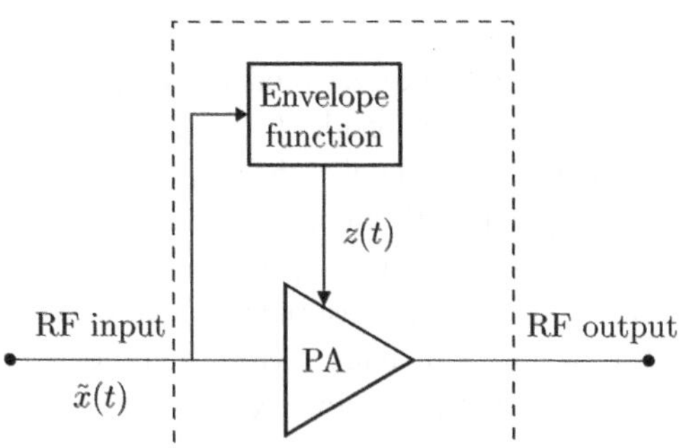

Figure 4.6 The envelope tracking power amplifier.

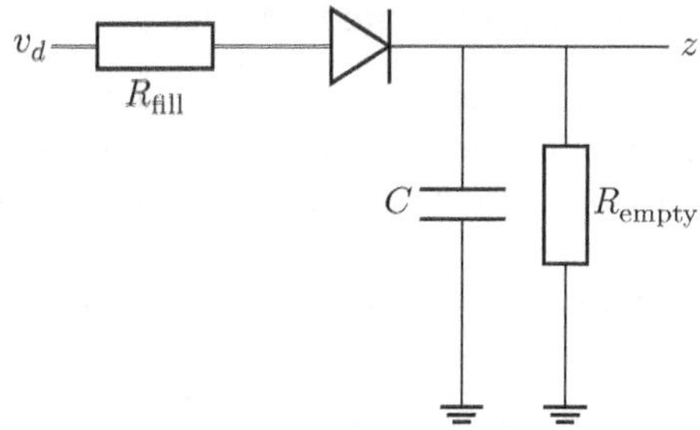

Figure 4.7 Simplified layout of the associated charge-trapping subnetwork in a single-FET amplifier.

devices (Jardel et al., 2007). A direct inspection applied to the simplified diagram of a one-trap model shown in Figure 4.7 reveals that this subnetwork basically works as an envelope detector, and the real-valued signal envelope generated at the output acts as a new internal variable that additionally controls the current source of the transistor. The analysis is performed in a similar way to the model in Section 4.4 by simply substituting the thermal variable θ for the new variable $z(t)$.

The concepts behind the two cases commented above can be integrated in the simple schematic of Figure 4.8, in which the internally generated waveform is represented acting as the second input to a bivariate Volterra system.

Referring to this figure, the discrete-time RF output of the amplifier when the bivariate block is driven by the RF signal $\tilde{x}(k)$ and the real-valued signal $z(k)$ can be expressed as the double Volterra series

$$\tilde{y}(k) = \hat{H}_{a,0}[\tilde{x}(k)] + \hat{H}_{b,0}[z(k)] + \sum_{n=1}^{N}\sum_{m=1}^{M}\sum_{\mathbf{q}_n=0}^{\mathbf{Q}_n}\sum_{\mathbf{q}'_m=0}^{\mathbf{Q}'_m} h_{n,m}[\mathbf{q}_n, \mathbf{q}'_m]\prod_{r=1}^{n}\tilde{x}(k-q_r)\prod_{s=1}^{m} z(k-q'_s), \tag{4.38}$$

where $\hat{H}_{a,0}[\cdot]$ and $\hat{H}_{b,0}[\cdot]$ represent conventional univariate Volterra series.

The output of the bivariate Volterra system has been partitioned in three groups: the first group is a univariate Volterra series dependent on the RF signal $\tilde{x}(k)$, the second group is another univariate Volterra series dependent on the internal variable $z(k)$, and the third group contains cross products of $\tilde{x}(k)$ by $z(k)$.

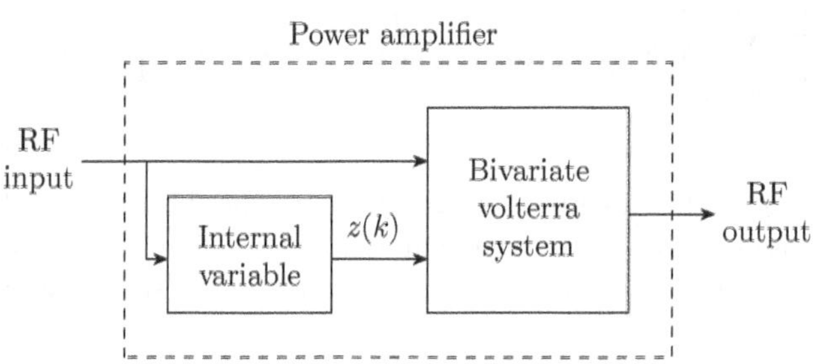

Figure 4.8 A bivariate Volterra approach to PAs with an internal variable generation.

4.5.1 The Bivariate FV Model

Since we are interested in the fundamental frequency zone, the baseband equivalent of the first group in (4.38) reduces to a model dependent on the complex envelope $x(k)$. Recalling that $z(k)$ is a baseband real-valued variable, the second group of terms do not contribute to the fundamental zone and can be neglected. Summing-up, the output complex envelope can be written as

$$y(k) = H_{a,0}[x(k)] + \sum_{m=1}^{M} \sum_{\mathbf{q}'_m=\mathbf{0}}^{\mathbf{Q}'_m} H_{a,m}[x(k)] \prod_{s=1}^{m} z(k - q'_s), \tag{4.39}$$

where $H_{a,0}[x(k)]$ is the conventional FV model (3.10) and the second group of terms is the result of multiplying the baseband Volterra model by monomials of the form $z(k - q'_1) \cdots z(k - q'_m)$. This perspective discloses non explored modeling alternatives based on the specific conventional Volterra model used for $H_{a,0}[x(k)]$, covering from the FV model, with a richer regressors set, to simpler models, with pruned sets of regressors. The relationship (4.39) is referred to as the bivariate FV model.

4.5.2 The Bivariate-CKV Model

Recalling that the supply voltage in an ET power amplifier or the internal variable generated by the charge-trapping mechanism is similar to the signal envelope, we can assume $z(k) \approx |x(k)|$ in a first approximation. Accordingly, the aforementioned bivariate FV model (4.39) includes the beneficial terms of the GMP model, among many other Volterra regressors. Since we are interested in a reduced set of regressors, a pruning procedure is appreciated, in particular by following the circuit knowledge approach.

Accepting the same line of reasoning based on circuit knowledge of the univariate CKV model to illustrate the present approach, the relationship (4.39) yields

$$\begin{aligned} y(k) = &\sum_{p=1}^{P} \sum_{\mathbf{q}_p=0}^{\mathbf{Q}_p} h_{2p-1}(\mathbf{q}_p) x(k - q_1) \prod_{r=2}^{p} |x(k - q_1 - q_r)|^2 \\ &+ \sum_{m=1}^{M} \sum_{\mathbf{q}'_m=\mathbf{0}}^{\mathbf{Q}'_m} \sum_{p=1}^{P} \sum_{\mathbf{q}_p=0}^{\mathbf{Q}_p} h'_{2p-1}(\mathbf{q}_p) x(k - q_1) \prod_{r=2}^{p} |x(k - q_1 - q_r)|^2 \prod_{s=1}^{m} |x(k - q'_s)|. \end{aligned} \tag{4.40}$$

Henceforth, the representation (4.40) is referred to as the bivariate-CKV model (Crespo-Cadenas et al., 2021). The first sum of this expression is made up of conventional odd-order Volterra terms.[1] On the contrary, for the second sum

1 Recall that the product $\prod_{r=2}^{p} |x(k - q_1 - q_r)|^2$ is understood to have the value 1 when $p = 1$.

we observe that, in the particular case that the indices satisfy $q_r = q_2$ for all $r \geq 2$, and $q'_s = q_1 + q_2$, the second sum incorporates terms $x(k - q_1)|x(k - q_1 - q_2)|^{2p-1}$, directly identified as the beneficial terms of the GMP model (4.2) if we use the notation $q_1 \rightarrow l$ and $q_2 \rightarrow m$.

4.5.3 Model for Concurrent Dual-Band Signal

When more bands are included in mobile communication systems, as it is the case in the fourth generation (4G) and fifth generation (5G) standards, often the power amplifier is operated with concurrent dual-band signals. In concurrent dual-band operation, the baseband signal can be split into two single-band signals, expressed as

$$x(k) = x_1(k)e^{-j\Omega k} + x_2(k)e^{j\Omega k}, \tag{4.41}$$

where $x_1(k)$ and $x_2(k)$ are the input lower band and upper band signals centered in the lower and upper zones, around $-\Omega$ and $+\Omega$, respectively. Likewise, the output signal is split into the lower band $y_1(k)$ and the upper band $y_2(k)$ output signals, with a spectrum as that displayed in Figure 4.9. With a dual-band signal, the output signal power spectral density shows an excessive regrowth compared to the case of a single-band signal, suggesting the need for a specific modeling.

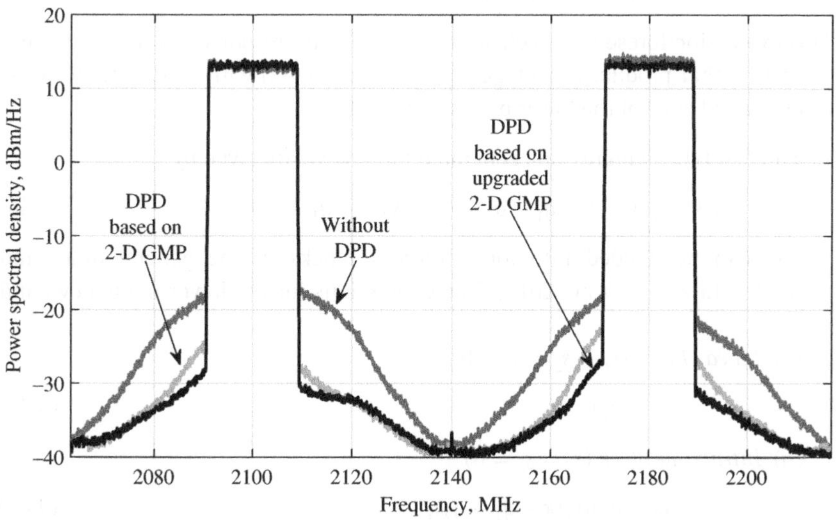

Figure 4.9 Power spectral density of the dual-band output signal and comparison with the linearized signal by applying a DPD based on the 2-D GMP model and other DPD based on an upgraded version.

The dual-band formulation can be derived from a standard Volterra model, for example, the bivariate-CKV model. According to the above discussion, the particular case of indices $q_r = q_2$ for all $r \geq 2$, and $q'_s = q_1 + q_2$, reduces the structure (4.40) to the GMP model (4.2) that has been considered among the best alternatives to dual-band signal formulation (Enzinger et al., 2018, Pérez-Hernández et al., 2021). Assuming this model, the regressors contain delayed versions of the term

$$x(k-q_1)|x(k-q_1-q_2)|^{2\mu}, \qquad \mu = 0, 1, 2, \ldots . \tag{4.42}$$

Focusing on the envelope signal, we can write

$$\begin{aligned} |x(k)|^{2\mu} &= \left[|x_1(k)e^{-j\Omega k} + x_2(k)e^{j\Omega k}|^2\right]^\mu \\ &= \left[|x_1(k)|^2 + |x_2(k)|^2 + x_1(k)x_2^*(k)e^{-j2\Omega k} + x_1^*(k)x_2(k)e^{j2\Omega k}\right]^\mu . \end{aligned} \tag{4.43}$$

Using the multinomial theorem

$$(a_1 + a_2 + \cdots + a_m)^\mu = \sum_{s_1+s_2+\cdots+s_m=\mu} \frac{\mu!}{s_1!s_2!\cdots s_m!} \prod_{t=1}^{m} a_t^{s_t}, \tag{4.44}$$

equation (4.43) can be written as

$$\begin{aligned} |x(k)|^{2\mu} = &\sum_{s_1+s_2+s_3+s_4=\mu} \frac{\mu!}{s_1!s_2!s_3!s_4!} |x_1(k)|^{2s_1} |x_2(k)|^{2s_2} \\ &\times x_1(k)^{s_3} x_2(k)^{*,s_3} x_1(k)^{*,s_4} x_2(k)^{s_4} e^{j2(s_4-s_3)\Omega k} . \end{aligned} \tag{4.45}$$

This expression foresees a spectrum for $|x(k)|^{2\mu}$ with responses in the even zones $0, \pm 2\Omega, \ldots$ that produces two types of regressors (4.42) with contribution in the fundamental zone of the lower band around $-\Omega$:

1. The first type of regressors, generated for $s_3 = s_4$, are given by

$$x_1(k-q_1)|x_1(k-q_1-q_2)|^{2s_1+2s_4}|x_2(k-q_1-q_2)|^{2s_2+2s_4} \tag{4.46}$$

and can be divided into four different blocks in the general dual-band model (Enzinger et al., 2018). The expressions for the lower frequency band are

(a) *Linear Block* ($s_1 = s_2 = s_4 = 0$)

$$x_1(k-q_1) \tag{4.47}$$

(b) *Intraband Block* ($s_2 = s_4 = 0$)

$$x_1(k-q_1)|x_1(k-q_1-q_2)|^{2s_1} \tag{4.48}$$

(c) *Crossband Block* ($s_1 = s_4 = 0$)

$$x_1(k-q_1)|x_2(k-q_1-q_2)|^{2s_2} \tag{4.49}$$

(d) *MP Mixed-Band Block* ($q_2 = 0$)

$$x_1(k-q_1)|x_1(k-q_1)|^{2s_1+2s_4}|x_2(k-q_1)|^{2s_2+2s_4} \tag{4.50}$$

2. The second type of regressors, generated for $s_3 = s_4 - 1$, can be written as

$$\begin{aligned} &x_2(k-q_1)x_1(k-q_1-q_2)\,|x_1(k-q_1-q_2)|^{2s_1+2s_4} \\ &\times x_2^*(k-q_1-q_2)|x_2(k-q_1-q_2)|^{2s_2+2s_4}. \end{aligned} \tag{4.51}$$

This second type of regressors forms a novel mixed-band block that represents an upgrade of the conventional dual-band model 2-D GMP (Bassam et al., 2011a,b). For the response inside the fundamental zone of the upper band around $+\Omega$, the corresponding regressors are obtained by swapping the signals $x_1(k)$ and $x_2(k)$ in the above expressions.

The DPD designed with the new mixed-band terms reported in Pérez-Hernández et al. (2021) improved the 2-D GMP ACPR from about −43.4 to −45.3 dBc in the first adjacent channel and from about −42.5 to −45.6 dBc in the second adjacent channel, with an EVM reduction from 1.7/2.1% to 1.3/1.4% for the lower and upper band, respectively.

4.6 Volterra Models for I/Q Modulators

An indispensable block in the transmitter of a wireless communications system is the I/Q modulator. Without an appropriate characterization of the I/Q modulator, distortions caused by the transmitter itself could be wrongly attributed to other circuits.

4.6.1 Two-Tone Test for I/Q Modulators

As it was introduced in Section 1.3, the ideally linear behavior of I/Q modulators can be affected by several types of impairments. The characterization of impairments in a quadrature modulator considering only a linear behavior has been studied in Faulkner et al. (1991), where a simple compensation algorithm was proposed to correct the carrier leakage, gain imbalance, and phase error. A formal study about the imperfections of quadrature modulators can be found in Cavers and Liao (1993, Cavers, 1997). In addition to these frequency-independent or static models, a frequency-dependent model and compensation technique for DC offset and gain/phase imbalance was presented in Ding et al. (2008). Despite their different approaches, all the aforementioned works are based on the use of simple linear analytical models for the I/Q modulator.

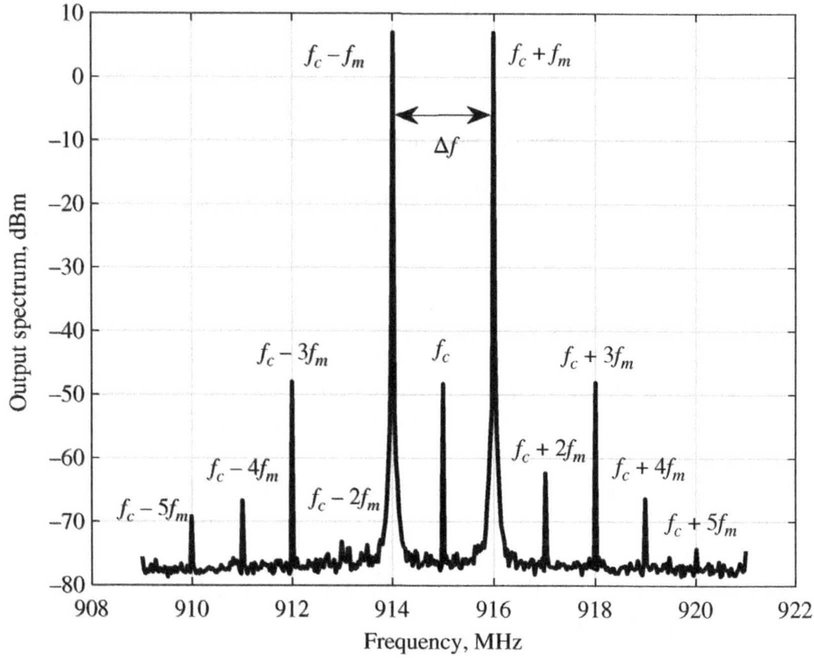

Figure 4.10 Spectrum at the output of an I/Q modulator for the generation of a two-tone signal.

However, a standard two-tone test of a communications modulator can demonstrate that these models do not account for all the actual distortion mechanisms. Let us consider a two-tone signal generated by a sinusoidal complex envelope with frequency $f_m = \frac{\Delta f}{2}$, where Δf is the tone spacing. As shown in Figure 4.10, a number of frequency components can be observed at the output of the I/Q modulator with frequencies $f_c \pm mf_m$, where $m \in \mathbb{N}$. The only spectral components that the linear model can explain are those at f_c and $f_c \pm f_m$, i.e., the carrier residue and the possibly asymmetrical tones. This is a severe limitation that indicates the need to adopt a nonlinear model.

Nonlinear behavior in I/Q modulators has been considered (Wisell, 2000, Gadringer et al., 2008, Cao et al., 2009), not only in a possible internal RF amplifier but also in both I and Q paths that could include in their hardware digital-to-analog converters (DACs), mixers, and amplifiers at baseband. On the contrary to the approach of (Wisell, 2000), where a fixed operation point was used to identify the parameters of the model, in (Gadringer et al., 2008, Cao et al., 2009) different modulator signal levels were taken into consideration. In Cao et al. (2009), a dual-input nonlinear model based on a Volterra representation was proposed. The inverse model was implemented as an I/Q imbalance pre-compensator.

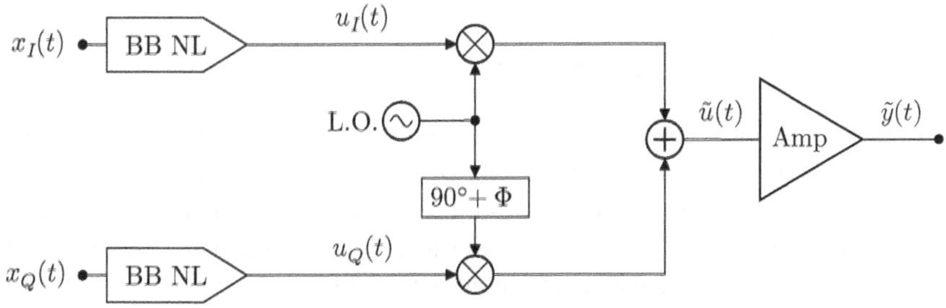

Figure 4.11 Nonlinear block model of an I/Q modulator.

The simple nonlinear block model shown in Figure 4.11 allows us to account for nonlinear distortions as well as gain/phase imbalance and carrier leakage by means of the nonlinear transfer functions of the I and Q paths, which can include a DC offset. In this part, we assume that the linear crosstalk is only produced by the phase quadrature error and neglect other causes. Note that, whereas the RF amplifier exhibits a bandpass nonlinear characteristic that is usually modeled by an odd-order nonlinear transfer function, DACs, and other elements in the I and Q paths are baseband nonlinear blocks generally modeled by nonlinear characteristics with both even and odd-order terms, and actual mixers are represented by nonlinear models with memory. Taking this into account, the crosstalk also contains nonlinear and memory effects.

Another observation made from the standard two-tone test can be a notable asymmetric behavior for the measured levels at $f_c + 2f_m$ and $f_c - 2f_m$ and a dependence on tone spacing, both related to memory effects. To explain it, a Volterra series analysis can be carried out by considering two sets of baseband nonlinear transfer functions, $H_{In/Qn}(\omega_1, \ldots, \omega_n)$, one per each I/Q path. Subscripts in H represent I or Q baseband model path and n-th order nonlinear transfer function, respectively. Consider a two-tone input signal

$$x_I(t) = x_Q(t) = \frac{A}{2}\left(e^{j\omega_m t} + e^{-j2\omega_m t}\right). \tag{4.52}$$

The second-order output signals for I and Q paths $u_{I2/Q2}(t)$ can be obtained in terms of the baseband nonlinear transfer functions

$$\begin{aligned} u_{I2}(t) = u_{Q2}(t) = \frac{A^2}{4}\big[&H_{I2/Q2}(\omega_m, -\omega_m) + H_{I2/Q2}(-\omega_m, \omega_m) \\ &+H_{I2/Q2}(-\omega_m, -\omega_m)e^{-j2\omega_m t} + H_{I2/Q2}(\omega_m, \omega_m)e^{j2\omega_m t}\big]. \end{aligned} \tag{4.53}$$

Then, the second-order output complex envelope is described by

$$\begin{aligned} y_2(t) = \frac{c_1 A^2}{4}\Big[&H_{I2}(\omega_m, -\omega_m) + H_{I2}(-\omega_m, \omega_m) \\ &+ H_{Q2}(\omega_m, -\omega_m)e^{j\pi/2} + H_{Q2}(-\omega_m, \omega_m)e^{j\pi/2} \end{aligned}$$

$$+ H_{I2}^*(\omega_m, \omega_m)e^{-j2\omega_m t} + H_{I2}(\omega_m, \omega_m)e^{j2\omega_m t}$$
$$+ H_{Q2}^*(\omega_m, \omega_m)e^{-j(2\omega_m t - \pi/2)} + H_{Q2}(\omega_m, \omega_m)e^{j(2\omega_m t + \pi/2)}\Big], \tag{4.54}$$

where c_1 is the linear coefficient of the RF amplifier, whose nonlinear distortions are neglected in this analysis for the sake of clarity. Note that the first two terms are the DC contribution of the second-order terms, or carrier residue in the real bandpass output signal. The output power level of the frequency components at $\omega_c \pm 2\omega_m$, referred to a real Z_0 load, can be expressed as

$$P_{\omega_c \pm 2\omega_m} = \frac{c_1^2 A^4}{16 \cdot 2Z_0} \left\{ |H_{I2}(\omega_m, \omega_m)|^2 + |H_{Q2}(\omega_m, \omega_m)|^2 \right.$$
$$\left. \pm 2|H_{I2}(\omega_m, \omega_m)||H_{Q2}(\omega_m, \omega_m)| \sin(\Delta\Psi) \right\}, \tag{4.55}$$

where the phase difference is defined as $\Delta\Psi = \angle H_{I2}(\omega_m, \omega_m) - \angle H_{Q2}(\omega_m, \omega_m)$.

The asymmetrical behavior of the frequency components at $f_c + 2f_m$ and $f_c - 2f_m$ predicted by equation (4.55) has been plotted in Figure 4.12 with solid and dashed lines, showing a good agreement with measurements. Therefore, though simple and with room for improvement, the nonlinear block model of Figure 4.11 seems

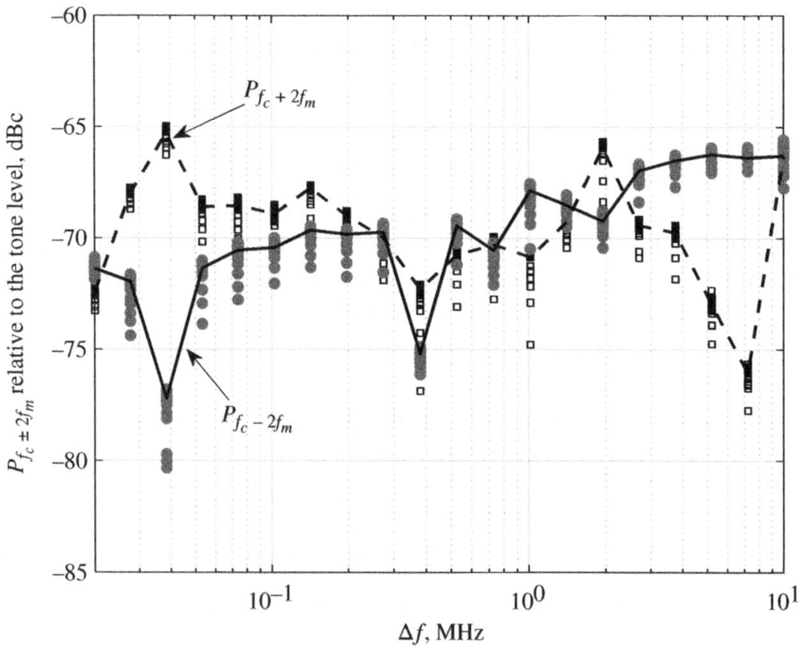

Figure 4.12 Asymmetries between the output components at $f_c + 2f_m$ and $f_c - 2f_m$ versus the tone spacing for different carrier power levels. Prediction based on a Volterra series analysis (solid and dashed lines) and measurement (marks).

a better representation of the behavior of I/Q modulators than a linear model with I/Q impairments.

4.6.2 Widely Nonlinear Approach for I/Q Modulators

The modulator transforms two discrete-time real-valued signals, $x_I(k)$ and $x_Q(k)$, in a complex-valued signal, $u(t) = u_I(t) + ju_Q(t)$, and the Volterra models examined above are not appropriate to explain this system.

An ideally perfect linear I/Q modulator transforms the input signal $x = x_I + jx_Q$ into the complex envelope of the output RF signal u as a simply linear system. That is to say, u is a proper signal that is uncorrelated with its complex conjugate, i.e., its complementary correlation is zero. However, as it has been stated, actual I/Q modulators present unbalanced gains, DC offsets and errors in the quadrature of the carriers of the two modulator branches. These imperfections make u an improper signal that explicitly depends on the input signal x and its complex conjugate x^* (Ding et al., 2003). According to the definitions in Adali et al. (2011), such a modulator can be considered a *widely linear* system. Specific signal processing techniques can be applied to complex-valued signals, such as Wirtinger calculus that allows keeping all computations for widely linear systems in the complex domain, in a similar way to the real-valued case (Wirtinger, 1927). Unfortunately, the modulator also produces nonlinear distortions and therefore it cannot be represented by a widely linear model. Based on the knowledge of the physical blocks that constitute the I/Q modulator, we are going to deduce a more realistic model involving a nonlinear function to transform $x(k)$ into $u(k)$, referred to as a *widely nonlinear* model.

As it is shown in Figure 4.13, the I branch of the I/Q modulator is basically a bivariate nonlinear (bi-NL) block with the in-phase signal $x_I(t)$ applied to one input and the local oscillator $A\cos\omega_c t$, applied to the second input. In the Q branch, the inputs are $x_Q(t)$ and $A\sin(\omega_c t - \phi)$. It is possible to analyze each branch by using a double Volterra series approach (Rice, 1973).

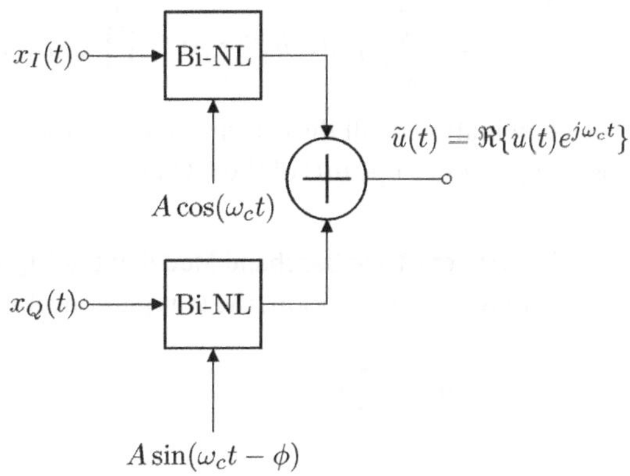

Figure 4.13 Equivalent model of an I/Q modulator with bivariate nonlinear (bi-NL) blocks.

4.6.2.1 Analysis of the I Branch

The output of the I branch contains three separate parts:

1. The output of a standard Volterra system driven by $x_I(t)$. Since this is a lowpass signal, its contribution to the fundamental band is negligible.

$$z_{I0}(t) = \sum_{n=1}^{\infty} \int h_{n0}^{(I)}(\tau_n) \prod_{r=1}^{n} x_I(t - \tau_r) d\tau_r. \tag{4.56}$$

2. The output of a Volterra system driven by the local oscillator signal. Its content in the fundamental band is a spectral line at ω_c, whose amplitude is denoted as $h_0^{(I)}$, a coefficient that does not depend on the baseband signal.

$$z_{0I}(t) = \sum_{m=1}^{\infty} A^m \int h_{0m}^{(I)}(\zeta_m) \prod_{s=1}^{m} \cos(\omega_c t - \omega_c \zeta_s) d\zeta_s. \tag{4.57}$$

3. The cross modulation between $x_I(t)$ and the local oscillator. Sorting those terms that contribute to the fundamental zone around ω_c, the output can be expressed as $\Re\{z_I(t)e^{j\omega_c t}\}$, where

$$z_I(t) = \sum_{n=1}^{\infty} \int h_n^{(I)}(\tau_n) \prod_{r=1}^{n} x_I(t - \tau_r) d\tau_r. \tag{4.58}$$

Note that the explicit dependence of the kernel $h_n^{(I)}(\tau_n)$ on ω_c and A is omitted, since we assume fixed levels for the carrier frequency and the local oscillator.

4.6.2.2 Analysis of the Q Branch

Proceeding in a similar way and replacing $\omega_c t$ by $(\omega_c t - \pi/2 - \phi)$, we can compute the corresponding components at the output of the Q branch bi-NL block. The distortion generated by $x_Q(t)$ in the fundamental zone is negligible and the amplitude of the spectral line is $-je^{-j\phi}h_0^{(Q)}$. The contribution of the mixed terms can be expressed as $\Re\{z_Q(t)e^{j\omega_c t}\}$, where

$$z_Q(t) = -j \sum_{n=1}^{\infty} \int d\tau_n h_n^{(Q)}(\tau_n) e^{-j\phi} \prod_{r=1}^{n} x_Q(t - \tau_r). \tag{4.59}$$

The combination of all these terms at the output of the I and Q branches produce the complex envelope $u(t)$ of the I/Q modulator RF output.

4.6.2.3 Discrete-Time Baseband Model of the I/Q Modulator

Using discrete-time notation, the complex envelope is given by the sum

$$u(k) = u_0 + \sum_{n=1}^{\infty} u_n(k), \tag{4.60}$$

where $u_0 = h_0^{(I)} + je^{-j\phi}h_0^{(Q)}$ is the carrier leakage and

$$u_n(k) = \sum_{\mathbf{q}_n=\mathbf{0}}^{\mathbf{Q}_n} \left[h_n^{(I)}(\mathbf{q}_n) \prod_{r=1}^{n} x_I(k-q_r) + jh_n^{(Q)}(\mathbf{q}_n)e^{-j\phi} \prod_{r=1}^{n} x_Q(k-q_r) \right]. \quad (4.61)$$

Defining the complex-valued input signal $x(k) = x_I(k) + jx_Q(k)$, we can write $x_I(k) = \frac{1}{2}[x(k) + x^*(k)]$ and $x_Q(k) = \frac{-j}{2}[x(k) - x^*(k)]$ to expand the products in (4.61). With the help of the following definition

$$h_{\mu,n-\mu}(\mathbf{q}_n) = \frac{1}{2^n} \binom{n}{\mu} \left[h_n^{(I)}(\mathbf{q}_n) + j^{n+2\mu+1} h_n^{(Q)}(\mathbf{q}_n) e^{-j\phi} \right], \quad (4.62)$$

the baseband model of the I/Q modulator can be written

$$u(k) = u_0 + \sum_{n=1}^{\infty} \left\{ \sum_{\mathbf{q}_n=\mathbf{0}}^{\mathbf{Q}_n} h_{n,0}(\mathbf{q}_n) \prod_{r=1}^{n} x(k-q_r) + \sum_{\mathbf{q}_n=\mathbf{0}}^{\mathbf{Q}_n} h_{0,n}(\mathbf{q}_n) \prod_{r'=1}^{n} x^*(k-q_{r'}) \right.$$
$$\left. + \sum_{\mu=1}^{n-1} \sum_{\mathbf{q}_n=\mathbf{0}}^{\mathbf{Q}_n} h_{n-\mu,\mu}(\mathbf{q}_n) \prod_{r=1}^{n-\mu} x(k-q_r) \prod_{r'=n-\mu+1}^{n} x^*(k-q_{r'}) \right\}. \quad (4.63)$$

Observe that, if we consider only linear imperfections in the modulator, the terms with $n \geq 2$ are neglected in (4.63) and the model is widely linear as in Ding et al. (2008, Zou et al., 2008).

4.6.2.4 Model Structure of a Transmitter in the Presence of I/Q Impairments

In a generic communication system, the baseband signal is used in the I/Q modulator to generate the RF signal, possibly with linear and nonlinear impairments, and the power amplifier delivers the signal at the desired power, adding further nonlinear distortions. The effects that the presence of I/Q impairments produce in the model structure of the complete transmitter are interesting to be discussed. To illustrate the relative importance of the kernels identified for a transmitter based on the complex Volterra series model of Section 3.7, we can define normalized kernels $|\hat{h}_{n-m,m}(\mathbf{q}_{n-m}, \mathbf{p}_m)|^2$ that can be interpreted as the power contribution of the corresponding regressor to the output signal. Graphs containing the ordered sequence of the normalized kernels are depicted in Figure 4.14 for an I/Q modulator under two scenarios: when it is operated with negligible nonlinear distortions for a reduced power level and without artificially-added impairments, and when it exhibits an increase of 6 dB in power level together with I/Q impairments, such as gain imbalance. The horizontal axis spans from index 0 and 1, corresponding to the DC term and the kernel associated with the $x(k)$ term, respectively, to higher indexes representing the nth-order kernel associated with the $(x^*(k))^n$ term. For the sake of clarity, only orders one to three are shown, and kernels with a level below −60 dB are assumed negligible.

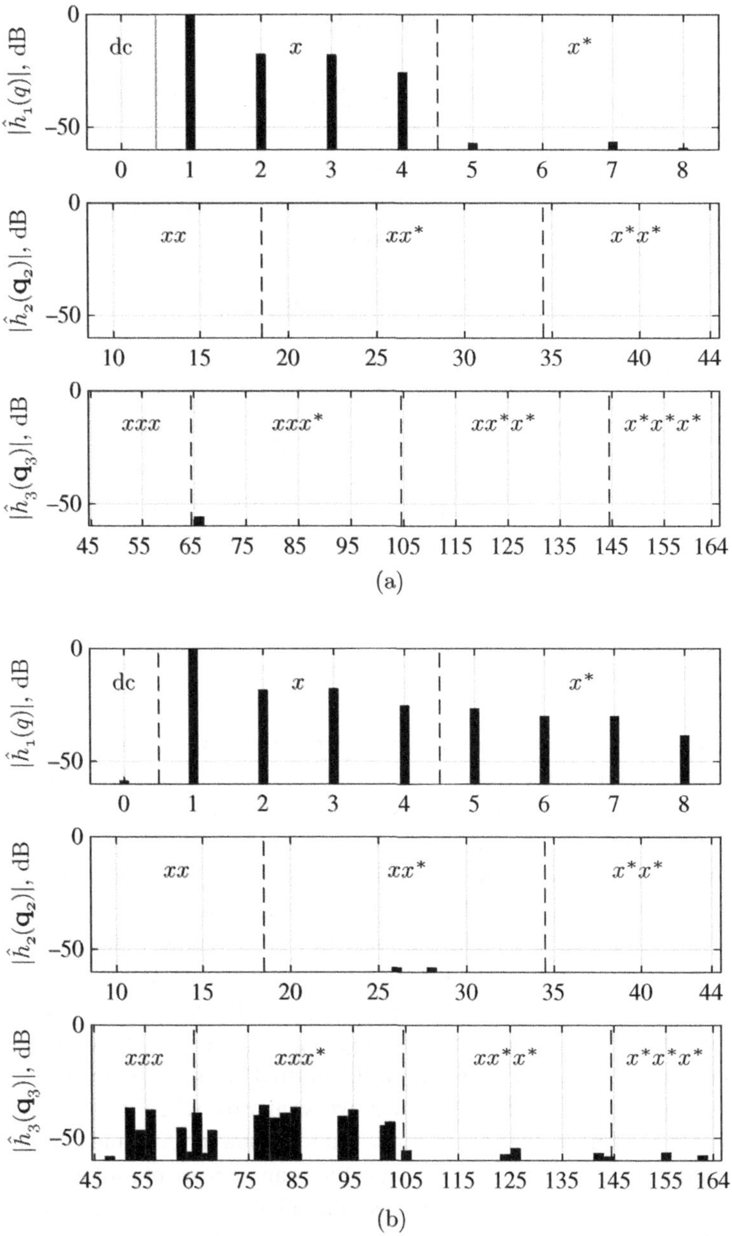

Figure 4.14 Normalized power of the regressors with respect to the memoryless linear term in dB for an I/Q modulator (a) without artificially-added impairments for reduced power level and (b) with a gain imbalance of −1 dB for high power level.

For the first order without I/Q impairments and nonlinear distortions, only regressors of the type x are present, with the maximum corresponding to the memoryless term (0-dB reference). On the other hand, with I/Q impairments the terms depending on the image signal x^* become non-negligible. The second-order kernels are negligible in general. For the third order, all the terms are below −60 dB without impairments except the memoryless coefficient associated with the regressors of the type xxx^*. However, the structure of the case with I/Q impairments is more complex, and all the types xxx, xxx^*, xx^*x^*, and $x^*x^*x^*$ are present, although with negligible values in some cases.

Bibliography

T. Adali, P.J. Schreier, and L.L. Scharf. Complex-valued signal processing: The proper way to deal with impropriety. *IEEE Transactions on Signal Processing*, 59(11):5101–5125, 2011. doi: 10.1109/TSP.2011.2162954.

S.A. Bassam, W. Chen, M. Helaoui, F.M. Ghannouchi, and Z. Feng. Linearization of concurrent dual-band power amplifier based on 2D-DPD technique. *IEEE Microwave and Wireless Components Letters*, 21(12):685–687, 2011a. doi: 10.1109/LMWC.2011.2170669.

S.A. Bassam, M. Helaoui, and F.M. Ghannouchi. 2-D digital predistortion (2-D-DPD) architecture for concurrent dual-band transmitters. *IEEE Transactions on Microwave Theory and Techniques*, 59(10):2547–2553, 2011b. doi: 10.1109/TMTT.2011.2163802.

H. Cao, A.S. Tehrani, C. Fager, T. Eriksson, and H. Zirath. I/Q imbalance compensation using a nonlinear modeling approach. *IEEE Transactions on Microwave Theory and Techniques*, 57(3):513–518, 2009. doi: 10.1109/TMTT.2008.2012305.

J.K. Cavers. The effect of quadrature modulator and demodulator errors on adaptive digital predistorters for amplifier linearization. *IEEE Transactions on Vehicular Technology*, 46(2):456–466, 1997. doi: 10.1109/25.580784.

J.K. Cavers and M.W. Liao. Adaptive compensation for imbalance and offset losses in direct conversion transceivers. *IEEE Transactions on Vehicular Technology*, 42(4):581–588, 1993. doi: 10.1109/25.260752.

C. Crespo-Cadenas, J. Reina-Tosina, and M.J. Madero-Ayora. Volterra series approach to behavioral modeling: Application to an FET amplifier. In *2006 Asia-Pacific Microwave Conference*, pages 445–448, 2006. doi: 10.1109/APMC.2006.4429459.

C. Crespo-Cadenas, J. Reina-Tosina, and M.J. Madero-Ayora. Volterra behavioral model for wideband RF amplifiers. *IEEE Transactions on Microwave Theory and Techniques*, 55(3):449–457, 2007. doi: 10.1109/TMTT.2006.890514.

C. Crespo-Cadenas, J. Reina-Tosina, and M.J. Madero-Ayora. Performance of an extended behavioral model for wideband amplifiers. In *2008 Asia-Pacific Microwave Conference*, pages 1–4, 2008. doi: 10.1109/APMC.2008.4958289.

C. Crespo-Cadenas, J. Reina-Tosina, and M.J. Madero-Ayora. Study of a power amplifier behavioral model with nonlinear thermal effects. In *The 40th European Microwave Conference*, pages 1138–1141, 2010a. doi: 10.23919/EUMC.2010.5616708.

C. Crespo-Cadenas, J. Reina-Tosina, M.J. Madero-Ayora, and J. Muñoz-Cruzado. A new approach to pruning Volterra models for power amplifiers. *IEEE Transactions on Signal Processing*, 58(4): 2113–2120, 2010b. doi: 10.1109/TSP.2009.2039815.

C. Crespo-Cadenas, J. Reina-Tosina, M.J. Madero-Ayora, and M. Allegue-Martínez. A Volterra-based procedure for multi-port and multi-zone GaN FET amplifier CAD simulation. *IEEE Transactions on Circuits and Systems I: Regular Papers*, 60(11):3022–3032, 2013. doi: 10.1109/TCSI.2013.2252691.

C. Crespo-Cadenas, M.J. Madero-Ayora, and J.A. Becerra. A bivariate Volterra series model for the design of power amplifier digital predistorters. *Sensors*, 21(17), 2021. doi: 10.3390/s21175897.

W. Dai, P. Roblin, and M. Frei. Distributed and multiple time-constant electro-thermal modeling and its impact on ACPR in RF predistortion. In *Conference, 2003. Fall 2003. 62nd ARFTG Microwave Measurements*, pages 89–98, 2003. doi: 10.1109/ARFTGF.2003.1459759.

L. Ding, Z. Ma, D.R. Morgan, M. Zierdt, and G.T. Zhou. Frequency-dependent modulator imbalance in predistortion linearization systems: modeling and compensation. In *The Thrity-Seventh Asilomar Conference on Signals, Systems & Computers, 2003*, volume 1, pages 688–692, 2003. doi: 10.1109/ACSSC.2003.1292002.

L. Ding, Z. Ma, D.R. Morgan, M. Zierdt, and G.T. Zhou. Compensation of frequency-dependent gain/phase imbalance in predistortion linearization systems. *IEEE Transactions on Circuits and Systems I: Regular Papers*, 55 (1):390–397, 2008. doi: 10.1109/TCSI.2007.910545.

H. Enzinger, K. Freiberger, and C. Vogel. Competitive linearity for envelope tracking: Dual-band crest factor reduction and 2D-vector-switched digital predistortion. *IEEE Microwave Magazine*, 19 (1):69–77, 2018. doi: 10.1109/MMM.2017.2759618.

M. Faulkner, T. Mattson, and W. Yates. Automatic adjustment of quadrature modulators. *Electronics Letters*, 27(3):214–216, 1991. doi: 10.1049/el:19910139.

F. Filicori and G. Vannini. Mathematical approach to large-signal modelling of electron devices. *Electronics Letters*, 27(4):357–359, 1991. doi: 10.1049/el:19910226.

M.E. Gadringer, C. Schuberth, and G. Magerl. Characterization and modeling of direct conversion transmitters. In *2008 38th European Microwave Conference*, pages 745–748, 2008. doi: 10.1109/EUMC.2008.4751560.

O. Jardel, F. De Groote, T. Reveyrand, J.-C. Jacquet, C. Charbonniaud, J.-P. Teyssier, D. Floriot, and R. Quere. An electrothermal model for AlGaN/GaN power HEMTs including trapping effects to improve large-signal simulation results on high VSWR. *IEEE Transactions on Microwave Theory and Techniques*, 55(12):2660–2669, 2007. doi: 10.1109/TMTT.2007.907141.

J. Kim and K. Konstantinou. Digital predistortion of wideband signals based on power amplifier model with memory. *Electronics Letters*, 37(23):1417–1418, 2001. doi: 10.1049/el:20010940.

H. Ku and J. S. Kenney. Behavioral modeling of nonlinear RF power amplifiers considering memory effects. *IEEE Transactions on Microwave Theory and Techniques*, 51(12):2495–2504, 2003. doi: 10.1109/TMTT.2003.820155.

D.R. Morgan, Z. Ma, J. Kim, M.G. Zierdt, and J. Pastalan. A generalized memory polynomial model for digital predistortion of RF power amplifiers. *IEEE Transactions on Signal Processing*, 54(10):3852–3860, 2006. doi: 10.1109/TSP.2006.879264.

A. Pérez-Hernández, J.A. Becerra, M.J. Madero-Ayora, and C. Crespo-Cadenas. An upgraded dual-band digital predistorter model for power amplifiers linearization. *IEEE Microwave and Wireless Components Letters*, 31(1):33–36, 2021. doi: 10.1109/LMWC.2020.3040101.

S.O. Rice. Volterra systems with more than one input port — distortion in a frequency converter. *The Bell System Technical Journal*, 52(8):1255–1270, 1973. doi: 10.1002/j.1538-7305.1973.tb02019.x.

W. Wirtinger. Zur formalen theorie der funktionen von mehr komplexen veranderlichen. *Mathematische Annalen*, 97:357–375, 1927.

D. Wisell. Identification and measurement of transmitter nonlinearities. In *56th ARFTG Conference Digest-Fall*, volume 38, pages 1–6, Boulder, CO, Nov. 2000.

D. Wisell and M. Isaksson. Derivation of a behavioral RF power amplifier model with low normalized mean-square error. In *2007 European Microwave Integrated Circuit Conference*, pages 485–488, 2007. doi: 10.1109/EMICC.2007.4412755.

A. Zhu, J.C. Pedro, and T.J. Brazil. Dynamic deviation reduction-based Volterra behavioral modeling of RF power amplifiers. *IEEE Transactions on Microwave Theory and Techniques*, 54(12):4323–4332, 2006. doi: 10.1109/TMTT.2006.883243.

A. Zhu, J.C. Pedro, and T.R. Cunha. Pruning the Volterra series for behavioral modeling of power amplifiers using physical knowledge. *IEEE Transactions on Microwave Theory and Techniques*, 55(5):813–821, 2007. doi: 10.1109/TMTT.2007.895155.

Y. Zou, M. Valkama, and M. Renfors. Digital compensation of I/Q imbalance effects in space-time coded transmit diversity systems. *IEEE Transactions on Signal Processing*, 56(6):2496–2508, 2008. doi: 10.1109/TSP.2007.916132.

5

Regression of Volterra Models

5.1 Introduction

The third and fourth chapters of this book presented the mathematical modeling of power amplifiers and wireless communication systems through the Volterra series. There exists a plethora of different model structures that are developed following different approaches with the intention of capturing the power amplifier behavior. Some of them are the model equivalent of a Swiss army knife, valid for general cases, while others are based on circuit knowledge that allows interpretation of the physical meaning of the regressors. The two main branches that develop model structures give rise to models based on either a priori or a posteriori information. Model structures based on a priori information use a generalist approach that includes those regressors that better represent the input–output relationship from a theoretical, heuristic, or physical point of view. So far in this book, the main reasoning was to consider the full Volterra as the most generic model, taking subsets of its regressors pool to attain pruned models. In contrast, based on the fact that not all basis functions within an a priori structure are significant, aposterioristic models use real measurements to determine and optimize their structure and are calculated by a certain sparse signal processing algorithm. A posteriori information can be used for an ulterior refinement of model structures, explicitly pruning not relevant regressors. This framework produces a paradigm shift — instead of seeking simple models, we opt for large models with sufficient regressor richness to be pruned. The latter techniques will be covered in detail in the next chapter.

This chapter is devoted to the introduction of the mathematical tools to calculate the regression of Volterra models, i.e., once the model structure is selected and fixed, the necessary steps are taken to attain their coefficients so that their behavior resembles the power amplifier's. Although the same principles can be applied to digital predistortion, the basic theory of regression is often presented

A Volterra Approach to Digital Predistortion: Sparse Identification and Estimation, First Edition.
Carlos Crespo-Cadenas, María José Madero-Ayora, and Juan A. Becerra.

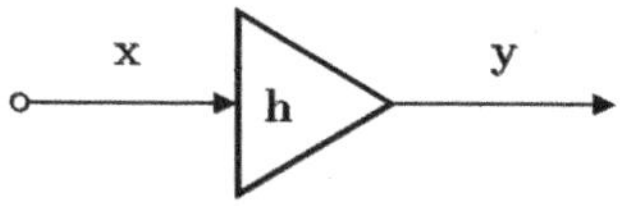

Figure 5.1 Relation between the amplifier's input signal **x**, its output **y** and the power amplifier behavior, represented by its coefficient vector **h**.

in a forward modeling scheme, since the same framework applies in digital predistortion (DPD) scenarios by just swapping the input and output signals.

5.2 Least Squares Algorithm

The least squares method is the cornerstone in the field of regression and is widely used in all areas of knowledge. The convenience of this standard approach to attain a solution of an overdetermined system lies in its closed-form single solution, which provides a ready-to-use formula that allows one to estimate the coefficients vector.

5.2.1 The Measurement Equation

The measurement equation, represented in Figure 5.1, relates the complex envelope output of the power amplifier

$$\mathbf{y} = [y(k), y(k+1), \dots, y(k+m-1)]^T \in \mathbb{C}^{m\times 1} \tag{5.1}$$

to its input

$$\mathbf{x} = [x(k), x(k+1), \dots, x(k+m-1)]^T \in \mathbb{C}^{m\times 1} \tag{5.2}$$

as

$$\mathbf{y} = \mathbf{X}\mathbf{h} + \epsilon, \tag{5.3}$$

where **x** and **y** are vectors that hold m consecutive samples of the input and output signals, respectively, k is the discrete-time index, $\mathbf{h} \in \mathbb{C}^{n\times 1}$ is the vector of model coefficients, **X** is the measurement matrix that holds n regressors, and $\boldsymbol{\epsilon}$ stands for the unavoidable error that the model cannot capture.

The matrix **X** is shaped by the model structure, which holds one regressor in each of its columns:

$$\mathbf{X} = \begin{bmatrix} \phi_1 & \phi_2 & \cdots & \phi_n \end{bmatrix} \in \mathbb{C}^{m\times n}, \tag{5.4}$$

where ϕ_i is the ith generic regressor of the model that generally consists of some kind of function of the input complex signal $x(k)$, commonly being combinations of the signal, its conjugate, and its envelope in Volterra series models.

For illustration purposes, consider a simple third-order univariate memory polynomial (MP) model (3.28) with memory depth of $Q = 1$, which results in four coefficients. Using the compact notation of $h_{n,q} = h_n(q)$, the model equation follows

$$y(k) = h_{1,0}x(k) + h_{1,1}x(k-1) + h_{3,0}x(k)|x(k)|^2 + h_{3,1}x(k-1)|x(k-1)|^2, \tag{5.5}$$

and taking the generic regressor notation,

$$y = h_{1,0}\phi_1 + h_{1,1}\phi_2 + h_{3,0}\phi_3 + h_{3,1}\phi_4. \tag{5.6}$$

A diligent reader will notice that there exists a map between the two dimensional notation of the MP and the above mentioned unidimensional notation. This mapping depends on the exact configuration of the model, letting i in ϕ_i be a generic consecutive index that denotes the regressor order of appearance.

This model structure admits the matrix form of (5.3). The measurement matrix follows

$$\mathbf{X}_{\text{MP}} = \begin{bmatrix} x(k) & x(k-1) & x(k)|x(k)|^2 & x(k-1)|x(k-1)|^2 \\ x(k+1) & x(k) & x(k+1)|x(k+1)|^2 & x(k)|x(k)|^2 \\ \vdots & \vdots & \vdots & \vdots \\ x(k+q) & x(k+q-1) & x(k+q)|x(k+q)|^2 & x(k+q-1)|x(k+q-1)|^2 \\ \vdots & \vdots & \vdots & \vdots \\ x(k+m-1) & x(k+m-2) & x(k+m-1)|x(k+m-1)|^2 & x(k+m-2)|x(k+m-2)|^2 \end{bmatrix}, \tag{5.7}$$

where its ith column represents the ith generic regressor ϕ_i. Similarly, its Volterra coefficient vector has the form of

$$\mathbf{h}_{\text{MP}} = \left[h_{1,0}\ h_{1,1}\ h_{3,0}\ h_{3,1}\right]^T. \tag{5.8}$$

Although an example of a MP model has been given, all the discussion within this chapter is focused on a general Volterra framework, and it does not depend on the underlying model structure.

The main purpose of the least squares method is to obtain an estimate $\hat{\mathbf{h}}$ of the Volterra kernel vector that minimizes the modeling error. Considering that the output of this system is the estimated output $\hat{\mathbf{y}}$, i.e.,

$$\hat{\mathbf{y}} = \mathbf{X}\hat{\mathbf{h}}, \tag{5.9}$$

the error between the estimation and the real output can be defined as

$$\mathbf{e} = \mathbf{y} - \hat{\mathbf{y}}. \tag{5.10}$$

Recall that Volterra series have the peculiarity that although they reproduce a nonlinear link between the input and the output of the system, the output is linear with respect to the regressors. Observe that the whole process can be intuitively considered as a kernel vector $\mathbf{h}$ that passes through a linear system with response $\mathbf{X}$ controlled by $\mathbf{x}$, as shown in Figure 5.2. This framework allows a

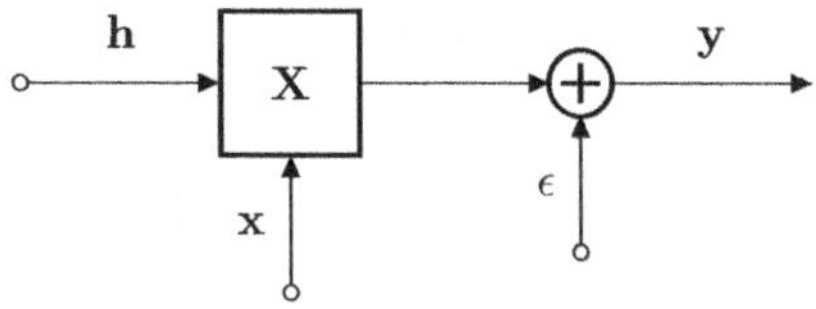

Figure 5.2 Block diagram of Volterra series represented as a measurement process where **h** is the vector of Volterra coefficients to be estimated. Note that the power amplifier input signal **x** shapes the system matrix **X**, whose input is the Volterra kernel vector **h**.

straightforward application of general signal processing techniques to regression problems in power amplifiers modeling. Some of these techniques will be shown in Chapter 6.

5.2.2 The Least Squares Method

The least squares method aims at attaining an estimator that reduces the quadratic error, known in the literature as the residual sum of squares (RSS), which is defined as the ℓ_2-norm of the modeling error, i.e.,

$$\text{RSS} = \|\mathbf{e}\|_2^2 = \|\mathbf{y} - \hat{\mathbf{y}}\|_2^2. \tag{5.11}$$

Please note that RSS and NMSE are closely related: RSS measures the error in natural units and NMSE represents the error scaled to the reference signal power, typically expressed in decibels when referring to behavioral modeling and linearization of power amplifiers. Aiming at reducing the difference between the system output $\mathbf{y}$ and the model output $\hat{\mathbf{y}}$, it is ensured that the model will behave similarly to the system. This process is depicted in Figure 5.3. Although this last comment is generally valid, the modeling performance of the least squares solution will also depend on the quality of the measurement matrix $\mathbf{X}$, providing a better solution in the RSS sense as long as it shows enough richness in its regressors and it is adequate to represent the physical characteristics of the system. In this sense, the measurement matrix quality is determined by the ability to select the appropriate model to represent the behavior of the nonlinear system.

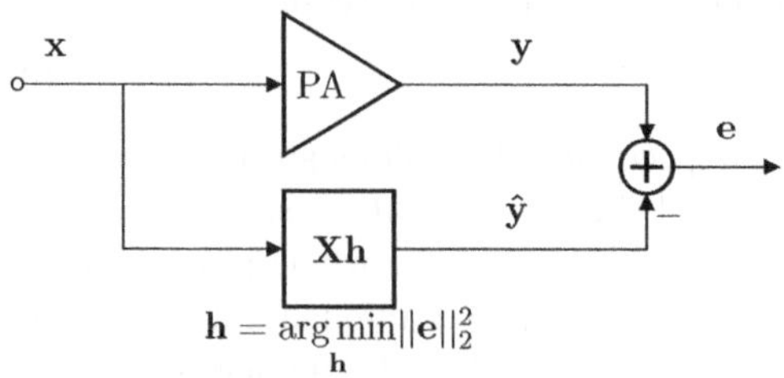

Figure 5.3 Modeling scheme representation. The objective of least squares is to minimize the error between the system and model outputs.

For example, a memoryless model is less appropriate than a model with memory when representing the actual behavior of a power amplifier.

The RSS can be expanded as

$$\mathrm{RSS} = (\mathbf{y} - \hat{\mathbf{y}})^H(\mathbf{y} - \hat{\mathbf{y}}). \tag{5.12}$$

From (5.9) we have

$$\mathrm{RSS} = (\mathbf{y} - \mathbf{X}\hat{\mathbf{h}})^H(\mathbf{y} - \mathbf{X}\hat{\mathbf{h}}). \tag{5.13}$$

Hence,

$$\begin{aligned}\mathrm{RSS} &= (\mathbf{y}^H - \hat{\mathbf{h}}^H\mathbf{X}^H)(\mathbf{y} - \mathbf{X}\hat{\mathbf{h}})\\ &= \mathbf{y}^H\mathbf{y} - \hat{\mathbf{h}}^H\mathbf{X}^H\mathbf{y} - \mathbf{y}^H\mathbf{X}\hat{\mathbf{h}} + \hat{\mathbf{h}}^H\mathbf{X}^H\mathbf{X}\hat{\mathbf{h}}.\end{aligned} \tag{5.14}$$

In order to find the value of $\mathbf{h}$ that produces a minimum in the RSS, its derivative with respect to the coefficients is taken and equals to zero,

$$\frac{d\,\|\mathbf{e}\|_2^2}{d\mathbf{h}} = -2\mathbf{X}^H\mathbf{y} + 2\mathbf{X}^H\mathbf{X}\hat{\mathbf{h}} = 0, \tag{5.15}$$

that can be rewritten as the set of equations known as the *normal equations*

$$\mathbf{X}^H\mathbf{X}\hat{\mathbf{h}} = \mathbf{X}^H\mathbf{y}. \tag{5.16}$$

The normal equations relate the estimated Volterra coefficients vector $\hat{\mathbf{h}}$ with the projection of the output signal $\mathbf{y}$ over the regressors of the measurement matrix, i.e., $\mathbf{X}^H\mathbf{y}$. Please note that the term $\mathbf{X}^H\mathbf{X}$ is mixing the components of the Volterra vector following the statistics between the model regressors.

Considering that the product $\mathbf{X}^H\mathbf{X}$ is a nonsingular matrix, the resulting least squares solution is then given by

$$\hat{\mathbf{h}} = (\mathbf{X}^H\mathbf{X})^{-1}\mathbf{X}^H\mathbf{y}, \tag{5.17}$$

and defining the Moore–Penrose pseudoinverse $\mathbf{X}^\dagger$ of $\mathbf{X}$ as

$$\mathbf{X}^\dagger = \left(\mathbf{X}^H\mathbf{X}\right)^{-1}\mathbf{X}^H, \tag{5.18}$$

the solution can take the form of

$$\hat{\mathbf{h}} = \mathbf{X}^\dagger\mathbf{y}. \tag{5.19}$$

The projection of the output signal over the pseudoinverse $\mathbf{X}^\dagger$ computes the LS solution.

The estimation of the output signal $\hat{\mathbf{y}}$ is attained through

$$\hat{\mathbf{y}} = \mathbf{X}\hat{\mathbf{h}} = \mathbf{X}\left(\mathbf{X}^H\mathbf{X}\right)^{-1}\mathbf{X}^H\mathbf{y}, \tag{5.20}$$

where

$$\mathcal{H} = \mathbf{X}\left(\mathbf{X}^H\mathbf{X}\right)^{-1}\mathbf{X}^H \tag{5.21}$$

is usually called the hat matrix (Draper and Smith, 1966; Hoaglin and Welsch, 1978) that converts values from the observed variable into estimations.

The normal equations (5.16) are an interesting intermediate result of the regression process. Consider, for example, an orthonormal basis set $\mathbf{X}_{\text{orth}}$, i.e., a situation in which, if we take a vector space interpretation of the basis functions, all the regressors are perpendicular and therefore fulfill the following relation

$$\mathbf{X}_{\text{orth}}^{H}\mathbf{X}_{\text{orth}} = \mathbf{I}. \tag{5.22}$$

Using this result in (5.16) reveals that

$$\hat{\mathbf{h}} = \mathbf{X}_{\text{orth}}^{H}\mathbf{y}, \tag{5.23}$$

thus, when the basis functions are independent, the Volterra kernel coefficient is calculated through the scalar product of its basis function and the system output. This scalar product consists of the correlation between both signals; therefore one can conclude that a least squares regression consists basically in finding the similarities between the regressors and the output signal, with the coefficient higher as their resemblance grows. The effect of a more general nonorthogonal basis function set is deeper analyzed in the next subsections.

5.2.3 The Autocorrelation and Crossvariance Matrices

5.2.3.1 Autocovariance Matrix

Internally, the regression is dominated by the statistics and correlations of the signals. The sample autocovariance matrix $\mathbf{R}_{\text{X}}$ can be defined as

$$\mathbf{R}_{\text{X}} = \mathbf{X}^{H}\mathbf{X} = \begin{bmatrix} \rho_{11} & \rho_{12} & \cdots & \rho_{1n} \\ \rho_{21} & \rho_{22} & \cdots & \rho_{2n} \\ \vdots & \vdots & \ddots & \vdots \\ \rho_{n1} & \rho_{n2} & \cdots & \rho_{nn} \end{bmatrix} \in \mathbb{C}^{n\times n}, \tag{5.24}$$

where the component crosscorrelation is defined as

$$\rho_{ij} = \phi_i^{H}\phi_j, \tag{5.25}$$

and the precision matrix $\mathbf{P}_{\text{X}}$ as its inverse, i.e.,

$$\mathbf{P}_{\text{X}} = \mathbf{R}_{\text{X}}^{-1}. \tag{5.26}$$

The autocovariance matrix accounts for the relations between the basis functions. Figure 5.4 shows a graphical representation of the absolute value of an autocovariance matrix of a generalized MP model. Darker shades represent high covariances, closer to 1, while lighter shades represent low covariances. The representation shows a clear structure of three blocks that coincide with the three parts in which the generalized memory polynomial (GMP) model structure shown in

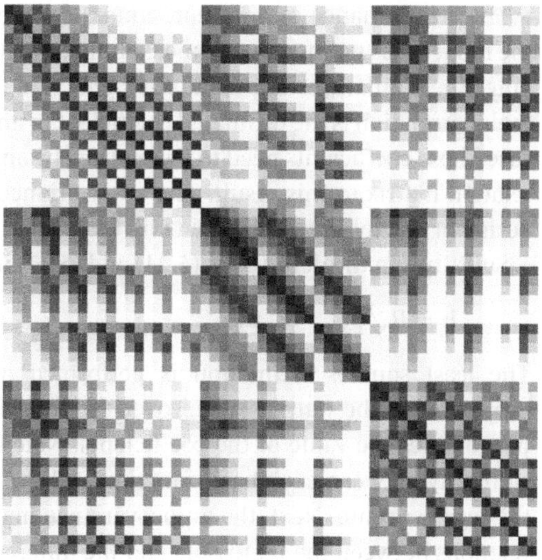

Figure 5.4 Representation of the autocovariance matrix absolute value of a generalized memory polynomial model.

equation (4.2) can be separated. The regressors are shown in the appearance order of the equation, i.e., the first block corresponds to diagonal regressors, the second block represents the delayed envelope basis functions, and the third one corresponds to the advanced envelope. The diagonal of the matrix exhibits the highest value possible, since the similarity between the basis functions and themselves is equal to one. This diagonal also represents the power of the basis functions, which is equal to one in every regressor since the measurement matrix is normalized.

Please note that the autocovariance—or covariance—matrix is coincident with the autocorrelation matrix, since, in general, the regressors are centered and their mean is zero.

5.2.3.2 Definition of Least Squares (LS) in Terms of the Crosscorrelation and Crossvariance Matrices

The statistical relations between the signal $\mathbf{y}$ and the Volterra kernels $\mathbf{X}$, seen as random variables, are characterized by their covariance. These statistics can be estimated by averaging the product of samples measured from both processes, allowing to use the sample crosscorrelation as an estimator of the crosscorrelation. The sample crossvariance vector $\mathbf{p}_{\mathbf{Xy}}$ takes the form of

$$\mathbf{p}_{\mathbf{Xy}} = \mathbf{X}^H\mathbf{y} = \begin{bmatrix}\gamma_1 & \gamma_2 & \cdots & \gamma_n\end{bmatrix}^T \in \mathbb{C}^{n\times 1}, \tag{5.27}$$

where the sample crosscorrelation between the ith regressor and the output obeys

$$\gamma_i = \phi_i^H\mathbf{y}. \tag{5.28}$$

The crossvariance matrix represents the projection of the output signal **y** over the model axes defined by the measurement matrix **X**. This particular result computes the similarities of the output signal and the model basis functions, where their correlation is the estimated coefficient in an orthonormal model. In a general case, these coefficients need to be ***unmixed*** by applying the inverse of the crosscorrelation matrix to this result. The crossvariance matrix is specially useful in the design of greedy algorithms that will be covered in the next chapter.

Substituting the definitions (5.24) and (5.27) in (5.17), the regression becomes

$$\hat{\mathbf{h}} = \mathbf{R}_{\mathbf{X}}^{-1}\mathbf{p}_{\mathbf{Xy}} = \mathbf{P}_{\mathbf{X}}\mathbf{p}_{\mathbf{Xy}}. \tag{5.29}$$

The least squares estimation is performed in two internal steps. First, the projection of the output signal over the model basis functions, represented by $\mathbf{p}_{\mathbf{Xy}}$, defines the value of the coefficients. Note that this projection measures the similarity between each basis function and the output no matter how correlated the regressors are. Next, the autocovariance matrix is inverted to account for the relations between the regressors and the differences in their norm. The precision matrix is then applied to the projection to decorrelate the coefficients.

The definitions provided in this section allow the reader to interpret the process of estimation through least squares. These variables will be later used in Chapter 6, where it will be shown how the coefficients selection algorithms use these variables to determine the relevance of each regressor.

5.2.4 Centering, Normalization, and Standardization

Centering, normalization, and standardization are operations that, in general, help for numerical issues in the regression and for easier interpretation of the results.

Centering is the operation of subtracting the average from a signal. Normalization is the process of scaling a basis function, generally dividing it by its ℓ_2 norm, while standardization accounts for centering and normalization. While this explanation is valid in a general regression framework, in the context of communication signals one can assume—or enforce—that the signals are centered. Nevertheless, since Volterra series deal with exponentiations to different orders, the norm of the basis functions are generally quite different. Thus, in the current context, it is desirable to apply normalization before the regression and denormalize the resulting Volterra coefficients vector.

The normalization process consists of scaling the basis functions so that they have a same-scale norm before the regression. The normalization matrix $\mathcal{N}$ holds the ℓ_2-norm of the regressors and is defined as

$$\mathcal{N} = \operatorname{diag}\left(\sqrt{\mathbf{R}_{\mathbf{X}}}\right), \tag{5.30}$$

where diag(·) returns a diagonal matrix whose entries are the elements in the diagonal of its argument. Defining the normalized regressors matrix as

$$\overline{\mathbf{X}} = \mathbf{X} \cdot \mathcal{N}^{-1} \tag{5.31}$$

and the normalized Volterra kernel vector

$$\overline{\mathbf{h}} = \mathcal{N} \cdot \mathbf{h}, \tag{5.32}$$

results in a regression that follows

$$\hat{\mathbf{h}} = \mathcal{N}^{-1}(\overline{\mathbf{X}}^H \overline{\mathbf{X}})^{-1} \overline{\mathbf{X}}^H \mathbf{y}, \tag{5.33}$$

and using (5.24),

$$\hat{\mathbf{h}} = \mathcal{N}^{-1} \mathbf{R}_{\overline{\mathbf{X}}}^{-1} \overline{\mathbf{X}}^H \mathbf{y}. \tag{5.34}$$

Generally, the normalization process produces a lower condition number in $\mathbf{R}_{\overline{\mathbf{X}}}$ than in $\mathbf{R}_{\mathbf{X}}$, thus enhancing the numerical properties of the estimator.

5.2.5 Performance Indicators

Before exemplifying the regression in the next section, we introduce some indicators necessary to assess the performance achieved with different model structures. The quality of the regression process can be measured through different figures of merit. Some of them were previously introduced in Chapter 3. The first and most straightforward quantity that determines performance is the *identification NMSE* (3.91), which represents the similarity of the model and power amplifier output samples over the set of samples used in the regression, i.e., the capacity of the model to represent the nonlinear function with samples seen previously.

As a rule of thumb, when dealing with not complex hardware situations and in direct modeling schemes, NMSEs in the range of −30 to −40 represent a poor performance. NMSEs ranging from −40 to −50 dB are representative of fair and good modeling capabilities, while errors below −50 dB are reached in cases where the fidelity is very high. It is important to remark that, since NMSE is a logarithmic function, reducing its value by 3 dB represents a decrease of one half of the error in a natural scale, therefore it is easier to enhance it working in the high-end range than when the NMSE is very low. These ranges are generally valid in direct modeling, i.e., capturing the behavior in the forward path of the power amplifier.

The second important figure of merit that measures model quality is *validation NMSE*. Validation error is calculated between the model and amplifier outputs for not-previously-seen samples. It represents the generalization capabilities of the constructed model. Generally, validation NMSE follows the trend of the identification NMSE and the identification NMSE is few decibels lower than the validation NMSE. A situation in which the validation NMSE is much higher than the identification NMSE—the NMSE being good enough to think that the model structure is sufficient for modeling the nonlinear behavior— is generally indicative of either numerical problems or overfitting. The overfitting phenomenon appears when the model is too aligned to the dataset, i.e., the model has used *too much information*—including noise—to calculate its coefficients and it is very precise with the identification samples but loses its generalization properties. Overfitting normally occurs when the model structure is not adequate or sufficiently rich, when the number of samples is low compared to the number of regressors, or when the number of regressors is too high. Overfitting can be mitigated with regularization techniques, covered next in this chapter.

Another variation of NMSE that is highly relevant in digital predistortion scenarios is the *linearization NMSE*. In this case, the reference signal is considered to be the ideal output, i.e., the scaled input signal to the system, and the error quantified is the one between the linearized output of the amplifier and the scaled input to the system.

The adjacent channel error power ratio (ACEPR) (3.94) is often also presented as an additional performance metric that allows to distinguish between two similar modeling scenarios. In some amplifiers the combination of NMSE and ACEPR determines the performance of the modeling while in other cases just the NMSE is able to indicate the performance. The difference between them is that NMSE is dominated by the in-band error and the ACEPR considers the corresponding adjacent channels (Landin et al., 2008; Ghannouchi and Hammi, 2009).

5.2.6 A Practical Regression

A practical example of a regression through Volterra series is presented next. The device under test is a class AB commercial power amplifier.

The first exercise is to identify the model with a set of measurements that consist of 360 000 consecutive samples of a signal compliant with the fifth generation new radio (5G-NR) standard. The first 10% segment,[1] i.e., samples numbered from 1 to 36 000 will be used for the regression. For this first example, the model to be used is the MP (4.1), for which the memory depth and nonlinear order will be iterated from 0 to 10 and 1 to 14, respectively. Figure 5.5(a) shows the evolution of the identification NMSE. The general trend in such a simple and manageable model is that the richer the basis set—evidenced by an increase in the memory depth or nonlinear order— the better fidelity between its output and the signal to model. The edge of the surface that corresponds to nonlinear order 1 ($P = 0$)

1 This illustration exercise is meant to be very simple, but in a real situation the identification set should be chosen following a strategy that ensures that it is representative of the overall signal. For example, taking a segment of samples that represent the whole signal range avoids divergence of the model when the validation is performed out of the identification range.

represents the modeling error for a number of taps indicated by the memory depth. This edge is about −30 dB, clearly suggesting the need for nonlinear modeling. The side of the figure that corresponds to memory depth equals to zero, in the left part of the figure, is decreasing with the nonlinear order, reaching to a steady point of about −45.5 dB. This line corresponds to a nonlinear memoryless model that can be enhanced with the addition of memory to the model. The bottom surface is represented by a set of combinations that provide better modeling capabilities. At one extreme, the point of 6th order and memory depth of two results in an identification NMSE of −49.8 dB, ranging to −52.9 dB for the model with nonlinear order of 14 and memory depth of 10. The trend of the curve is always decreasing when the nonlinear order and memory depth increase, indicating that the inclusion of regressors is beneficial. The slope of this curve is different in each direction, making it more important to include parameters that cause a greater decrease in the curve. The fact that some regressors are much more relevant to the modeling process than others will be discussed in detail in the next chapter.

Once the model has been identified with the first 10% of the samples, its output with the remaining 90% was calculated and compared to the power amplifier output. Note that although in other fields it is a common trend to use the majority of the signal for training and the remaining to evaluate the fitting performance, in this case of power amplifiers behavioral modeling, the first 10% segment was enough to avoid overfitting and the computational complexity of the matrix inversion was not so demanding. This validation NMSE is plotted in Figure 5.5(b), where it is noticeable how the trend of both quantities are similar. Now, looking closely at the vertical axes, while the identification NMSE falls in the bottom part of the surface to near −53 dB, the validation NMSE almost reaches the same value with a decrease of less than 0.5 dB. This difference in identification and validation NMSEs is very likely to happen, but since the validation NMSE exhibits a good performance, we can conclude that the model is able to generalize its behavior to unseen data samples. Please note that, for proper modeling, the unseen sample data should exhibit similar characteristics to those used for the modeling.

The comparison between different model performances is somehow complex. While it is clear that models with higher nonlinear order and memory depth perform better in terms of modeling error, their main drawback is the increase in complexity. Complexity is measured in the literature in many different ways, such as running time, number of parameters, or number of floating point operations (FLOPs) (Tehrani et al., 2010). The first and most basic step in examining the complexity of a model is to take the number of coefficients as its computational complexity, leading to a higher computational complexity for models with a corresponding higher number of coefficients or regressors. Following the same experiment, Figure 5.6 shows the identification NMSE against the number of coefficients for the linear, third-, fifth- and seventh- order

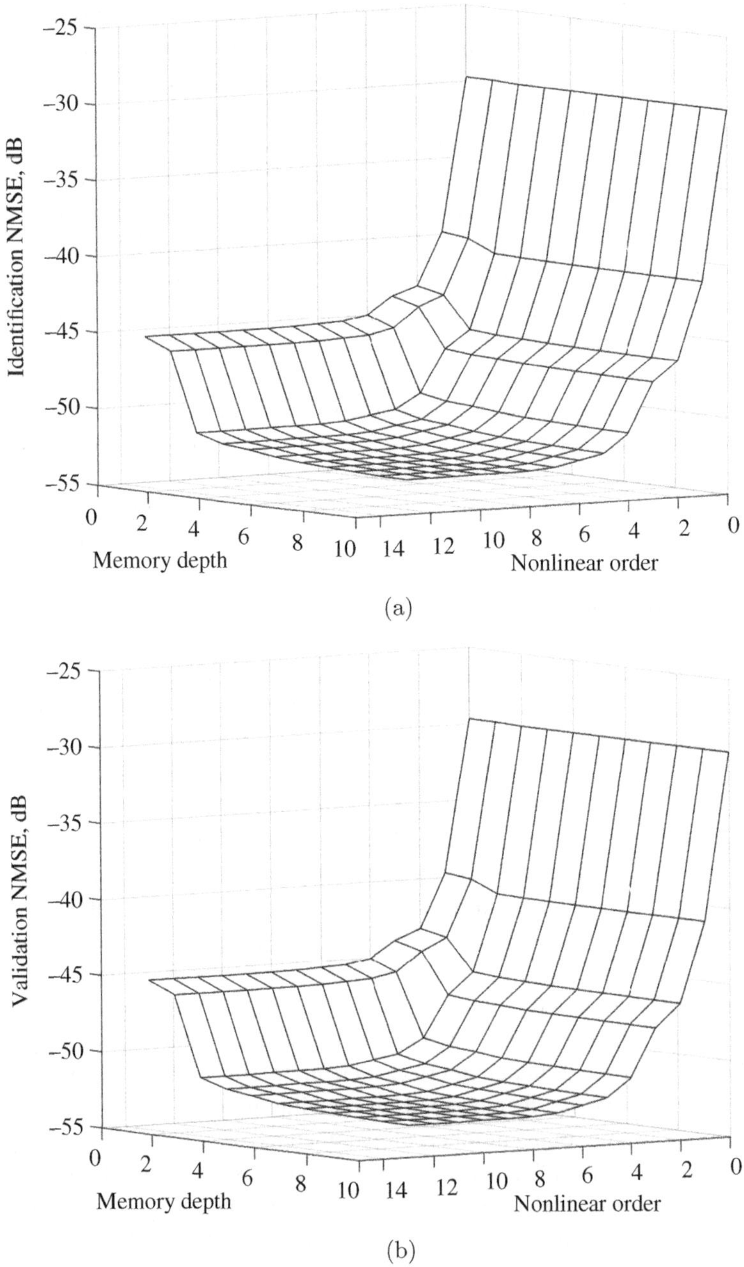

Figure 5.5 Identification NMSE (a) and validation NMSE (b) for a memory polynomial model in a sweep of its configuration parameters (memory depth and nonlinear order).

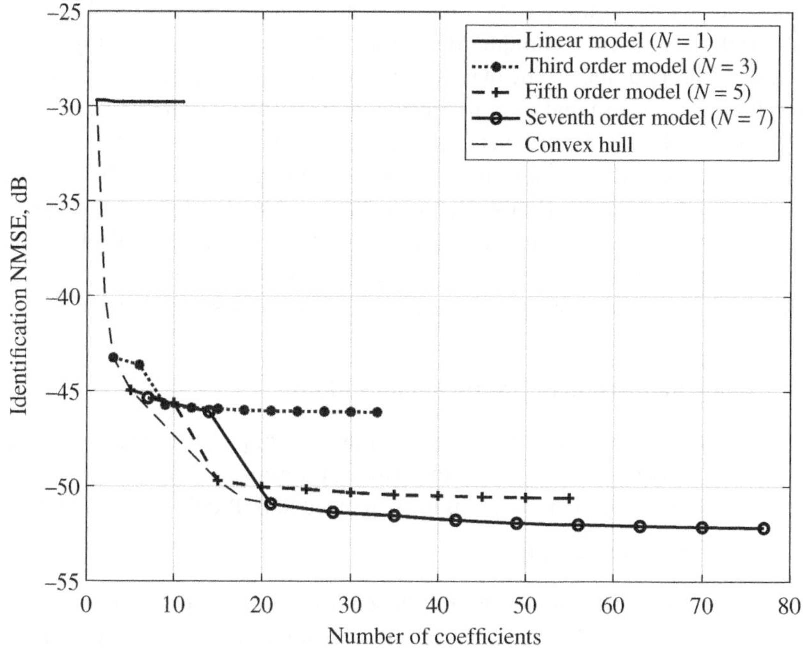

Figure 5.6 Identification NMSE for a memory polynomial model versus number of coefficients.

MP models. This simple representation shows, again, how increasing the memory depth for a given nonlinear order enhances the modeling error up to a certain point, leading to the need for increasing the nonlinear order to strengthen the modeling capabilities. In this example, the linear model shows a poor error that is outperformed by nonlinear configurations. The third nonlinear order increases its number of coefficients in multiples of 3, corresponding to the regressors $x(k-q)$, $x(k-q)|x(k-q)|$, and $x(k-q)|x(k-q)|^2$ for each value of an incremental q. The same behavior is observed for the fifth and seventh orders, increasing the number of coefficients in multiples of 5 and 7, respectively. In this scenario, the points that belong to the convex hull or convex envelope represent the highest performance in NMSE per coefficient.

5.3 Regularization

The least squares method is well-known because of its features, such as providing a single solution that leads to the minimum of the identification error. Nevertheless, there are special cases in which the regression does not perform properly and it is necessary to apply some kind of regularization (Becerra et al., 2022). Regularization is a set of techniques that enhance the regression, generally affecting the condition number of the matrix to be inverted. In this group of regularization techniques, one can find several strategies that lead to, for example, overfitting mitigation or reducing the number of active coefficients in the estimation.

Examples of regularization in the literature are (Liu et al., 2021), where Ridge regression was applied as a means of reducing the number of samples in the regression, (Gilabert et al., 2020) that provided a general framework of DPD including regularization, and (Guan and Zhu, 2012) that proposed a low-complexity implementation of the least squares method based on 1-bit Ridge regression.

5.3.1 Ridge Regression (ℓ_2-Norm Minimization)

Ridge regression, also known as ℓ_2-norm minimization, performs the estimation by enforcing a second condition in addition to the minimization of the error norm. In this case, the variable to minimize is the penalized RSS (PRSS), that is defined as

$$\text{PRSS} = \| \mathbf{y} - \hat{\mathbf{y}} \|_2^2 + \lambda \, \| \hat{\mathbf{h}} \|_2^2 . \tag{5.35}$$

The novelty against the previously introduced RSS minimization is the appearance of a penalty factor that depends on the norm of the coefficients $\| \hat{\mathbf{h}} \|_2^2$ and is controlled by the regularization factor λ. This regularization factor is a trade-off between the error minimization and the variance of the estimator. It has the ability of reducing the absolute value of the estimators, therefore reducing the dependence on noise in the regression process.

The PRSS is expanded as

$$\begin{aligned} \text{PRSS} &= (\mathbf{y} - \hat{\mathbf{y}})^H(\mathbf{y} - \hat{\mathbf{y}}) + \lambda \hat{\mathbf{h}}^H \hat{\mathbf{h}} \\ &= \mathbf{y}^H\mathbf{y} - \hat{\mathbf{h}}^H \mathbf{X}^H \mathbf{y} - \mathbf{y}^H \mathbf{X} \hat{\mathbf{h}} + \hat{\mathbf{h}}^H \mathbf{X}^H \mathbf{X} \hat{\mathbf{h}} + \lambda \hat{\mathbf{h}}^H \hat{\mathbf{h}}, \end{aligned} \tag{5.36}$$

and taking its derivative with respect to the coefficients value,

$$\frac{d\text{PRSS}}{d\mathbf{h}} = -2\mathbf{X}^H\mathbf{y} + 2\mathbf{X}^H\mathbf{X}\hat{\mathbf{h}} + 2\lambda\hat{\mathbf{h}} = 0. \tag{5.37}$$

The equivalent to the normal equations in (5.16) take the form of

$$\mathbf{X}^H\mathbf{y} = (\mathbf{X}^H\mathbf{X} + \lambda\mathbf{I})\hat{\mathbf{h}}, \tag{5.38}$$

where it is clear that the effect of the Ridge regression is to add a diagonal matrix to the autocorrelation matrix before the inversion. This addition increases the singular values of the matrix $\mathbf{X}^H\mathbf{X} + \lambda\mathbf{I}$, therefore increasing the condition number.

Finally, when the inversion is performed, the estimation of the coefficients can be attained through

$$\hat{\mathbf{h}} = (\mathbf{X}^H\mathbf{X} + \lambda\mathbf{I})^{-1}\mathbf{X}^H\mathbf{y}. \tag{5.39}$$

Ridge regression is affected by the scale of the regressors, i.e., their power. Therefore, it is advisable to normalize the basis functions prior to the Ridge regression (James et al., 2013).

The effect of the Ridge regression is an attenuation of the absolute value of the coefficients, tending to zero when $\lambda \to \infty$. In the other extreme, when $\lambda = 0$, no regularization is applied and the regression becomes least squares. This effect is shown in Figure 5.7(a), where the absolute value of the coefficients of a polynomial memoryless model is represented against a sweep of λ values.

The benefits of regularization are shown in Figure 5.7(b). In this case, the identification and validation NMSEs are plotted against the regularization parameter. It can be seen how bias is traded by variance of the estimator, showing lower identification NMSE values at low values of the regularization parameter while at its other range extreme, the ℓ_2 norm of the estimators is low, having a high identification error. Regarding validation NMSE, the interesting feature of regularization is that there exist values of the regularization parameter in which this error is lower than in the least squares case. For example, in this figure, the least squares NMSE corresponds to values of $\lambda = 10^{-10}$, while for the value of $\lambda = 10^{-3}$ the validation NMSE of the Ridge regression outperforms that of the least squares. The selection of the regularization factor value depends on the specific nature of the problem and it is left to be taken by the model designer. There are some references in the literature that present a proposal for fixing this value, for example (Guan and Zhu, 2012) for applications in the field of power amplifiers behavioral modeling.

5.3.2 LASSO (ℓ_1-Norm Minimization)

The least absolute shrinkage and selection operator (LASSO) is a regularization technique that includes variable selection. Compared to Ridge regression, instead of minimizing the PRSS that includes an ℓ_2 penalty on the model coefficients, it minimizes a penalty based on the ℓ_1-norm of the estimator.

$$\hat{\mathbf{h}} = \min_{\mathbf{h}} \left\{ \|\mathbf{y} - \hat{\mathbf{y}}\|_2^2 + \lambda_{\text{LASSO}} \|\mathbf{h}\|_1 \right\}. \tag{5.40}$$

Although the difference between Ridge and LASSO might seem subtle, the inclusion of the ℓ_1 norm forces the regression to produce estimators whose sparsity can be controlled by tuning the regularization parameter λ_{LASSO}.

The main numerical issue of LASSO is that its loss function is not differentiable, therefore a closed formula to attain the estimator does not exist. Nevertheless, it establishes a general sparse regression framework. A wide set of techniques exists in the literature to perform an ℓ_1 regularization, having different approaches or assumptions on the data. Next chapter will cover some of these techniques applied to the sparse regression of Volterra models.

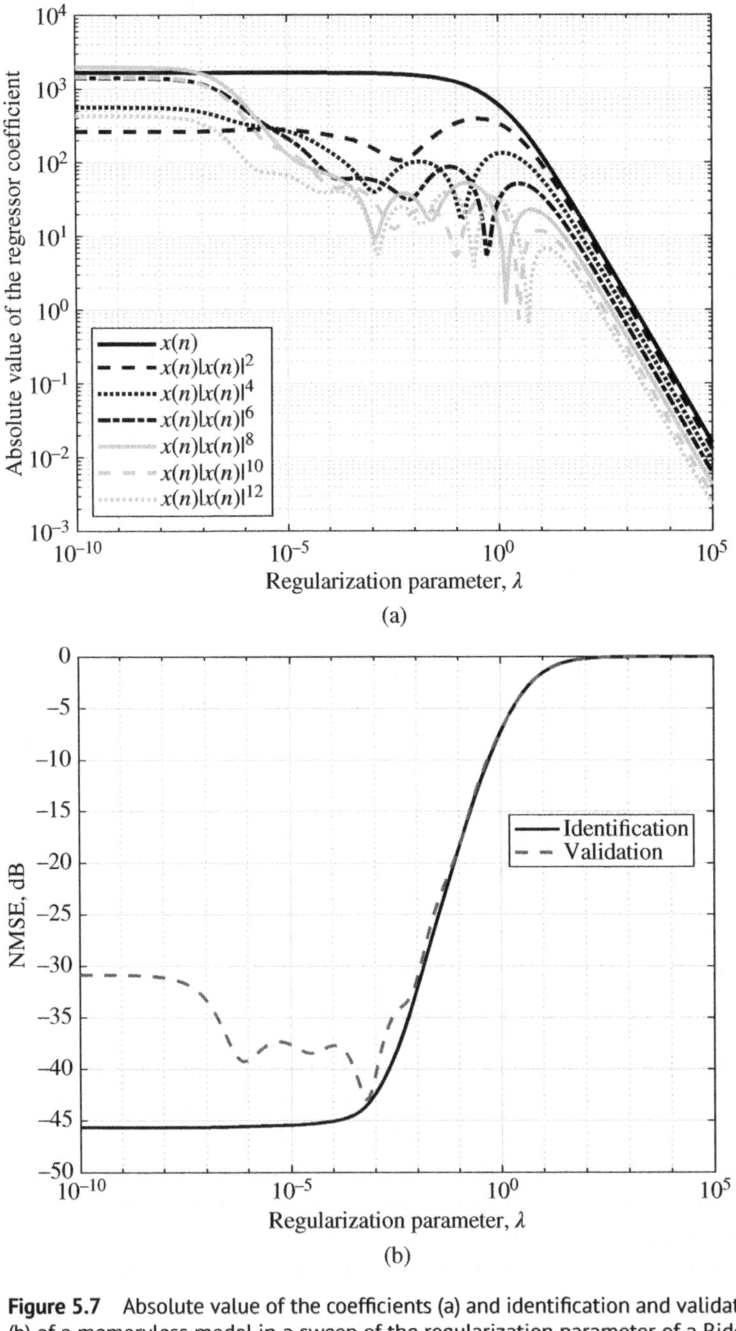

Figure 5.7 Absolute value of the coefficients (a) and identification and validation NMSEs (b) of a memoryless model in a sweep of the regularization parameter of a Ridge regression.

5.4 Adaptive Optimization and Iterative Regression

Adaptive optimization is a group of techniques in which the kernel vector estimation is updated dynamically in time. It has the advantage of being more robust to the change of system conditions. This family of algorithms is able to follow the effects that may appear with a change of any variable like signal type, power level, or temperature, with the drawback of constantly updating the system model. The scheme is very similar to least squares, but taking into account only a number of the most recent samples of each signal.

5.4.1 Steepest Descent

The steepest descent takes the same definitions of (5.3) with an explicit dependence of the time index k. For the derivation of the algorithm, we depart from the measurement process equation that, this time, takes the form

$$\hat{\mathbf{y}}(k) = \mathbf{X}(k)\mathbf{h}(k), \tag{5.41}$$

where $\hat{\mathbf{y}}(k) \in \mathbb{C}^{m\times 1}$ is the estimated output, $\mathbf{h}(k) \in \mathbb{C}^{n\times 1}$ is the Volterra vector with the n coefficients of the model, and $\mathbf{X}(k) \in \mathbb{C}^{m\times n}$ is the measurement matrix, whose n columns corresponds to the regressors of the model formed with a section of the last m samples of the input signal

$$\mathbf{x}(k) = [x(k), x(k+1), \dots, x(k+m-1)]^T \in \mathbb{C}^{m\times 1}. \tag{5.42}$$

The optimization error is defined as

$$\mathbf{e}(k) = \hat{\mathbf{y}}(k) - \mathbf{y}(k), \tag{5.43}$$

or equivalently,

$$\mathbf{e}(k) = \mathbf{X}(k)\mathbf{h}(k) - \mathbf{y}(k). \tag{5.44}$$

These elements are represented in a block diagram in Figure 5.8.

Considering the ℓ_2-norm as the cost function to minimize,

$$J(k) = E\left[\|\mathbf{e}(k)\|^2\right], \tag{5.45}$$

that is,

$$J(k) = E\left[\left(\mathbf{h}^H(k)\mathbf{X}^H(k) - \mathbf{y}^H(k)\right)(\mathbf{X}(k)\mathbf{h}(k) - \mathbf{y}(k))\right], \tag{5.46}$$

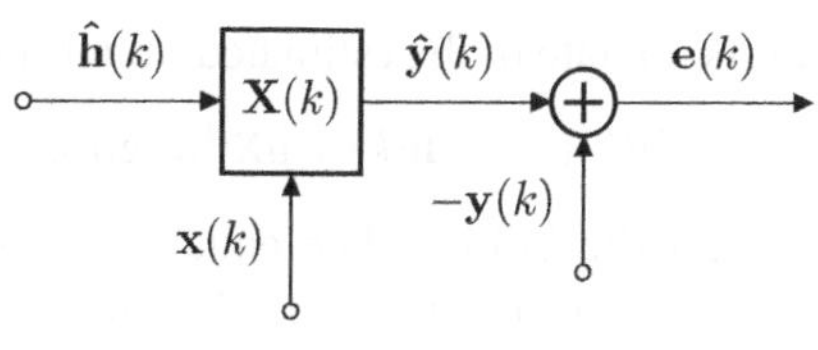

Figure 5.8 Block diagram of the steepest descent main variables.

and handling the terms considering the random nature of $\mathbf{X}(k)$ and $\mathbf{y}(k)$,

$$\begin{aligned} J(k) = {} & E\left[\mathbf{y}^H(k)\mathbf{y}(k)\right] \\ & - E\left[\mathbf{y}^H(k)\mathbf{X}(k)\right]\mathbf{h}(k) \\ & - \mathbf{h}^H(k)E\left[\mathbf{X}^H(k)\mathbf{y}(k)\right] \\ & + \mathbf{h}^H(k)E\left[\mathbf{X}^H(k)\mathbf{X}(k)\right]\mathbf{h}(k). \end{aligned} \tag{5.47}$$

The final cost function remains as

$$J(k) = \sigma_y^2 - \mathbf{p}^H\mathbf{h}(k) - \mathbf{h}^H(k)\mathbf{p} + \mathbf{h}^H(k)\mathbf{R}\mathbf{h}(k), \tag{5.48}$$

where $\mathbf{p}$ is the crosscorrelation between $\mathbf{X}(k)$ and $\mathbf{y}^H(k)$ and $\mathbf{R}$ is the correlation matrix of $\mathbf{X}(k)$. If we set the direction of the kernel vector update to its gradient

$$-\nabla J(k) = 2\mathbf{p} - 2\mathbf{R}\mathbf{h}(k), \tag{5.49}$$

the update equation of the estimated values is

$$\mathbf{h}(k+1) = \mathbf{h}(k) + \frac{1}{2}\mu\left[-\nabla J(k)\right], \tag{5.50}$$

or

$$\mathbf{h}(k+1) = \mathbf{h}(k) + \mu\left[\mathbf{p} - \mathbf{R}\mathbf{h}(k)\right], \tag{5.51}$$

where μ is the step size that controls the amount of correction in the gradient direction applied over the model coefficients at each iteration.

5.4.2 The Least Mean Squares (LMS) Algorithm

The *a priori* information required by the steepest descent algorithm is not always known. One possible approach for this issue is to use estimated values for the cross-correlation $\mathbf{p}$ and the correlation matrix $\mathbf{R}$. The resulting algorithm is known as least mean squares (LMS), developed in the 60s by Professor Bernard Widrow and his first PhD student Ted Hoff (Widrow, 2005). Using the following instantaneous estimates

$$\begin{aligned} \mathbf{R}_{\mathbf{X}}(k) &= \mathbf{X}^H(k)\mathbf{X}(k), \\ \mathbf{p}_{\mathbf{X}\mathbf{y}}(k) &= \mathbf{X}^H(k)\mathbf{y}(k). \end{aligned} \tag{5.52}$$

and using the direction of the update

$$-\nabla J(k) = -2\mathbf{X}^H(k)\mathbf{y}(k) + 2\mathbf{X}^H(k)\mathbf{X}(k)\mathbf{h}(k) = 2\mathbf{X}^H(k)\mathbf{e}(k) \tag{5.53}$$

in the update of the estimated Volterra vector

$$\mathbf{h}(k+1) = \mathbf{h}(k) + \mu\mathbf{X}^H(k)\mathbf{e}(k), \tag{5.54}$$

we get the LMS update equation. This update is extremely simple and it has the property of averaging the large variance that the instantaneous estimates may have. Due to the simplicity and good performance of this algorithm, it is considered a standard benchmark against which other algorithms are compared.

The LMS update equation performs a least squares regression in a recursive manner. Instead of calculating the estimator at once, it also allows the signals to evolve over time and to adjust the estimator consequently. This scheme is specially convenient in time-varying scenarios, where the classical least squares estimation would not use previous information to recalculate the Volterra vector.

Bibliography

J.A. Becerra, M.J. Madero-Ayora, E. Marqués-Valderrama, M. Nogales, and C. Crespo-Cadenas. Preconditioning the regression of power amplifier behavioral models and digital predistorters. In *2022 IEEE Topical Conference on RF/Microwave Power Amplifiers for Radio and Wireless Applications (PAWR)*, pages 58–61, 2022. doi: 10.1109/PAWR53092.2022.9719703.

N.R. Draper and H. Smith. *Applied Regression Analysis*. Wiley series in probability and mathematical statistics. Wiley, New York, 1966. ISBN 0471221708.

F.M. Ghannouchi and O. Hammi. Behavioral modeling and predistortion. *IEEE Microwave Magazine*, 10(7):52–64, 2009. doi: 10.1109/MMM.2009.934516.

P.L. Gilabert, R.N. Braithwaite, and G. Montoro. Beyond the Moore-Penrose inverse: Strategies for the estimation of digital predistortion linearization parameters. *IEEE Microwave Magazine*, 21(12):34–46, 2020. doi: 10.1109/MMM.2020.3023220.

L. Guan and A. Zhu. Optimized low-complexity implementation of least squares based model extraction for digital predistortion of RF power amplifiers. *IEEE Transactions on Microwave Theory and Techniques*, 60(3):594–603, 2012. doi: 10.1109/TMTT.2011.2182656.

D.C. Hoaglin and R.E. Welsch. The hat matrix in regression and ANOVA. *The American Statistician*, 32(1):17–22, 1978. doi: 10.2307/2683469.

G. James, D. Witten, T. Hastie, and R. Tibshirani. *An Introduction to Statistical Learning: with Applications in R*. Springer, New Yok, 2013. ISBN 9781461471387.

P. Landin, M. Isaksson, and P. Handel. Comparison of evaluation criteria for power amplifier behavioral modeling. In *2008 IEEE MTT-S International Microwave Symposium Digest*, pages 1441–1444, 2008. doi: 10.1109/MWSYM.2008.4633050.

Y. Liu, X. Xia, Q. Xu, W. Pan, W. Ma, S. Shao, and Y. Tang. Relaxing requirements on training samples in digital predistortion by using Ridge regression. *IEEE Microwave and Wireless Components Letters*, 31(6): 616–619, 2021. doi: 10.1109/LMWC.2021.3067346.

A.S. Tehrani, H. Cao, S. Afsardoost, T. Eriksson, M. Isaksson, and C. Fager. A comparative analysis of the complexity/accuracy tradeoff in power amplifier behavioral models. *IEEE Transactions on Microwave Theory and Techniques*, 58(6):1510–1520, 2010. doi: 10.1109/TMTT.2010.2047920.

B. Widrow. Thinking about thinking: the discovery of the LMS algorithm. *IEEE Signal Processing Magazine*, 22(1):100–106, 2005. doi: 10.1109/MSP.2005.1407720.

6

Sparse Machine Learning

6.1 Introduction

Traditional procedures use aprioristic models assuming that all their regressors are pertinent. Can any of these be discarded while keeping their modeling performance? Compressed sensing techniques are a set of interesting algorithms that achieve a reduction in the number of necessary elements that properly define a signal. While the most general definition of the term *compressed sensing* leads to selecting which samples are relevant to reproduce the signal after a sampling process, these techniques have been applied in different frameworks and disciplines to essentially extract the most meaningful parts of high dimensional problems. Since Volterra models are characterized by a number of coefficients that explodes with configuration parameters (that is nonlinear order and memory depth), compressed sensing techniques become a perfect match to reduce this problematic high number of regressors. This chapter covers a general review of coefficients selection by particularizing the compressed sensing framework to the regression of high dimensional Volterra-based models. The main objective of these approaches to coefficients selection is to attain a *sparse* model, that is, a model in which not all coefficients are active but still shows a good level of performance.

The content that is presented next is a change of paradigm with respect to the previous chapters of this book. Chapters three and four presented Volterra series and the motivation of having a simple model that is able to be regressed without overfitting or high computational complexity. Conversely, once we introduce the capabilities of selecting which basis functions of the model are most relevant, the initial model is preferred to be rich in its components—evidenced by a high number of coefficients before pruning—so we ensure that the desired pruned model structure is mixed with irrelevant terms that can be discarded, provided that the necessary terms were present in the original model.

A Volterra Approach to Digital Predistortion: Sparse Identification and Estimation, First Edition.
Carlos Crespo-Cadenas, María José Madero-Ayora, and Juan A. Becerra.

6.2 Thresholding

The first section of this chapter presents the mathematical tools that can be used to perform sparse estimation. The most intuitive way of thinking of a sparse model is through the application of a threshold. This is the nonlinear operation of setting to zero those elements of a vector that are below a certain quantity called threshold. Mathematically, it can be defined as

$$\eta(a,\theta) = \begin{cases} 0, & |a| < \theta, \\ a, & |a| \geq \theta, \end{cases} \tag{6.1}$$

where a is its argument, θ is the threshold value, and $|\cdot|$ stands for the absolute value of its argument.

When the threshold operation is applied to the Volterra kernel vector $\mathbf{h}$, the highest-value elements are retained. If we assume that the regressors are weighted by the values of the Volterra coefficients, it seems appropriate to prune those that have less impact on the output. Figure 6.1 shows the scenario where all the basis functions in $\mathbf{X}$ are represented by some proportion in $\mathbf{y}$. This reasoning is valid under certain circumstances that are not fulfilled in regressors generated by Volterra models. First, it is assumed that only the coefficients magnitudes impact the output, but generally the basis functions are very different in magnitude since they operate polynomially over the input signal. Also, claiming that the coefficient itself is able to represent the proportion of the basis function on the output does not take into account the correlations between the regressors, which are also high in Volterra models.

A similar approach to the actual operation of model pruning is through the support set S. The support set is a list of indices that indicate the active regressors within a model. Defining $\mathbf{X}_S$ as the operation of column selection of matrix $\mathbf{X}$, the least squares regression of the support set follows

$$\hat{\mathbf{h}}_S = \left(\mathbf{X}_S^H \mathbf{X}_S\right)^{-1} \mathbf{X}_S^H \mathbf{y}, \tag{6.2}$$

with the rest of elements of the coefficient vector equal to zero, i.e., $\hat{\mathbf{h}}_{\bar{S}} = \mathbf{0}$.

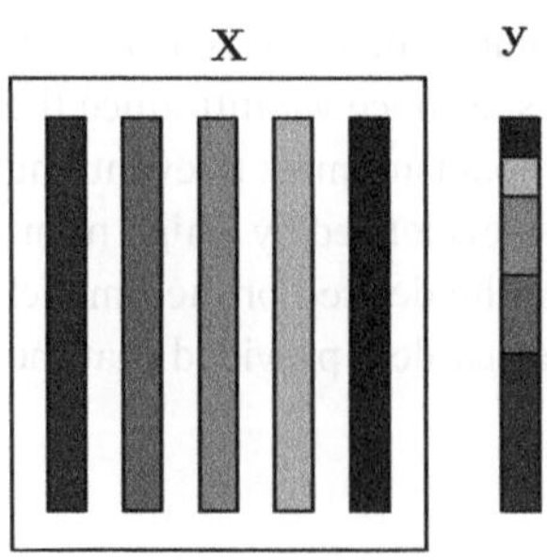

Figure 6.1 General scheme of coefficients selection techniques. Different shades represent the correlation between each regressor and the output without considering the correlation between regressors, with the bar length indicating the correlation magnitude.

Other important operations that sparsify a vector are:

- The nonlinear projection $H_k(\cdot)$. This operation sets all but the largest k elements of its argument to zero.
- The support operation, denoted as supp(·), that returns the indices of the elements of the argument that are not equal to zero. In conjunction with the nonlinear projection, an algorithm is able to gather the indices of the regressors that should be retained or discarded in the pruning process.

The operations that have been presented in this section are combined following different strategies to form a specific technique.

6.3 Local Search: Hill Climbing

Hill climbing techniques are optimization algorithms that belong to the family of local search. These iterative algorithms work with the concept of neighbor models, which are defined as those models that are in one of the next positions in the model configuration parameter space. For example, if we consider that a memory polynomial model represented by MP(N, Q) is defined by its hyperparameters memory depth Q and nonlinear order N, a possible set of its neighbors could be those that add or subtract a quantity to at least one of its hyperparameters i.e., MP($N+1, Q$), MP($N, Q+1$), MP($N+1, Q+1$), MP($N-1, Q$), MP($N, Q-1$), MP($N-1, Q-1$), MP($N+1, Q-1$), MP($N-1, Q+1$). The concept of neighbouring is specifically designed in each algorithm implementation, always looking at reducing the need for performing too many operations.

When the hill climbing search is implemented to select a model, in each iteration it checks all the neighbors against some cost function like the normalized mean squared error (NMSE) or the Bayesian information criterion (BIC), later introduced in Section 6.5.2. If any of the neighbors decreases or increases—depending on how the cost function is defined—the value of the cost function, the algorithm moves to that model and iterates again. Examples of the use of these techniques in the field of power amplifier modeling and predistortion can be found in Wang et al. (2018) and Barry et al. (2021).

6.4 Greedy Pursuits

A greedy algorithm is a technique that is able to solve a problem by applying some local decision (Eldar and Kutyniok, 2012; Becerra et al., 2019). In the field of model pruning, greedy techniques will select the most relevant coefficients by applying some consideration on the statistics of the model elements.

Algorithm 6.1: Greedy Pursuits Framework

Input: $\mathbf{X} \in \mathbb{C}^{m \times n}, \mathbf{y} \in \mathbb{C}^{m}$

Output: $S^{(\text{end})}, \hat{\mathbf{h}}^{(\text{end})}$

1: **for** $t = 1$ till stopping criterion is met **do**

2: Select elements and include them in $S^{(t)}$ based on some algorithm-dependent criteria that prioritizes the most relevant regressors.

3: Update the model vector: $\hat{\mathbf{h}}_{S^{(t)}} = \mathbf{X}^{\dagger}_{S^{(t)}}\mathbf{y}$.

4: Update the estimate: $\hat{\mathbf{y}}^{(t)} = \mathbf{X}_{S^{(t)}}\hat{\mathbf{h}}$.

5: Update the residual error: $\mathbf{r}^{(t)} = \mathbf{y} - \hat{\mathbf{y}}^{(t)}$.

6: **end for**

The general framework of greedy pursuits is presented in Algorithm 6.1. The inputs are generally the measurement matrix $\mathbf{X}$ and the output signal $\mathbf{y}$. Other input parameters may appear depending on the specific algorithm. Then, greedy techniques follow an iterative fashion to select elements and add them to the support set $S^{(t)}$ at iteration t. The residual estimate at iteration t is calculated with

$$\mathbf{r}^{(t)} = \mathbf{y} - \hat{\mathbf{y}}^{(t)}, \tag{6.3}$$

where $\hat{\mathbf{y}}^{(t)}$ is the output estimation at iteration t. The process is repeated until convergence is decided through some criterion or all the components have been analyzed. Strategies for stopping criteria are analyzed in Section 6.5.2. The output of the algorithm is the support set at the end of the execution, $S^{(\text{end})}$ and the estimation of the Volterra kernel vector $\hat{\mathbf{h}}^{(\text{end})}$.

6.4.1 Orthogonal Matching Pursuit (OMP)

The orthogonal matching pursuit (OMP) (Reina-Tosina et al., 2015) is a greedy algorithm that performs coefficients selection with the characteristic that the residual error is orthogonal to the selected coefficients. Its pseudocode is presented in Algorithm 6.2. The selection is performed by calculating the correlation $\mathbf{g}$ between the normalized basis functions and the residual error in step 3 through

$$\mathbf{g}^{(t)}_{\{i\}} \xleftarrow{\forall i} \frac{\mathbf{X}^{H}_{\{i\}}}{\|\mathbf{X}_{\{i\}}\|_2}\mathbf{r}^{(t-1)}, \tag{6.4}$$

where i is the generic index to sweep every regressor, the normalization of the basis functions is performed by dividing by their ℓ_2 norm, and $\mathbf{X}_{\{i\}}$ stands for the selection of the ith column of matrix $\mathbf{X}$, as defined in Section 6.2. This correlation measures the proportion of the residual error that can be modeled with each regressor, therefore the logic behind the decision is to minimize the modeling error after the coefficients selection.

Algorithm 6.2: Orthogonal Matching Pursuit (OMP) Pseudocode

Input: $\mathbf{X}, \mathbf{y}$

Output: $S^{(\text{end})}, \hat{\mathbf{h}}^{(\text{end})}$

1: *Initialization* : $\mathbf{r}^{(0)} \leftarrow \mathbf{y}, S^{(0)} \leftarrow \emptyset$

2: **for** $t = 1$ till stopping criterion is met **do**

3: $\quad \mathbf{g}^{(t)}_{\{i\}} \overset{\forall i}{\longleftarrow} \dfrac{\mathbf{X}^{H}_{\{i\}}}{\|\mathbf{X}_{\{i\}}\|_2}\mathbf{r}^{(t-1)}$

4: $\quad s^{(t)} \leftarrow \text{supp}\left(H_1\left(\left|\mathbf{g}^{(t)}\right|\right)\right)$

5: $\quad S^{(t)} \leftarrow S^{(t-1)} \cup s^{(t)}$

6: $\quad \hat{\mathbf{h}}_{S^{(t)}} \leftarrow \mathbf{X}^{\dagger}_{S^{(t)}}\mathbf{y}$

7: $\quad \hat{\mathbf{y}}^{(t)} \leftarrow \mathbf{X}_{S^{(t)}}\hat{\mathbf{h}}_{S^{(t)}}$

8: $\quad \mathbf{r}^{(t)} \leftarrow \mathbf{y} - \hat{\mathbf{y}}^{(t)}$

9: **end for**

The coefficient that has more resemblance with the residual error is selected through the threshold operations in step 4:

$$s^{(t)} \leftarrow \text{supp}\left(H_1\left(\left|\mathbf{g}^{(t)}\right|\right)\right) \tag{6.5}$$

and included in the support set in step 5 with

$$S^{(t)} \leftarrow S^{(t-1)} \cup s^{(t)}. \tag{6.6}$$

Recall that $\text{supp}(\cdot)$ and $H_1(\cdot)$ stand for support operation and nonlinear projection, respectively, as defined in Section 6.2.

At each iteration t of the algorithm, a model with t active components is regressed in step 6. Then, the estimated modeled output and the corresponding residual error are calculated in steps 7 and 8 to be used in the next algorithm iteration.

The OMP is characterized by its simplicity and a fair pruning performance in Volterra-based models. The operations that are performed in loop are quite simple, being the pseudoinverse calculation of step 6, the one that exhibits the highest computational complexity. Although the OMP is not optimal from a statistical point of view since it does not take into account the high correlation between the basis functions of the Volterra models, its output offers a reasonable performance with the benefit of a low computational complexity and simplicity of operations.

6.4.2 Principal Component Analysis (PCA)

The principal component analysis (PCA) is a whitening technique that allows to describe a matrix as a set of orthogonal vectors. It is specially useful in multidimensional environments in which it is necessary to reduce the dimensionality of the problem and, therefore, explain the output variable with a subset of predictors of the original pool set. The idea is conceptually simple: those regressors that exhibit a higher variance are supposed to contain more information, and consequently will be prioritized in the selection process. The new orthogonal set of coordinates will be constructed iteratively, selecting the components in decreasing order of variance and orthogonalizing them with respect to the already-selected regressors. Since in a Volterra-model scenario the correlation between the basis functions is very high, it is expected that a subset of the full-model regressors is able to model the output of the system with an equivalent performance—or even better when it comes to analyzing the numerical properties of the regression.

The principal component method is commonly explained with the singular value decomposition (SVD) of the matrix $\mathbf{X}$ as

$$\mathbf{X} = \mathbf{U}\boldsymbol{\Sigma}\mathbf{W}^T, \tag{6.7}$$

where $\boldsymbol{\Sigma}$ is a diagonal matrix that holds the singular values of $\mathbf{X}$, in its diagonal, $\mathbf{U}$ holds the left singular vectors of $\mathbf{X}$, and $\mathbf{W}$ contains the right singular vectors of $\mathbf{X}$. Although (6.7) is the classical introduction of PCA that has been extensively studied and can be widely found in the literature (James et al., 2013), in this text we introduce the calculation of PCA as an iterative algorithm following the selection fashion of greedy algorithms. The classical PCA technique performs regression in an orthogonal domain, resulting in a lack of interpretability of the coefficient values. Here, we adapt its formulation to perform the regression of the selected coefficients in the Volterra domain. Also, the formulation presented here is compatible with that of the OMP and the doubly orthogonal matching pursuit (DOMP), therefore it is advisable to review those explanations as their steps are very similar.

Algorithm 6.3 shows a pseudocode of the PCA technique applied to regress a Volterra model. Initialization is performed in step 1 and the variance of the regressors is gathered from the diagonal of the autocovariance matrix $\mathbf{R_Z}$ (5.24) in step 3, following

$$\mathbf{g}_{\{i\}}^{(t)} \xleftarrow{\forall i} \text{diag}(\mathbf{R_Z}). \tag{6.8}$$

Note that this operation can be performed with either $\mathbf{R_Z}$ or $\mathbf{R_X}$, as long as it is ensured that the same regressor is not selected twice. Next, the component with the maximum variance is merged into the support set $S^{(t)}$ and the rest of regressors are orthogonalized with respect to the selected one through a Gram–Schmidt procedure in step 7. This step is discussed in more detail in the next subsection. Matrix $\mathbf{Z}$ holds the orthogonal regressors attained by the PCA technique. The regressors in the support set are used to regress the model through the Moore–Penrose pseudoinverse in step 8. Examples of the use of PCA and its derivations in the literature for power amplifiers behavioral modeling are (Gilabert et al., 2013; López-Bueno et al., 2018).

Algorithm 6.3: Principal Component Analysis (PCA) Pseudocode

Input: $\mathbf{X}, \mathbf{y}$

Output: $S^{(\text{end})}, \hat{\mathbf{h}}^{(\text{end})}$

1: *Initialization* : $S^{(0)} \leftarrow \emptyset, \mathbf{Z}^{(0)} \leftarrow \mathbf{X}$
2: **for** $t = 1$ till stopping criterion is met **do**
3: $\quad \mathbf{g}^{(t)}_{\{i\}} \overset{\forall i}{\leftarrow} \text{diag}(\mathbf{R}_{\mathbf{Z}})$
4: $\quad s^{(t)} \leftarrow \text{supp}\left(H_1\left(|\mathbf{g}^{(t)}|\right)\right)$
5: $\quad S^{(t)} \leftarrow S^{(t-1)} \cup s^{(t)}$
6: $\quad \mathbf{p}^{(t)} \leftarrow \mathbf{Z}^{(t-1)^H}_{\{i^{(t)}\}} \mathbf{Z}^{(t-1)}$
7: $\quad \mathbf{Z}^{(t)} \leftarrow \mathbf{Z}^{(t-1)} - \mathbf{p}^{(t)} \otimes \mathbf{Z}^{(t-1)}_{\{i^{(t)}\}}$
8: $\quad \hat{\mathbf{h}}_{S^{(t)}} \leftarrow \mathbf{X}^{\dagger}_{S^{(t)}} \mathbf{y}$
9: $\quad \hat{\mathbf{y}}^{(t)} \leftarrow \mathbf{X}_{S^{(t)}} \hat{\mathbf{h}}$
10: **end for**

6.4.3 Doubly Orthogonal Matching Pursuit (DOMP)

Volterra basis functions are highly correlated. This effect is produced by the way they are constructed: each regressor is the result of multiplication of delayed versions of the signal and its envelope. The high correlation produces a mixing effect that affects the selection process. Since each regressor has presence in others, when one is selected, that part of others is indirectly selected too. After that, when the correlation between the regressors and the residual error is calculated, the remaining parts of the regressors are not correctly scaled. This can be observed in Figure 6.2. In Figure 6.2(a), the first regressor has been selected, and its corresponding part of other regressors have been filled in white. If the correlation between these basis functions and the residual was performed—as it is the case in the OMP—the remainings of each regressor would not have the correct scale, therefore producing an incorrect selection order. In this case, the OMP would select the first, third, and second regressors shaded in the plot. The DOMP adds an orthogonalization process after the selection, as it is shown in Figure 6.2(b). The orthogonalization and normalization processes allow to correctly scale the basis functions after selection, providing the correct selection order of first, second, and third regressors in this example.

The application of orthogonalization to selection processes widely appears in the literature. The modified Gram–Schmidt (MGS) technique (Chen et al., 1989), also known as orthogonal least squares (OLS) (Maymon and Eldar, 2015; Hashemi and Vikalo, 2016, 2018), and OMP were first fused into DOMP (Becerra et al., 2018) for the application of behavioral model pruning.

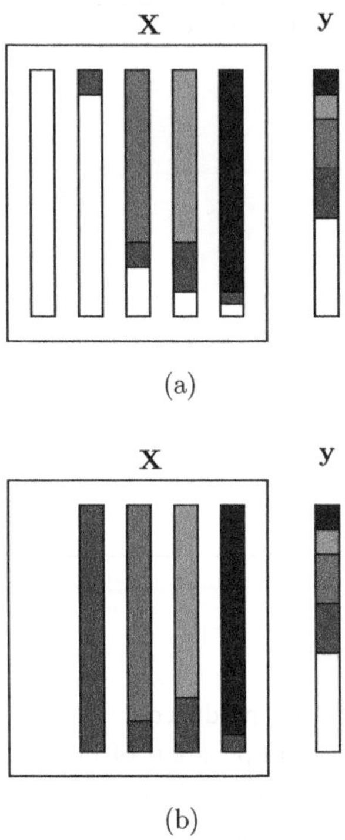

Figure 6.2 Coefficients selection in a scenario with high correlation between the measurement matrix columns. The different shades represent correlated parts between the regressors and the output. Once the first regressor has been selected, orthogonalization (a) and normalization (b) are applied.

Several variations of the original DOMP technique can be also found in the literature. The DOMP was evolved to its simplified sparse parameter identification (SSPI) version that showed less computational complexity by transforming the pseudoinverse calculation into a recursive operation. A reduced complexity DOMP (RC-DOMP) was presented in Becerra et al. (2020b), in which the operations are performed in the correlation space and through the covariance and crosscorrelation matrices, avoiding to perform operations in the time domain and therefore further reducing its complexity.

The pseudocode of the DOMP is provided in Algorithm 6.4. The operation is very similar to that of the OMP with the difference that the DOMP keeps track of the information in two subspaces simultaneously: the Volterra space defined by **X** and an orthonormal space **Z**. The orthonormal space is the result of applying orthogonalization to the measurement matrix **X** in the order of the support set.

In the algorithm initialization, the residual error is set to the signal to model, the support set is empty and $\mathbf{Z} = \mathbf{X}$, since no regressors have been selected yet. The selection is performed by maximizing the correlation between the columns of the orthonormal space and the residual error, and the index of the basis function that fulfils this condition is added to the support set. Then, the not-yet-selected

Algorithm 6.4: Doubly Orthogonal Matching Pursuit (DOMP) Pseudocode

Input: $\mathbf{X}, \mathbf{y}$

Output: $S^{(\text{end})}, \hat{\mathbf{h}}^{(\text{end})}$

1: *Initialization* : $\mathbf{r}^{(0)} \leftarrow \mathbf{y}, S^{(0)} \leftarrow \emptyset, \mathbf{Z}^{(0)} \leftarrow \mathbf{X}$
2: **for** $t = 1$ till stopping criterion is met **do**
3: $\quad \mathbf{g}_{\{i\}}^{(t)} \overset{\forall i}{\leftarrow} \dfrac{\mathbf{Z}_{\{i\}}^H}{\|\mathbf{Z}_{\{i\}}\|_2} \mathbf{r}^{(t-1)}$
4: $\quad s^{(t)} \leftarrow \text{supp}\left(H_1\left(|\mathbf{g}^{(t)}|\right)\right)$
5: $\quad S^{(t)} \leftarrow S^{(t-1)} \cup s^{(t)}$
6: $\quad \mathbf{p}^{(t)} \leftarrow \mathbf{Z}_{\{i^{(t)}\}}^{(t-1)H} \mathbf{Z}^{(t-1)}$
7: $\quad \mathbf{Z}^{(t)} \leftarrow \mathbf{Z}^{(t-1)} - \mathbf{p}^{(t)} \otimes \mathbf{Z}_{\{i^{(t)}\}}^{(t-1)}$
8: $\quad \hat{\mathbf{h}}_{S^{(t)}} \leftarrow \mathbf{X}_{S^{(t)}}^{\dagger} \mathbf{y}$
9: $\quad \hat{\mathbf{y}}^{(t)} \leftarrow \mathbf{X}_{S^{(t)}} \hat{\mathbf{h}}$
10: $\quad \mathbf{r}^{(t)} \leftarrow \mathbf{y} - \hat{\mathbf{y}}^{(t)}$
11: **end for**

columns of $\mathbf{Z}$ are orthogonalized with respect to the selected basis function in operations 6 and 7 of the algorithm. The first step calculates the projection $\mathbf{p}$ of the selected basis function $\mathbf{Z}_{\{i^{(t)}\}}$ in the rest of the subspace,

$$\mathbf{p}^{(t)} \leftarrow \mathbf{Z}_{\{i^{(t)}\}}^{(t-1)H} \mathbf{Z}^{(t-1)}. \tag{6.9}$$

Next, the regressor is scaled by the projection and subtracted from the rest of the basis

$$\mathbf{Z}^{(t)} \leftarrow \mathbf{Z}^{(t-1)} - \mathbf{p}^{(t)} \otimes \mathbf{Z}_{\{i^{(t)}\}}^{(t-1)}, \tag{6.10}$$

where $\otimes$ stands for the Kronecker product. The estimation of the Volterra kernel vector $\hat{\mathbf{h}}$, output signal $\hat{\mathbf{y}}$, and the residual update are performed in the same fashion as in the OMP.

6.5 Stopping Criteria

6.5.1 Custom Target

The techniques presented previously generally provide an ordered set of regressors in order of increasing importance. Some of them are defined without stopping criteria while others like the sparse Bayesian pursuit (SBP) internally include a strategy to fix the number of coefficients. What is the optimum number of coefficients? The choice of the number of coefficients can be seen as a stopping criterion, since, in this context, the regressors are selected atom by atom, and by fixing the number of components following some criterion, the algorithm can be halted without further using computation time to generate data that will not be used. On the other hand, the set of stopping criteria presented in this section can be interpreted as a means of comparing different models' performance that allow prioritizing one over the others.

The most intuitive stopping criterion is to set a target NMSE. Once this NMSE is achieved by the technique, the number of coefficients at that iteration is considered to be enough for the experiment and it is fixed. This approach has several drawbacks: first, since it does not consider the number of coefficients to stop the algorithm, it can lead to models with too many coefficients. When the target NMSE threshold is too low, the technique may lead to having too many basis functions or all of them, if the model is not rich enough.

6.5.2 Bayesian Information Criterion

The BIC is a tool for model selection. It performs as a cost function that allows model evaluation with the preference of a lower BIC. The BIC follows

$$\mathrm{BIC} = 2m \ln \hat{\sigma}_{\mathbf{e}}^2 + 2n \ln 2m, \tag{6.11}$$

where m is the number of training samples, $\hat{\sigma}_{\mathbf{e}}^2$ is the error variance, and n is the number of components in the model. The BIC equation (6.11) is composed of two parts: an error and a penalty. On the one hand, the error is generally decreasing with the number of coefficients since the error variance $\hat{\sigma}_{\mathbf{e}}^2$ decreases as more regressors are added to the model. On the other hand, the penalty increases with the number of regressors n. When the BIC is applied to the result of a greedy algorithm (Reina-Tosina et al., 2015; Becerra et al., 2020a) (that provides a sorted list of regressors or a set of models with incremental number of components), its minimum can be used to determine the optimum model. Figure 6.3 shows a classical evolution of the error and the BIC. Since the models are incremental in the horizontal axis, every new component added to it enriches the support set and the error decreases. The penalty allows to find an optimum number of coefficients that in this case equals 70.

The BIC allows several transformations such as an explicit dependence of the NMSE (Crespo-Cadenas et al., 2017), being the optimum number of coefficients given by BIC

$$n_{\mathrm{BIC}} = \arg\min_{n} \left[\mathrm{NMSE} + \frac{n}{m} 10 \log_{10} 2m \right]. \tag{6.12}$$

Depending on the scenario, the BIC may choose too many components. In some situations, it is preferable to use a modified BIC with the following cost function

$$\mathrm{BIC}(\alpha) = m \ln \hat{\sigma}_{\mathbf{e}}^2 + \alpha n \ln 2m, \tag{6.13}$$

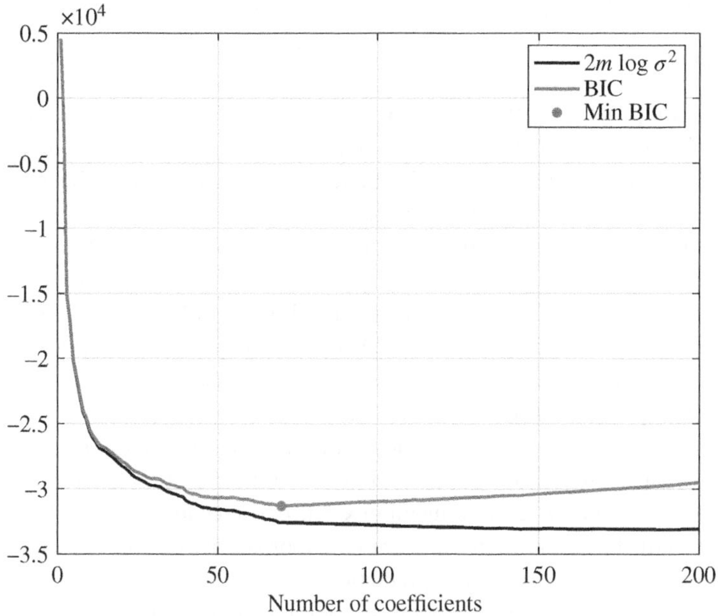

Figure 6.3 Example of the BIC execution over the output of a greedy algorithm.

α being a trade-off factor that allows for increasing the penalty proportion. Two examples of this way of using the BIC are Wang et al. (2018) and the practical sparse regression that will be discussed in Section 6.7.

6.6 Sparse Bayesian Learning

The techniques that have been presented so far in this chapter belong to a class of algorithms that optimize the solution with respect to some local criteria. In this section, we introduce a comprehensive Bayesian treatment, which includes not only a novel sparse regression based on statistical analysis of the data to select active regressors but also provides other important features. For instance, it establishes a particular stopping criterion, or makes possible to check if a selected regressor is actually active and remove it if not relevant. Because of this, the general Bayesian analysis deserves the detailed examination this section will devote to it.

Referring to the standard setup of Figure 5.1, we aspire to model the amplifier with the objective of making accurate predictions of the acquired output samples $\mathbf{y}$ for unobserved values of the input signal $\mathbf{x}$. In equation (5.3), it is assumed that the acquired output are samples of the model with additive noise:

$$\mathbf{y} = \mathbf{X}\mathbf{h} + \epsilon, \tag{6.14}$$

where ϵ is a noise process assumed to be mean-zero Gaussian with variance σ^2. Thus, the model output is a complex proper Gaussian random vector with mean $\mathbf{y}_d = \mathbf{X}\mathbf{h}$ and variance σ^2. The presence of this additive noise implies that the principal modeling challenge is to avoid overfitting.

Given a set of input samples and the corresponding acquired output, in this section we present the modification of the Bayesian treatment of a generalised real-valued linear model, the relevance vector machine (RVM) (Tipping, 1999; Faul and Tipping, 2001; Tipping and Faul, 2003), to a complex-valued approach to infer model coefficients and make predictions in the case of amplifiers and predistorters (Crespo-Cadenas et al., 2021, 2022).

The objective is a model of the dependency of the amplifier output on the input resulting from a supervised learning employing a given training set composed of an acquired output **y** and the corresponding input **x**. Since in a second acquisition the measurement would be different, in a probabilistic approach we desire to estimate the parameters in order to improve our prediction.

The present Bayesian approach assumes a joint complex-valued Gaussian distribution over the coefficients **h** and the vector of the acquired data **y**. Considering the statistical independence of the acquired samples, the probability of the evidence given the parameters **h** and σ^2, that is, the likelihood of the complete training data set, can be written as

$$p(\mathbf{y}|\mathbf{h}, \sigma^2) = \frac{1}{(\pi\sigma^2)^m} e^{-\frac{1}{\sigma^2}\|\mathbf{y}-\mathbf{X}\mathbf{h}\|^2}. \tag{6.15}$$

To avoid overfitting, here we adopt a zero-mean Gaussian prior probability distribution over the coefficients **h**, given by

$$p(\mathbf{h}|\boldsymbol{\alpha}) = \prod_{i=1}^{n} \frac{1}{\pi\alpha_i^{-1}} e^{-\alpha_i|h_i|^2}, \tag{6.16}$$

where the regression coefficients $h_1, \dots, h_n$, are assumed mutually independent with respective variances α_i^{-1}, $i = 1, \dots, n$. If we define the diagonal precision matrix of the n *hyperparameters* vector $\boldsymbol{\alpha}$ as $\mathbf{A} = \text{diag}(\alpha_1, \dots, \alpha_n)$, we can write the prior distribution over **h** as

$$p(\mathbf{h}|\boldsymbol{\alpha}) = \frac{1}{\pi^n|\mathbf{A}|^{-1}} e^{-\mathbf{h}^H\mathbf{A}\mathbf{h}}. \tag{6.17}$$

Furthermore, we define uniform hyperprior distributions over $\boldsymbol{\alpha}$ that allow some of the $\boldsymbol{\alpha}$ variables to result in high values, and the associated coefficients to be concentrated in a zero value, giving the result of a pruned coefficients set.

Following the Bayes rule, the deduction of the posterior distribution over the coefficients yields the expression

$$p(\mathbf{h}|\mathbf{y},\boldsymbol{\alpha},\sigma^2) = \frac{1}{\pi^{n+1}\det\boldsymbol{\Sigma}} e^{-(\mathbf{h}-\mu)^H\boldsymbol{\Sigma}^{-1}(\mathbf{h}-\mu)}, \tag{6.18}$$

where the *posterior precision* and *mean of the coefficients* are respectively given by

$$\boldsymbol{\Sigma}^{-1} = \beta\mathbf{X}^H\mathbf{X} + \mathbf{A} \tag{6.19}$$

and

$$\hat{\mathbf{h}} \equiv \mu = \beta\boldsymbol{\Sigma}\mathbf{X}^H\mathbf{y}, \tag{6.20}$$

with $\beta = \sigma^{-2}$. Notice that equation (6.19) includes conventional *aprioristic* estimators. The simplest case is for an initial vague information concerning the variance of parameters $\mathbf{h}$, for which $\mathbf{A} \to \mathbf{0}$ and the posterior precision matrix tends to one of the least squares estimator, i.e., $\boldsymbol{\Sigma}^{-1} \to \beta\mathbf{X}^H\mathbf{X}$. Another common assumption could be $\mathbf{A} = \lambda\mathbf{I}$, which provides the classical Ridge parameter estimator.

A significant powerfulness of the present Bayesian proposal is given by its capability to incorporate the extra information provided by the measurements in order to increase the precision of the coefficients and optimally update the posterior precision matrix $\boldsymbol{\Sigma}^{-1}$ in an additive manner.

The sparse Bayesian learning (SBL) in the case of complex-valued systems is formulated as the maximization with respect to $\boldsymbol{\alpha}$ of the marginal likelihood, given by

$$p(\mathbf{y}|\boldsymbol{\alpha},\sigma^2) = \frac{1}{\pi^m|\sigma^2\mathbf{I} + \mathbf{X}\mathbf{A}^{-1}\mathbf{X}^H|} e^{-\mathbf{y}^H(\sigma^2\mathbf{I}+\mathbf{X}\mathbf{A}^{-1}\mathbf{X}^H)^{-1}\mathbf{y}}. \tag{6.21}$$

It is equivalent and more convenient to maximize the logarithm of the marginal likelihood, $\ln p(\mathbf{y}|\boldsymbol{\alpha},\beta)$, which gives the objective function

$$\mathcal{L}(\boldsymbol{\alpha}) = -\ln|\sigma^2\mathbf{I} + \mathbf{X}\mathbf{A}^{-1}\mathbf{X}^H| - \mathbf{y}^H(\sigma^2\mathbf{I} + \mathbf{X}\mathbf{A}^{-1}\mathbf{X}^H)^{-1}\mathbf{y}. \tag{6.22}$$

The $m \times m$ matrix

$$\mathbf{C} = \sigma^2\mathbf{I} + \mathbf{X}\mathbf{A}^{-1}\mathbf{X}^H \tag{6.23}$$

refers to the covariance matrix of the measurement vector that may be written as a function of the posterior coefficients covariance $\boldsymbol{\Sigma} = (\sigma^{-2}\mathbf{X}^H\mathbf{X} + \mathbf{A})^{-1}$, a $n \times n$ matrix.

The particular strategy of the proposed sparse Bayesian learning is the maximization of $\mathcal{L}(\boldsymbol{\alpha})$ considering the contribution of a single hyperparameter α_i. This splitting is possible if the matrix $\mathbf{C}$ is decomposed as:

$$\mathbf{C} = \sigma^2\mathbf{I} + \sum_{m\neq i}\alpha_m^{-1}\phi_m\phi_m^H + \alpha_i^{-1}\phi_i\phi_i^H = \mathbf{C}_{-i} + \alpha_i^{-1}\phi_i\phi_i^H, \tag{6.24}$$

where $\mathbf{C}_{-i}$ is $\mathbf{C}$ with the contribution of the regressor ϕ_i removed. Using the inversion identity

$$\mathbf{C}^{-1} = \mathbf{C}_{-i}^{-1} - \frac{\mathbf{C}_{-i}^{-1}\phi_i\phi_i^H\mathbf{C}_{-i}^{-1}}{\alpha_i + \phi_i^H\mathbf{C}_{-i}^{-1}\phi_i} \tag{6.25}$$

and substituting in (6.22), the objective function $\mathcal{L}(\boldsymbol{\alpha})$ additively decouples as

$$\mathcal{L}(\boldsymbol{\alpha}) = \mathcal{L}(\boldsymbol{\alpha}_{-i}) + \ell(\alpha_i), \tag{6.26}$$

where $\mathcal{L}(\boldsymbol{\alpha}_{-i})$ denotes the value of the objective function when it is evaluated without the contribution of regressor ϕ_i, and $\ell(\alpha_i)$ refers to the increment obtained in the objective function due to the incorporation of this regressor. The corresponding increment is given by

$$\ell(\alpha_i) = \frac{|q_i|^2}{s_i + \alpha_i} - \ln\left(1 + \frac{s_i}{\alpha_i}\right), \tag{6.27}$$

where we have used the definitions of *sparsity factor*

$$s_i = \phi_i^H\mathbf{C}_{-i}^{-1}\phi_i \tag{6.28}$$

and *quality factor*

$$q_i = \phi_i^H\mathbf{C}_{-i}^{-1}\mathbf{y}. \tag{6.29}$$

The sparsity factor s_i is a measure of the extent that regressor ϕ_i overlaps those already present in the model. The quality factor q_i is complex-valued in this approach and its magnitude gives a measure of how well ϕ_i increases the marginal likelihood by helping to explain the data. The objective function $\mathcal{L}(\boldsymbol{\alpha})$ has a unique maximum with respect to α_i at

$$\alpha_i = \frac{s_i^2}{|q_i|^2 - s_i} \qquad \text{if } |q_i|^2 > s_i, \tag{6.30}$$

or at $\alpha_i = \infty$ otherwise. During the regressor pursuit, many α_i tend to infinity, meaning that these coefficients are peaked at zero, i.e., $h_i = 0$ and the corresponding regressors are not included in the active set.

6.6.1 Sparse Bayesian Pursuit (SBP)

The search procedure here is a complex-valued reformulation of the learning method in Tipping and Faul (2003) that also features the capability of coefficients selection. The full stock of regressors of the initially selected model is the potential set and σ^2 is assigned some sensible value, for example $\sigma^2 = \frac{1}{m}||\mathbf{y}||^2 \times 10^{-6}$ to start the pursuit with an empty active set of regressors, thus $\mathbf{C}_{-i} = \sigma^2$ in (6.24).

The values of s_i and q_i are computed with equations (6.28) and (6.29), and the potential regressor ϕ_i that maximizes $\mathcal{L}(\boldsymbol{\alpha})$, or equivalently maximizes

$$\ell(\alpha_i) = \frac{|q_i|^2 - s_i}{s_i} + \ln \frac{s_i}{|q_i|^2}, \tag{6.31}$$

is incorporated to the set of active regressors.

As in the greedy pursuits, this step is equivalent to selecting the regressor with the greatest projection $|\phi_i^H \mathbf{y}|$, but unlike OMP and DOMP, this procedure includes an additional condition so that the potential regressors not satisfying $|\phi_i^H \mathbf{y}|^2 > \sigma^2$ in (6.30) are deleted with $\alpha_i = \infty$. This can be interpreted as the fact that the regressors with projection below the noise level are not only not included in the active set, but are no longer considered regressors of the potential set.

In each iteration, the active set is increased by one regressor to which is assigned an updated α_i computed with equation (6.30) and the number of potential regressors is (hopefully) reduced. Likewise, the posterior covariance $\boldsymbol{\Sigma}$ and mean $\boldsymbol{\mu}$ of the coefficients, which are scalars in the first iteration, are computed along with the updated values of s_i and q_i for all potential regressors.

Sequentially, the SBP retrieves the regressor that maximizes the marginal likelihood, then the posterior covariance and mean of the coefficients are updated using the following formulas

$$\tilde{\boldsymbol{\Sigma}} = \begin{bmatrix} \boldsymbol{\Sigma} + \beta^2 \boldsymbol{\Sigma}_{ii} \boldsymbol{\Sigma} \mathbf{X}^H \phi_i \phi_i^H \mathbf{X} \boldsymbol{\Sigma} & -\beta \boldsymbol{\Sigma}_{ii} \boldsymbol{\Sigma} \mathbf{X}^H \phi_i \\ -\beta \boldsymbol{\Sigma}_{ii} (\boldsymbol{\Sigma} \mathbf{X}^H \phi_i)^H & \boldsymbol{\Sigma}_{ii} \end{bmatrix}, \tag{6.32}$$

$$\tilde{\boldsymbol{\mu}} = \begin{bmatrix} \boldsymbol{\mu} - \mu_i \beta \boldsymbol{\Sigma} \mathbf{X}^H \phi_i \\ \mu_i \end{bmatrix}, \tag{6.33}$$

where $\tilde{\boldsymbol{\Sigma}}$ and $\tilde{\boldsymbol{\mu}}$ are the updated posterior covariance and mean of the coefficients, $\boldsymbol{\Sigma}_{ii} = (\alpha_i + s_i)^{-1}$, $\mu_i = \boldsymbol{\Sigma}_{ii} q_i$ and $\mathbf{e}_i = \phi_i - \beta \mathbf{X} \boldsymbol{\Sigma} \mathbf{X}^H \phi_i$. The active set is increased until the candidate regressor set is exhausted or a stopping criterion is met. The pseudocode of this SBP is shown in Algorithm 6.5.

Enunciated in terms of the posterior coefficients covariance $\boldsymbol{\Sigma} = (\beta \mathbf{X}^H \mathbf{X} + \mathbf{A})^{-1}$, the estimation of the coefficients (6.20) is formulated as

$$\hat{\mathbf{h}} = (\mathbf{X}^H \mathbf{X} + \sigma^2 \mathbf{A})^{-1} \mathbf{X}^H \mathbf{y}, \tag{6.34}$$

which can be compared to (5.39) to illustrate that this method can be understood as a beneficial regularization procedure. Given that $\boldsymbol{\Sigma}$ is a $n \times n$ matrix, a second advantage is observed in terms of memory requirements compared to the direct use of the $m \times n$ regressors matrix $\mathbf{X}$ in a sparse system with $n < m$. Finally, the coefficients are estimated with a sequential algorithm avoiding the expensive computation of the regressor matrix pseudoinverse.

Examples of the SBP performance are the learning curves plotted in Figure 6.4 corresponding to the identification process of two generalized memory

Algorithm 6.5: Sparse Bayesian Pursuit (SBP) Pseudocode

Input: $\mathbf{X} \in \mathbb{C}^{m \times n}, \mathbf{y} \in \mathbb{C}^m$

Output: $\mathbf{h} \in \mathbb{C}^{n_a}$, $n_a \ll n$

1: *Initialization*: Initialize σ^2 to some sensible value (for example, $\sigma^2 = \frac{1}{m}||\mathbf{y}||^2 \times 10^{-6}$), $\mathbf{C}_{-i}^{(0)} \leftarrow \sigma^2$

2: Compute the values of s_i and q_i for all potential regressors ϕ_i, and remove from the potential set the regressors that do not fulfil the requirement $|q_i|^2 > s_i$.
$s_i = \phi_i^H \mathbf{C}_{-i}^{-1} \phi_i \quad \text{and} \quad q_i = \phi_i^H \mathbf{C}_{-i}^{-1} \mathbf{y}.$

3: Compute α_i and $\ell(\alpha_i)$ for all regressors.

$$\alpha_i = \frac{s_i^2}{|q_i|^2 - s_i} \qquad \text{if } |q_i|^2 > s_i,$$

$$\ell(\alpha_i) = \frac{|q_i|^2 - s_i}{s_i} + \ln \frac{s_i}{|q_i|^2}.$$

Move the potential regressor that maximizes $\ell(\alpha_i)$ to the active set.

4: Compute the updated values of Σ and μ (which are scalars initially).

$$\tilde{\Sigma} = \begin{bmatrix} \Sigma + \beta^2 \Sigma_{ii} \Sigma \mathbf{X}^H \phi_i \phi_i^H \mathbf{X} \Sigma & -\beta \Sigma_{ii} \Sigma \mathbf{X}^H \phi_i \\ -\beta \Sigma_{ii} (\Sigma \mathbf{X}^H \phi_i)^H & \Sigma_{ii} \end{bmatrix},$$

$$\tilde{\mu} = \begin{bmatrix} \mu - \mu_i \beta \Sigma \mathbf{X}^H \phi_i \\ \mu_i \end{bmatrix}.$$

Update $\mathbf{C}_{-i}$ along with s_i and q_i for all potential regressors.

5: Go to 2 until the *stopping criterion* is met, e.g., a target NMSE is reached, or the potential set is empty.

polynomial (GMP) models of 5th- and 9th-order for a class AB power amplifier. The experimental test bench will be described in Chapter 7. The initial set of candidate regressors of the 5th-order model is composed of 231 potential regressors and is reduced to a set of 77 of active regressors resulting in a NMSE performance of −52.2 dB. In the case of the 9th-order model, from an initial set of 451 regressors the SBP identifies 106 active regressors reaching a NMSE of −53.6 dB. If the stopping criterion is relaxed to halt the algorithm when the target NMSE is −51 dB, it is possible to reach a further reduction of the active sets to 19 and 16 regressors, respectively.

Notice that the 9th-order GMP model with a richer initial set yields a better NMSE with only 16 coefficients compared to the 5th-order GMP model, which needs 19 coefficients. This is explained by recalling that the 5th-order model is lacking regressors with higher nonlinear order and the algorithm tries to compensate for the missing regressors by adding less effective ones of lower nonlinear order, thus demanding a total of 19 regressors.

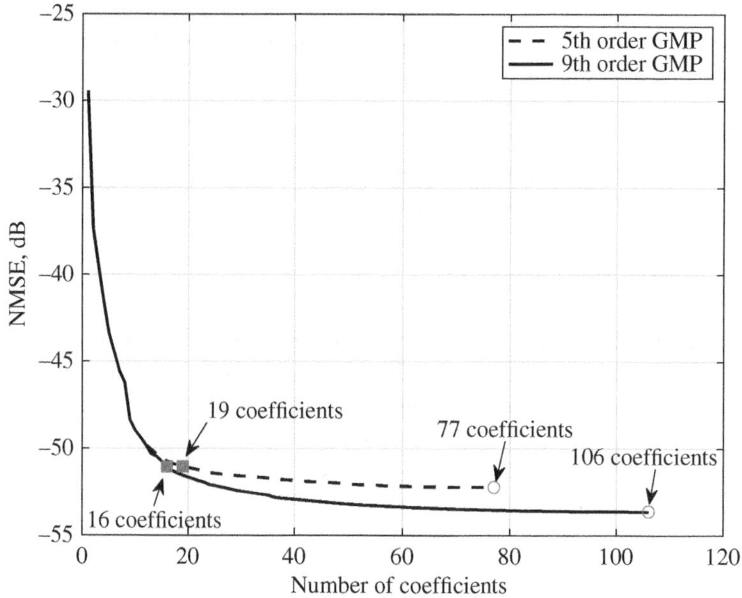

Figure 6.4 Performance of the SBP identification procedure in NMSE versus number of coefficients using two GMP models with 5th- and 9th-nonlinear order. The full active sets (circles) and the active sets that guarantee an NMSE = −51 dB (squares) are highlighted.

6.6.2 Deselecting Regressors

The Bayesian treatment also makes it possible to deselect inactive regressors in a given initial set. After a new regressor has been incorporated to the active set during the identification SBP procedure, it is reasonable to question if all regressors remain actually active or if some of them can be deselected otherwise. The algorithm to test the active set can be initiated with the n_a active regressors, the corresponding regressors matrix $\mathbf{X}_0 \in \mathbb{C}^{m \times n_a}$, and the *a priori* precision matrix $\mathbf{A}_0 \in \mathbb{C}^{n_a \times n_a}$, and it can be observed that the covariance matrix can be decomposed as

$$\mathbf{C}_0 = \sigma^2 \mathbf{I} + \mathbf{X}_0 \mathbf{A}_0^{-1} \mathbf{X}_0^H = \sigma^2 \mathbf{I} + \sum_{k=1}^{n_a} \alpha_k^{-1} \phi_k \phi_k^H. \tag{6.35}$$

If the regressor ϕ_j is deselected, the new covariance can be written as

$$\mathbf{C} = \sigma^2 \mathbf{I} + \sum_{k \neq j} \alpha_k^{-1} \phi_k \phi_k^H, \tag{6.36}$$

or

$$\mathbf{C} = \mathbf{C}_0 - \alpha_j^{-1} \phi_j \phi_j^H, \tag{6.37}$$

where $\mathbf{C}$ is $\mathbf{C}_0$ with the contribution of the regressor j (and also the dependence on α_j) removed. Proceeding in a similar way as in the previous SBP algorithm, the objective function after deselection $\mathcal{L}(\boldsymbol{\alpha})$ is written as

$$\mathcal{L}(\boldsymbol{\alpha}) = \mathcal{L}(\boldsymbol{\alpha}_0) + \ell(\alpha_j), \tag{6.38}$$

where $\mathcal{L}(\boldsymbol{\alpha}_0)$ is the objective function before deselection of regressor ϕ_j, and the increase of the objective function after deselection is

$$\ell(\alpha_j) = \frac{|q_j|^2}{s_j - \alpha_j} - \ln\left(1 - \frac{s_j}{\alpha_j}\right). \tag{6.39}$$

The regressor that maximizes the objective function (or the marginal likelihood) after deselection, is discarded from the active set, and the posterior covariance and mean of the coefficients are updated using the formulas

$$\tilde{\boldsymbol{\Sigma}} = \boldsymbol{\Sigma} - \frac{1}{\boldsymbol{\Sigma}_{jj}}\boldsymbol{\Sigma}_j\boldsymbol{\Sigma}_j^H, \tag{6.40}$$

$$\tilde{\boldsymbol{\mu}} = \boldsymbol{\mu} - \frac{\mu_j}{\boldsymbol{\Sigma}_{jj}}\boldsymbol{\Sigma}_j, \tag{6.41}$$

and the appropriate row and/or column j is removed from $\tilde{\boldsymbol{\Sigma}}$ and $\tilde{\boldsymbol{\mu}}$. $\boldsymbol{\Sigma}_j$ is the jth column of $\boldsymbol{\Sigma}$.

Unlike standard greedy pursuit techniques, the Bayesian treatment provides this beneficial deselection algorithm that enables to check if any regressor of the set identified by the SBP procedure is actually active. Referring to Figure 6.4, the deselection procedure was carried out for the active sets of 77 and 106 coefficients of the two models, with the result that no regressors had to be removed if the respective accuracy must be preserved. This is an indication that each regressor selected by the SBP algorithm is consistently active even if new candidate regressors are included in the set. However, some circumstances can be foreseen for which deselection of regressors would be convenient to reconfigure the pruned model. One possible example is an amplifier suffering a bias point modification leading to a less nonlinear operation and a consequent application of the deselection subroutine would be positive to remove higher order regressors. Perhaps it is more interesting in a different situation in which the power level of the input signal varies, but not the system itself. Therefore, the more likely model learned with the input signal at a given power level is also valid for other points with lower levels and the regressor selection routine is not necessary but only the coefficients re-estimation, referred to as *model reconfiguration*.

6.6.2.1 Reconfiguring an Amplifier Model

The details of the reconfiguration can be illustrated with a basic model identified for the amplifier discussed above. The SBP algorithm was applied employing a

5th-order bivariate circuit knowledge Volterra (CKV) model, defined in (4.37), to take advantage of a richer set of candidate regressors. For the amplifier tested at an output power of 27.4 dBm, the active set is constituted by 19 regressors.

Assuming a power-varying scenario, instead of performing again the complete SBP algorithm in a point-by-point basis, which means to repeat the identification of the active regressors and the estimation of the coefficients, a computationally more efficient procedure is to reuse the already identified active set, saving the searching steps, and to repeat only the coefficients re-estimation subroutine for the corresponding 19 coefficients. The efficiency improvement offered by the re-estimation subroutine with respect to the full SBP algorithm can be evaluated by comparing the respective execution times of each iteration displayed in Figure 6.5. The re-estimation (line with asterisks) execution time is nearly 80 times faster than that of the complete SBP algorithm, with active regressor selection and coefficients estimation (dashed line with circles).

A simplified procedure can be summarized as follows. Initially use the SBP algorithm to identify the active regressors and estimate the coefficients at the maximum operating power level. Since the regressors are arranged according to its likelihood order, they can be deselected systematically and the coefficients re-estimated if the power decreases.

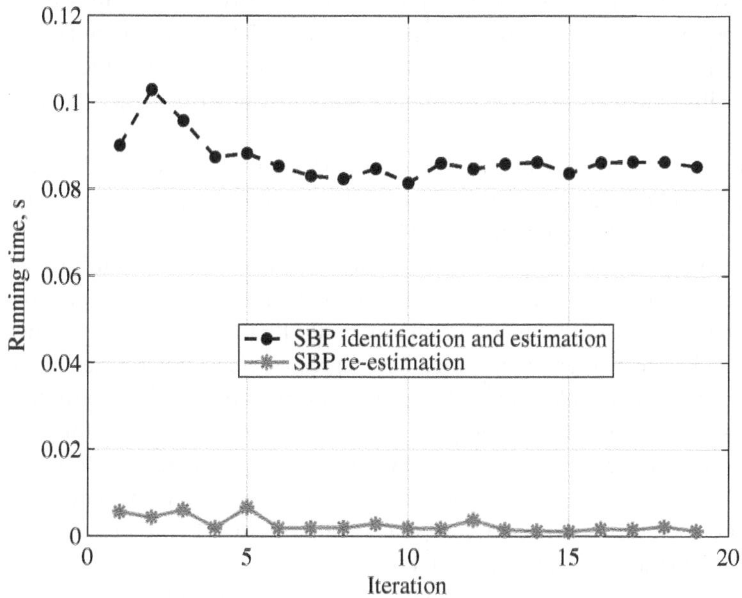

Figure 6.5 Comparison of the execution time for the complete identification and re-estimation procedures.

The problem is different if the amplifier is driven by a signal with an increased power level and the active set becomes insufficient to reach the NMSE objective. Therefore, the SBP pursuit algorithm can be repeated considering a new potential set with hopefully significant extra regressors attached to the active set.

6.6.3 Bayesian Upgrading

In amplifiers modeling and predistorters design, the engineer is in front of a dilemma to decide how much computational cost is acceptable for a given precision. On the one hand, to avoid an oversized regressors stock, the Volterra model nonlinear order and memory length are truncated and, since an important feature of a model is a reduced regressors set to diminish the computational cost of the coefficients estimation accomplished directly by the least squares procedure, reduced ad hoc or circuit knowledge-based models, perhaps with suboptimal structures, have been proposed in pioneering publications. On the other hand, a more complete and richer set is advantageous when greedy and the SBP techniques are applied to search for each one of the relevant actual active regressors.

The upgrade strategy presented here is proposed for situations in which the need for reducing the demand of computational resources available in standard processors limits the structure of the initial set of potential regressors, and consequently the final performance of the identified active set. To improve the modeling performance of such an incomplete active model, the search procedure is repeated involving a new potential set assembled by augmenting the sparse active set with new regressors having higher nonlinear order or larger memory length (Crespo-Cadenas et al., 2022).

Assuming an incomplete active set with a measurement matrix $\mathbf{X}_a \in \mathbb{C}^{m \times n_a}$, the present Bayesian upgrading approach proceeds by attaching a new stock of candidate regressors with measurement matrix $\mathbf{X}_b \in \mathbb{C}^{m \times n_b}$.

One possible procedure to search the upgraded active set can be outlined as follows. Determine α_i and $\ell(\alpha_i)$ using (6.30) and (6.31) for all potential regressors and update the active set by including the regressor that maximizes $\ell(\alpha_i)$. Once the active set has been augmented with a new regressor, check if it is necessary to remove a regressor of the old active set by proceeding with the deselection routine described in Section 6.6.2, i.e., compute the increase of the objective function after deselection of the regressor ϕ_j given by equation (6.39) and remove from the active set the regressor that maximizes $\ell(\alpha_j)$. Update the posterior covariance and mean of the coefficients using the formulas (6.40) and (6.41). The procedure is repeated for all regressors of the potential set.

A more direct line of action is to compose a potential set with regressors of the initial active set and the new candidate stock and perform the SBP algorithm 6.5 with the new measurement matrix $\mathbf{X} = [\mathbf{X}_a\, \mathbf{X}_b]$. In this way, the algorithm reorders the regressors according to their likelihood, with the result that perhaps some new candidate regressors are added to the updated active set, while at the same time some regressors of the old active set are removed.

6.6.3.1 SBL Comparison to Other Techniques

The formally deduced SBL approach involves the SBP procedure for the search of active regressors by following a marginal likelihood maximization order. It is worth noticing that the SBL method is a corpus of procedures that includes also other routines: removing irrelevant regressors, model upgrading, and model reconfiguration algorithms.

It is interesting to contrast the SBL search algorithm SBP with other approaches, such as the greedy OMP and DOMP techniques. Although a more complete comparison of these techniques will be detailed in Chapter 7 for a case of power amplifier linearization, let us state here that the experimental results demonstrate for SBP a NMSE performance similar to that of the DOMP algorithm and clearly superior with respect to the OMP algorithm.

Although the SBL procedure shares with other techniques the benefits of regularization, one important difference is that the precision matrix $\mathbf{A}$ follows directly from the application of the SBP algorithm. This is unlike other procedures, such as the Ridge regression, where the value of the regularization parameter λ in the estimation (5.39) is fixed by the designer depending on the specific nature of each problem.

Another particular feature of the SBL procedure is the inclusion of a novel Bayesian stopping criterion based on emptying the set of potential regressors, and consequently is distinct from the BIC defined in the previous section.

As a concluding statement, let us remark that, given an initial set of candidate regressors, the set selected by the SBL method can be objectively denoted as the most likely reduced-order model.

6.6.3.2 Two Cases of Model Upgrading

To illustrate the advantage of the Bayesian upgrading algorithm, we can consider two practical examples. As a first case, let us consider the thirteenth-order full Volterra model with a memory length of $Q = 10$ samples, for which it would be necessary to handle 19 617 regressors, a quantity that may exceed the average computer capabilities. Following the upgrading strategy, the SBP algorithm can be first applied to a model with only 3 samples of memory length, a structure with a manageable raw stock of 248 regressors. Notwithstanding that it is possible to assemble a sparse active set, it is not optimum because of the limited richness of the initial stock of regressors. Therefore, a first stage of model upgrading is possible by combining this suboptimal active set and a new candidate stock constituted by third-order regressors now with a memory length of 10 samples. Next, the

SBP algorithm is executed with this richer set of candidate regressors. Successive upgrading stages can be repeated with new candidate stocks, e.g., 5th-order regressors with 10 sample memory, 7th-order regressors with 10 sample memory, and so on. Notice that the final sparse model is constituted by the more likely regressors out of 19 617.

The second case is applicable to amplifiers operated near saturation, where Volterra models with practical nonlinear order truncation diverge and are unable to deliver an acceptable accuracy. Considering that the Volterra model is linear with respect to the coefficients, the kernels can be split in two components and the model expressed as the sum of two Volterra models. If the first model is selected memoryless and its nonlinear order is not truncated, it can be viewed as the Taylor series of a regressor $\phi_0(k)$ and the output can be written as

$$\mathbf{y} = h_0\phi_0 + \sum_{i=1}^{n} h_i\phi_i, \tag{6.42}$$

where ϕ_0 is a new candidate regressor in addition to the conventional regressors ϕ_i. The criterion adopted to select $\phi_0(k)$ can be a function that maximizes its correlation with the output $y(k)$. Observe that ϕ_0 may not be denoted as a conventional Volterra regressor, but it is the result of a Volterra series approach, with the benefit that this new regressor overcomes the inherent instability of a truncated polynomial in a compression region operation. If we denote $\mathbf{X}_b$ the measurement matrix of the second model, the SBP algorithm can be performed over the new stock of candidate regressors with the measurement matrix $\mathbf{X} = [\phi_0\,\mathbf{X}_b]$.

6.7 A Practical Sparse Regression

This section is devoted to extend Section 5.2.6 with the inclusion of the sparsity in a practical example of regression. Section 6.6 also showed a practical regression using the novel SBL algorithm associated to Bayesian techniques. A more detailed discussion and comparisons with other techniques will be performed in Chapter 7 for a linearization scenario. In this exercise, we particularize for the OMP and DOMP techniques to highlight the basic concepts of sparse regression.

In this experiment, the model under test is a GMP with the following configuration:

- **Part A (time-aligned term, corresponding to the MP):** Nonlinear order 13 and memory depth 2.
- **Parts B and C (leading and lagging terms):** Nonlinear order 7 and memory depth 5.

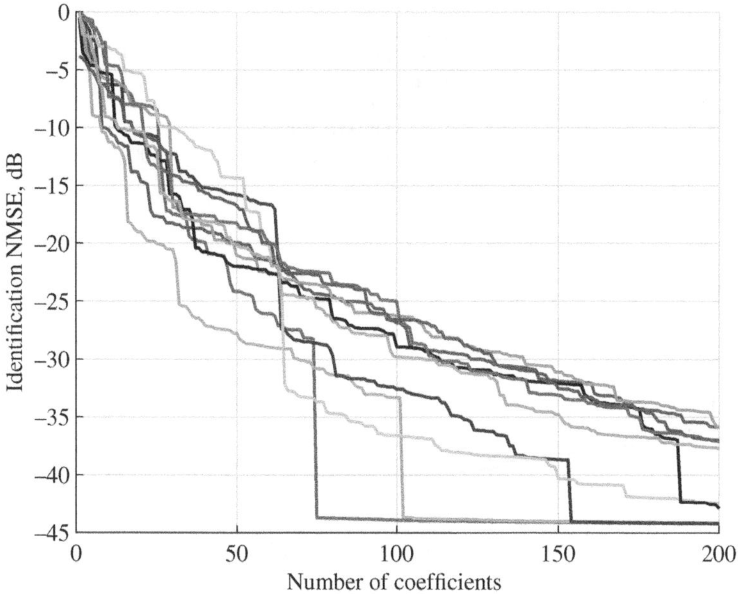

Figure 6.6 Evolution of the NMSE in models with components taken randomly from the initial pool set of basis functions.

The model generates a pool set of 405 regressors, and achieves an identification NMSE of −44.3 dB.

As a motivation exercise, 200 model regressors have been selected in random order to model the amplifier. The random selection of 200 regressors from the initial set of 405 has been repeated 10 times, and the resulting identification NMSE against the number of coefficients in random order is plotted in Figure 6.6. The general trend of the NMSE is to decrease with the number of coefficients, as it can be expected since the basis becomes richer and its capabilities to model the output increase. It is also expectable that all the NMSE evolutions reach the point of −44.3 dB, that corresponds to the complete model, when the number of selected coefficients equals the model number of coefficients. The interesting result of this experiment is that, although the trend is similar, taking coefficients at a random order produces a specific path in the NMSE versus number of coefficients space. Since the general aim of coefficients selection techniques is to choose those regressors that are more relevant to model the output, paths that drop quickly in the first stages of the selection process are preferred.

Next, the OMP and DOMP are executed over the measurements to attain a sorted list of regressors in increasing order of relevance. The result is shown in Figure 6.7. The trend of both techniques is also decreasing, but the enhancement

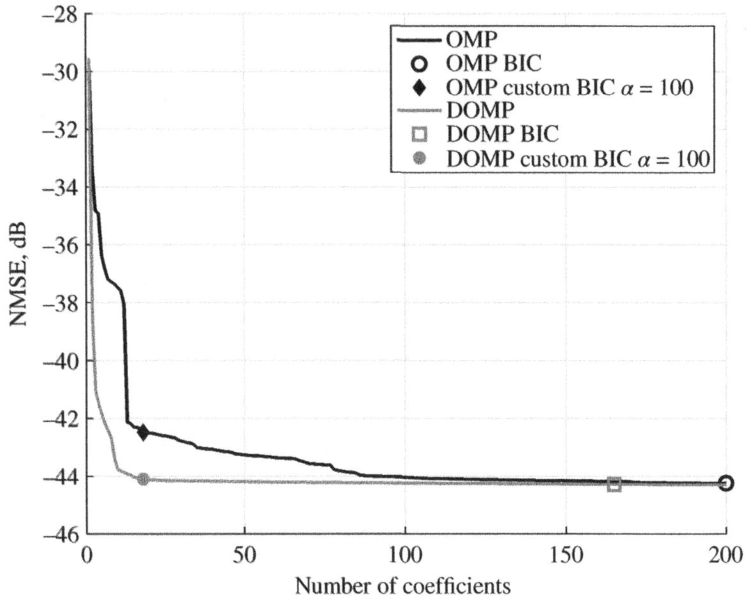

Figure 6.7 Evolution of NMSE in the OMP and DOMP techniques along with the optimal number of coefficients given by the BIC.

with respect to the random selection is noticeable. The OMP technique is able to provide a sparse solution with a fair performance while the DOMP is able to lower the identification NMSE, specially in the first iterations of the algorithm. The overall effect is evidenced by the optimum number of coefficients identified by the BIC in each case, being 200 for an NMSE of −44.3 dB and 165 for an NMSE of −44.3 dB for the OMP and DOMP cases, respectively. This BIC execution is an example of a situation where the custom BIC (6.13) allows to move the optimum model to the left when an optimum trade-off factor α is selected. Since the curve reaches a horizontal steady state, the number of coefficients of the model attained through the BIC can be reduced to 18 coefficients for both OMP and DOMP if the custom BIC is employed.

Why is there such a difference between the two techniques under comparison? Table 6.1 shows the first 30 regressors selected by each technique. As it was discussed in the previous chapter, those regressors of lower memory and lower order evidence a high impact in modeling the output. Both algorithms choose the linear memoryless regressor $x(n)$ in the first place. Now, looking at the memoryless regressors, the OMP chooses the 4th order, 11th order, and 2nd order in positions 1, 3, and 6, respectively. The DOMP selects 2nd-, 3rd-, 4th-, and 5th-order memoryless regressors in positions 2, 3, 5, and 11, respectively. These results reveal how the DOMP can prioritize relevant regressors over less important ones.

Table 6.1 Regressors selected by the OMP and DOMP techniques in the 30 first iterations.

	OMP Regressor	NMSE (dB)	DOMP Regressor	NMSE (dB)
1	$x(n)$	−29.55	$x(n)$	−29.55
2	$x(n)\|x(n)\|^3$	−33.40	$x(n)\|x(n)\|$	−38.58
3	$x(n)\|x(n)\|^8$	−34.81	$x(n)\|x(n)\|^3$	−41.07
4	$x(n-5)\|x(n-8)\|$	−34.91	$x(n-5)\|x(n)\|$	−41.52
5	$x(n)\|x(n)\|^{13}$	−36.36	$x(n-1)$	−41.87
6	$x(n-2)\|x(n+1)\|$	−36.84	$x(n-2)$	−42.19
7	$x(n-5)\|x(n)\|^3$	−37.19	$x(n)\|x(n+1)\|$	−42.45
8	$x(n-2)\|x(n-2)\|^{13}$	−37.28	$x(n)\|x(n)\|^6$	−42.69
9	$x(n-5)\|x(n-9)\|^4$	−37.35	$x(n)\|x(n)\|^2$	−43.39
10	$x(n-3)\|x(n+2)\|^3$	−37.47	$x(n)\|x(n)\|^4$	−43.77
11	$x(n)\|x(n+4)\|^4$	−37.58	$x(n)\|x(n)\|^5$	−43.83
12	$x(n-2)$	−38.03	$x(n)\|x(n)\|^7$	−43.88
13	$x(n)\|x(n)\|$	−42.14	$x(n)\|x(n-1)\|$	−43.92
14	$x(n-2)\|x(n-7)\|^3$	−42.18	$x(n-1)\|x(n)\|$	−43.96
15	$x(n)\|x(n-4)\|^2$	−42.30	$x(n-4)\|x(n)\|$	−44.04
16	$x^*(n-1)$	−42.31	$x(n-3)\|x(n)\|$	−44.07
17	$x(n-3)\|x(n-8)\|^6$	−42.36	$x^*(n-1)$	−44.09
18	$x(n-5)\|x(n)\|$	−42.49	$x(n)\|x(n)\|^8$	−44.11
19	$x(n-5)\|x(n-10)\|^6$	−42.50	$x(n-1)\|x(n+1)\|^4$	−44.12
20	$x(n)\|x(n+5)\|^6$	−42.51	$x(n-1)\|x(n)\|^5$	−44.13
21	$x(n-5)\|x(n-4)\|^4$	−42.52	$x(n)\|x(n)\|^9$	−44.14
22	$x(n-1)\|x(n+3)\|^2$	−42.55	$x(n)\|x(n)\|^{10}$	−44.14
23	$x(n)\|x(n+5)\|^2$	−42.57	$x(n)\|x(n-5)\|$	−44.15
24	$x(n-5)\|x(n)\|^6$	−42.60	$x(n-1)\|x(n-4)\|$	−44.15
25	$x(n-5)\|x(n-10)\|$	−42.61	$x(n)\|x(n)\|^{11}$	−44.15
26	$x(n-4)\|x(n-9)\|^2$	−42.63	$x(n-5)\|x(n-10)\|$	−44.16
27	$x(n)\|x(n+1)\|^5$	−42.65	$x(n-2)\|x(n)\|$	−44.16
28	$x(n-2)\|x(n+3)\|^6$	−42.67	$x(n-1)\|x(n-3)\|$	−44.16
29	$x(n)\|x(n-1)\|^5$	−42.75	$x(n-2)\|x(n+1)\|^3$	−44.16
30	$x(n)\|x(n-3)\|^6$	−42.79	$x^*(n)$	−44.16

The analysis of the linear regressors with memory reveals that the OMP does not prioritize any of them in the first 30 positions, while the DOMP selects $x(n-1)$ and $x(n-2)$ in the 6th and 7th positions of its selection.

The conclusion of this exercise is that by considering the correlation between the regressors in the selection process, the DOMP is able to better identify the basis functions that lower the identification NMSE.

Bibliography

A. Barry, W. Li, J.A. Becerra, and P.L. Gilabert. Comparison of feature selection techniques for power amplifier behavioral modeling and digital predistortion linearization. *Sensors*, 21(17), 2021. doi: 10.3390/s21175772.

J.A. Becerra, M.J. Madero-Ayora, R.G. Noguer, and C. Crespo-Cadenas. On the optimum number of coefficients of sparse digital predistorters: A Bayesian approach. *IEEE Microwave and Wireless Components Letters*, 30 (12):1117–1120, 2020a. doi: 10.1109/LMWC.2020.3027878.

J.A. Becerra, M.J. Madero-Ayora, J. Reina-Tosina, and C. Crespo-Cadenas. Sparse identification of Volterra models for power amplifiers without pseudoinverse computation. *IEEE Transactions on Microwave Theory and Techniques*, 68(11):4570–4578, 2020b. doi: 10.1109/TMTT.2020.3016967.

J.A. Becerra, M.J. Madero-Ayora, J. Reina-Tosina, C. Crespo-Cadenas, J. García-Frías, and G. Arce. A doubly orthogonal matching pursuit algorithm for sparse predistortion of power amplifiers. *IEEE Microwave and Wireless Components Letters*, 28(8):726–728, 2018. doi: 10.1109/LMWC.2018.2845947.

J.A. Becerra, M.J. Madero-Ayora, and C. Crespo-Cadenas. Comparative analysis of greedy pursuits for the order reduction of wideband digital predistorters. *IEEE Transactions on Microwave Theory and Techniques*, 67 (9):3575–3585, 2019. doi: 10.1109/TMTT.2019.2928290.

S. Chen, S.A. Billings, and W. Luo. Orthogonal least squares methods and their application to non-linear system identification. *International Journal of Control*, 50(5):1873–1896, 1989. doi: 10.1080/00207178908953472.

C. Crespo-Cadenas, M.J. Madero-Ayora, J. Reina-Tosina, and J.A. Becerra-González. Transmitter linearization adaptable to power-varying operation. *IEEE Transactions on Microwave Theory and Techniques*, 65(10): 3624–3632, 2017. doi: 10.1109/TMTT.2017.2742951.

C. Crespo-Cadenas, M.J. Madero-Ayora, J.A. Becerra, and S. Cruces. A fast sparse Bayesian pursuit approach for power amplifier linearization. In *2021 IEEE MTT-S International Wireless Symposium (IWS)*, pages 1–3, 2021. doi: 10.1109/IWS52775.2021.9499382.

C. Crespo-Cadenas, M.J. Madero-Ayora, J.A. Becerra, and S. Cruces. A sparse-Bayesian approach for the design of robust digital predistorters under power-varying operation. *IEEE Transactions on Microwave Theory and Techniques*, 70(9):4218–4230, 2022. doi: 10.1109/TMTT.2022.3157586.

Y.C. Eldar and G. Kutyniok. *Compressed Sensing: Theory and Applications*. Cambridge University Press, Cambridge, UK, 2012. ISBN 9781107005587.

A. Faul and M. Tipping. Analysis of sparse Bayesian learning. In T. Dietterich, S. Becker, and Z. Ghahramani, editors, *Advances in Neural Information Processing Systems*, volume 14. MIT Press, 2001.

P.L. Gilabert, G. Montoro, D. López, N. Bartzoudis, E. Bertran, M. Payaró, and A. Hourtane. Order reduction of wideband digital predistorters using principal component analysis. In *2013 IEEE MTT-S International Microwave Symposium Digest (MTT)*, pages 1–7, 2013. doi: 10.1109/MWSYM.2013.6697687.

A. Hashemi and H. Vikalo. Sparse linear regression via generalized orthogonal least squares. In *2016 IEEE Global Conference on Signal and Information Processing (GlobalSIP)*, pages 1305–1309, 2016. doi: 10.1109/GlobalSIP.2016.7906052.

A. Hashemi and H. Vikalo. Accelerated orthogonal least squares for large-scale sparse reconstruction. *Digital Signal Processing*, 82:91–105, 2018. doi: 10.1016/j.dsp.2018.07.010.

G. James, D. Witten, T. Hastie, and R. Tibshirani. *An Introduction to Statistical Learning: with Applications in R*. Springer, New Yok, 2013. ISBN 9781461471387.

D. López-Bueno, Q.A. Pham, G. Montoro, and P.L. Gilabert. Independent digital predistortion parameters estimation using adaptive principal component analysis. *IEEE Transactions on Microwave Theory and Techniques*, 66(12):5771–5779, 2018. doi: 10.1109/TMTT.2018.2870420.

S. Maymon and Y.C. Eldar. The Viterbi algorithm for subset selection. *IEEE Signal Processing Letters*, 22(5):524–528, 2015. doi: 10.1109/LSP.2014.2360881.

J. Reina-Tosina, M. Allegue-Martínez, C. Crespo-Cadenas, C. Yu, and S. Cruces. Behavioral modeling and predistortion of power amplifiers under sparsity hypothesis. *IEEE Transactions on Microwave Theory and Techniques*, 63(2): 745–753, 2015. doi: 10.1109/TMTT.2014.2387852.

M. Tipping. The Relevance Vector Machine. In S. Solla, T. Leen, and K. Müller, editors, *Advances in Neural Information Processing Systems*, volume 12. MIT Press, 1999.

M.E. Tipping and A.C. Faul. Fast marginal likelihood maximisation for sparse Bayesian models. In C.M. Bishop and B.J. Frey, editors, *Proceedings of the Ninth International Workshop on Artificial Intelligence and Statistics*, volume R4 of *Proceedings of Machine Learning Research*, pages 276–283. PMLR, 2003.

S. Wang, M.A. Hussein, O. Venard, and G. Baudoin. A novel algorithm for determining the structure of digital predistortion models. *IEEE Transactions on Vehicular Technology*, 67(8):7326–7340, 2018. doi: 10.1109/TVT.2018.2833283.

7

Transmitter Linearization with Digital Predistorters

7.1 Introduction

The evolution of wireless communications is motivated by an increasing need of higher bit rates to be transmitted over scarce spectrum resources. These requirements push for the development of new and more efficient modulation schemes such as the orthogonal frequency division multiplexing (OFDM), which allows to transmit information with variable bandwidths. With the transition from continuous phase modulations and code division multiple access technologies, that respectively architected the second and third generations of mobile communications, to OFDM, the transmitted signals markedly show peaks in the time domain. These peaks result in a high peak-to-average power ratio (PAPR), that affects the selection of the operating point of the power amplifier.

The power amplifier exhibits two operational zones: the linear behavior is shown at the lower end of input powers and the saturation zone appears at high input power ranges. Coincidentally, the power amplifier is more power efficient at high input powers. This situation is known as the nonlinearity versus efficiency trade-off (Lavrador et al., 2010). In other words, the power amplifier is efficient in power when it adds nonlinear distortions to the signal. As it was reviewed in previous chapters of this book, nonlinearity creates both in-band and out-of-band distortions, that generally do not comply with the standards' requirements.

The solutions to the aforementioned scenario are known as linearization techniques (Katz, 2001). In this chapter, we focus on the linearization of power amplifiers through digital predistortion. The concept is intuitively straightforward: the intention is to build a system, known as digital predistorter (DPD), that transforms the input signal to the power amplifier into something that makes the overall system behave linearly. It can be thought of as a system that is able to invert the not-desiderable behavior of the active device.

A Volterra Approach to Digital Predistortion: Sparse Identification and Estimation, First Edition.
Carlos Crespo-Cadenas, María José Madero-Ayora, and Juan A. Becerra.

7.2 Digital Predistortion

The objective of the DPD is to revert the nonlinear behavior of the power amplifier. This concept can be understood as it is shown in Figure 7.1, where the power amplifier is represented by an input-output relation that compresses at high input powers. The DPD is designed to represent the inverse characteristic of the power amplifier in a way that both the DPD and the PA in cascade exhibit a linear behavior.

Previous chapters of this book have shown how to model a nonlinear system with the use of Volterra series. Without going into detail of which model is more suitable and all the artifacts that have been presented previously, we are in the position of stating that the DPD will be adequately described by a Volterra series since its behavior is nonlinear with memory.

The first approach to this problem normally poses a counter intuitive question: How can the DPD be designed knowing that it affects the power amplifier behavior? Conceptually, we just need the pre-inverse system of the power amplifier, but since the power amplifier will change its behavior with and without DPD, we will not be able to find the power amplifier input-output function with DPD if we do not have the DPD in place. There is something of a chicken-and-egg situation here that can be solved considering that the pre-inverse of a system is identical to its post-inverse (Schetzen, 1976). In other words, in the noiseless case, designing a system that transforms the output of the power amplifier into its input is equivalent to the process of designing a DPD.

There exist multiple ways to perform digital predistortion. Amongst them, indirect learning architecture (ILA) and direct learning architecture (DLA) are the

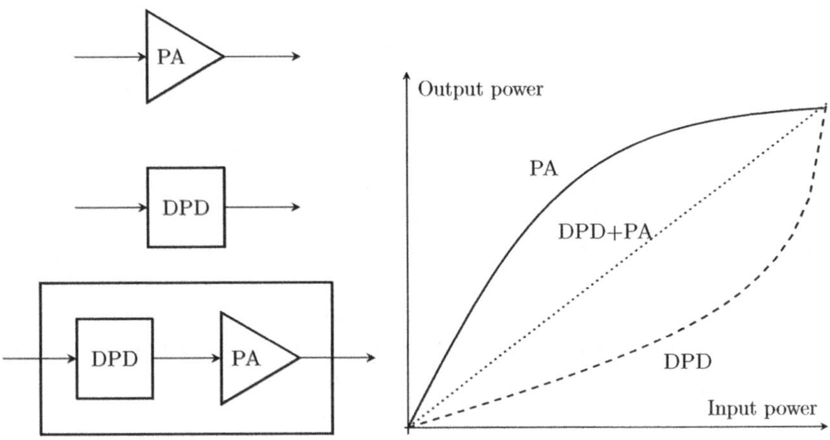

Figure 7.1 General scheme of digital predistortion.

most used. ILA is characterized by being very intuitive, since the process consists in the direct calculation of the inverse function of the power amplifier. In contrast, DLA encloses DPD and PA in a closed loop, and by means of an iterative control technique, the DPD coefficients are calculated iteratively. There is a third technique, closely related to DLA, that can be used to calculate the predistortion signal in a non-parametric fashion, i.e., there is no model and there are no coefficients associated to that model. These techniques are reviewed in the next Sections.

7.3 Indirect Learning Architecture

The ILA is a process that calculates the inverse function of the power amplifier (Ding et al., 2004). This process is performed in two stages, as shown in Figure 7.2. First, the DPD model is regressed as a post-inverse system, i.e., we are aiming at calculating a system that transforms the output of the power amplifier—with a corresponding scale—into its input. Considering that the input signal samples $u(k)$ are staked as

$$\mathbf{u} = [u(k), u(k+1), \dots, u(k+m-1)]^T \in \mathbb{C}^{m\times 1} \tag{7.1}$$

and the corresponding output signal follows

$$\mathbf{y} = [y(k), y(k+1), \dots, y(k+m-1)]^T \in \mathbb{C}^{m\times 1}, \tag{7.2}$$

the regression equation in this scheme becomes

$$\mathbf{w} = \mathbf{Y}^{\dagger}\mathbf{u}, \tag{7.3}$$

where $\mathbf{w}$ are the DPD coefficients and $\mathbf{Y}^{\dagger}$ is a pseudoinverse of $\mathbf{Y}$, as defined in (5.18), which holds the model regressors generated with the signal $\mathbf{y}/G_0$, where the normalization gain G_0 will be the gain of the linearized amplifier. In the DPD identification, the input signal to the PA equals the input signal to the system, $\mathbf{x} = \mathbf{u}$. When the DPD is active, $\mathbf{x}$ is the predistorted input signal.

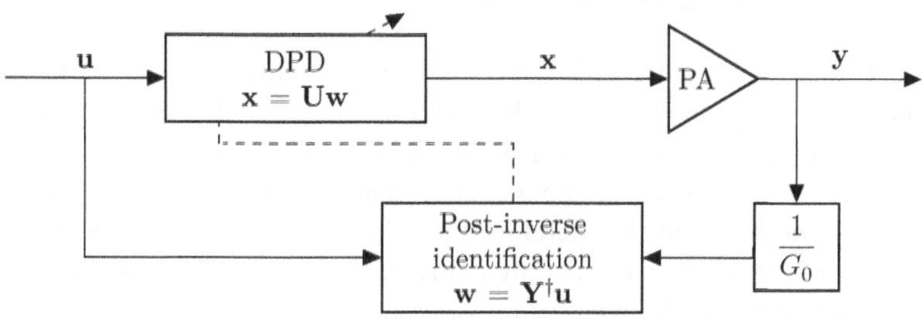

Figure 7.2 General scheme of indirect learning architecture.

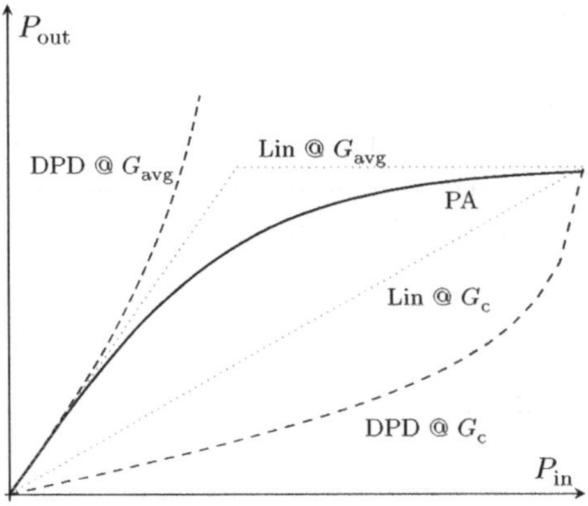

Figure 7.3 Output versus input relationships with linear and compressed gain in the DPD design.

The selection of the gain is crucial for the system performance, since it determines the backoff that the DPD will produce (Zhu et al., 2008). There are two typical gains that can be applied at this step: the average gain G_{lin} and the compressed gain G_c. The average gain is understood as the gain that is produced in the linear zone of the PA, and it will not produce any backoff in the system output. The compressed gain is the gain that the PA produces at high input powers, and it is lower than the average gain. The compressed gain will introduce a backoff in the DPD system of the difference between the linear and the compressed gains. The choice of these two possible gains is represented in Figure 7.3.

Once the coefficients are calculated, the PA input signal follows

$$\mathbf{x} = \mathbf{U}\mathbf{w}, \tag{7.4}$$

where $\mathbf{U}$ is calculated with the same model as (7.3) but with the system input signal $\mathbf{u}$ as the regressor generator and $\mathbf{w}$ is the coefficients vector calculated in (7.3). The ILA can be repeated by regressing the PA input and output signals with DPD following (7.3).

7.4 Direct Learning Architecture

Direct Learning Architecture (DLA) is a training architecture for adaptive DPD (Zhou and DeBrunner, 2007). The general scheme of this technique is presented in Figure 7.4. The signal $\mathbf{u}$ is the input signal to the system and is the signal that is used to build the Volterra model measurement matrix $\mathbf{U}$. This scheme is model agnostic, therefore the following explanations are valid independently of the configuration of the matrix $\mathbf{U}$. The DPD is able to generate the predistortion signal $\mathbf{x}$ following

$$\mathbf{x} = \mathbf{u} - \mathbf{U}\mathbf{w}, \tag{7.5}$$

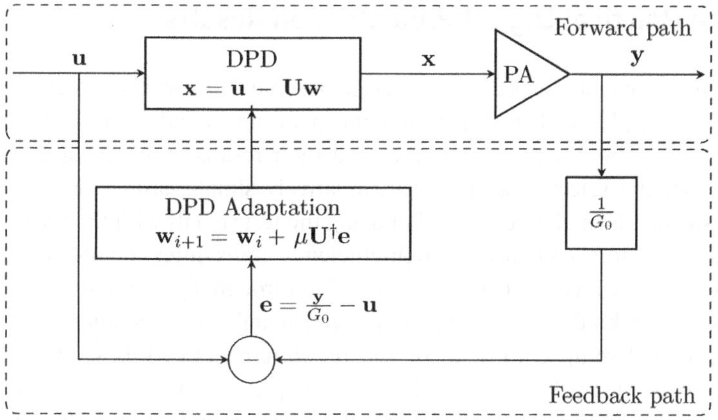

Figure 7.4 General scheme of direct learning architecture.

where $\mathbf{w}$ is the DPD coefficients vector and has a dimension that matches the number of regressors of $\mathbf{U}$. The signal $\mathbf{x}$ is set as the input of the power amplifier that provides at its output the predistorted signal $\mathbf{y}$, although $\mathbf{y}$ can also be the signal without DPD when the coefficients vector is zero, $\mathbf{w} = \mathbf{0}$. These aforementioned operations are included in the forward path, that is characterized by real-time processing.

The feedback path contains the DPD adaptation process, and it can be executed at a lower rate than the forward path. The feedback path rate depends mainly on the technical capabilities of the measurement platform and the desired update rate of the DPD. The DPD adaptation is performed in two steps. First, the error $\mathbf{e}$ between the normalized output $\mathbf{y}$ and the input $\mathbf{u}$ signals is calculated with

$$\mathbf{e} = \frac{\mathbf{y}}{G_0} - \mathbf{u}, \tag{7.6}$$

where G_0 is the desired gain of the system. The discussion of the desired gain value performed in the ILA is also valid in the DLA.

The DPD adaptation is performed with the formula

$$\mathbf{w}_{i+1} = \mathbf{w}_i + \mu \mathbf{U}^{\dagger} \mathbf{e}, \tag{7.7}$$

where i is the iteration index of the DPD that indicates how many times it has been updated, μ is the adaptation parameter and $\mathbf{U}^{\dagger}$ is a pseudoinverse matrix of $\mathbf{U}$. The adaptation parameter μ controls the speed of convergence and it varies in a range from 0 to 1. When μ is close to 0, the convergence speed is low, but the errors in convergence state are low. On the other hand, when μ grows to 1, the convergence speed is higher but the system will produce a higher error and it may lead to system instability. The DPD adaptation process (7.7) is attained by considering the error as the cost function and performing an LMS process (Yu and Zhu, 2015).

7.5 Some Practical Digital Predistortion Results

In this section, the practical aspects of the design of some DPDs will be presented. The provided examples with real experimental data are intended to illustrate the implications of the different techniques and their results in a didactic way. Because of that, the selected scenarios are relatively simple cases where the nonlinear distortion is moderate, instead of using the test signals with the broadest bandwidth or more complicated high-efficiency transmitter architectures. However, the explained techniques for linearization through digital predistortion are applicable to any kind of communications input signal and transmitters.

The experimental setup employed for the measurements of this section is the one shown in Figure 7.5, also available in Chapter 3. The equipment are controlled through LAN and GPIB (General-Purpose Instrumentation Bus) by using Standard Commands for Programmable Instruments (SCPI) from a PC, giving support to the acquisition of the datasets and also to the implementation of the digital predistortion scheme.

The probing signal is created by a vector signal generator (VSG) with built-in arbitrary waveform generator. We must recall that, in order to observe nonlinear distortions, it is necessary to have a measurement bandwidth available of, at least, three to five times the bandwidth occupied by the linear signal or the channel. The greater the oversampling factor, that is, the ratio between the measurement

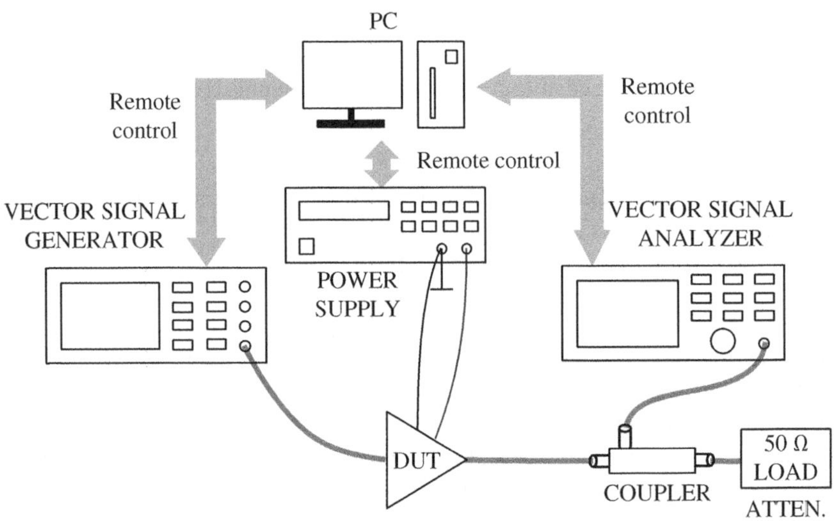

Figure 7.5 Framework of experimental equipment for the implementation of digital predistortion schemes.

bandwidth and the channel, the more accurate can the nonlinear modeling be and more adjacent channels can be observed in the frequency domain. The generation of the predistorted signal, nonlinear by nature, sets the same requirement of an oversampling rate of, at least, 3–5 in the transmit path. In that sense, the allowed maximum sampling rate of the vector signal generator imposes a limitation to the maximum signal bandwidth that can be linearized.

Many times, the vector signal generator needs to be followed by a preamplifier block to be able to drive the power amplifier under test in a mildly nonlinear operation. This is not only due to the limited power available at the output of the vector signal generator, but also seeks to keep the behavior of the modulator sufficiently linear by setting a low power level in it, thus minimizing its internal nonlinear operation.

All the experimental results of this section are provided for a commercial class AB power amplifier, operated at a center frequency of 3.6 GHz. The output of the power amplifier under test is fed to a vector signal analyzer (VSA) through a directional coupler and an attenuator to avoid introducing in the measurement undesired distortions from the equipment. In the VSA, the RF output signal is down-converted to baseband and acquired with an appropriate sampling frequency. The measurement dynamic range can be optimized by setting the equipment to average a certain number of acquisitions of the output signal. Please, note that this requires special care in the synchronization between the vector signal generator and the VSA by means of using an external reference signal of the generator as a clock reference for the analyzer and by the use of triggers.

The acquired signal can also be post-processed in the PC to make it adequate for the digital predistortion process. The usual post-processing consists of the following operations:

- Normalization: scaling of the signals in order to compensate the attenuations and gains of the measurement chain.
- Time alignment: fine synchronization of the input and output signals in time.
- Partition into several datasets: for the identification of the DPDs structure, a segment of consecutive samples is necessary with a limited length to reduce the computational complexity and avoid overfitting. It is recommended to chose this segment including the sample with the highest absolute value at the output. Afterwards, the complete signal is applied to the predistorter block in order to get the predistorted input signal and validate its performance.

7.5.1 Case 1: Basic Digital Predistorter with Indirect Learning Architecture

The first case is a DPD implemented following an indirect learning scheme. In this case, the test signal was generated with an OFDM format, 16-QAM symbols over all the subcarriers, and 15 MHz bandwidth. A common characteristic of OFDM signals is their high peak-to-average power ratio (PAPR). The aim to operate the power amplifier at high efficiency leads to more notable nonlinear impairments with high PAPR values. For this experiment, the input signal exhibited a PAPR of 9.8 dB and the datasets were composed of more than 360000 samples acquired with an oversampling factor of 6.

A standard ILA was used to design the DPD, exploiting that the post-distorter function shall be the same as the predistorter. Therefore, the basis functions were chosen to reduce the error between $x(k)$, as the desired output, and the signal $y(k)/G_c$, as the post-distorter input, with G_c representing the compressed gain, which is the target gain of the linearized power amplifier. In this case, the reason for this is that we seek the best linearization performance. If the small signal or the average gain of the amplifier were considered, the hard-saturation behavior of the amplifier would appear to some extent, and the linearization performance would be degraded.

As mentioned in Section 7.2, the objective of the DPD is to revert the nonlinear behavior of the power amplifier. In that sense, the DPD is also a nonlinear system that can be modeled by using the Volterra-based models discussed throughout Chapters 2–4. Therefore, the subjacent model structure of a DPD can be the same as that of a power amplifier, although the respectively identified nonlinear models will differ. In this example, the structure of the DPD corresponded to a GMP model following (4.2), with the following setting: part A presented thirteenth order and a maximum delay $Q = 15$ for the linear term, $Q = 10$ for the second order, $Q = 5$ for orders 3 to 7, and $Q = 1$ for the higher orders; parts B and C containing the non-diagonal branches just included odd-order terms up to the seventh-order with maximum values of $L = 1$ and $M = 1$ in both cases. This full basis model had a total number of 81 coefficients, which can be described as relatively high taking into consideration that the power amplifier does not present a demanding nonlinear distortion characteristic. A standard least squares regression was performed for the estimation of the coefficients, considering for identification about 5500 samples that produces a signal length to number of coefficients ratio of about 70.

Figures 7.6–7.8 show the dynamic AM–AM and AM–PM characteristics of the unlinearized power amplifier and for the amplifier's output when the predistorted signal is applied. It can be observed that both characteristics are successfully linearized with the DPD. Figure 7.7, representing the AM–AM characteristic in the format of the instantaneous gain versus the input level, deserves a special comment. The commercial amplifier under test presents a small-signal gain of 14.4 dB and a compressed gain for the peak level of the input signal of 12.7 dB, which corresponds to a gain compression of 1.7 dB. Since the target gain selected for the indirect learning scheme was the compressed gain, the linearized gain follows a

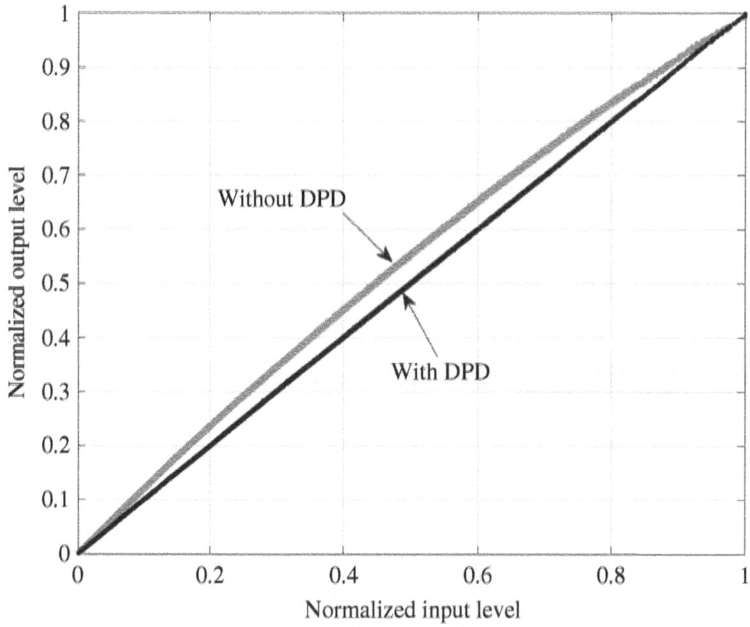

Figure 7.6 Normalized AM–AM characteristic of a commercial power amplifier, with and without linearization following an indirect learning architecture scheme and considering the compressed gain.

flat characteristic coinciding with the value of the compressed gain. This produces a reduction in the output power level of the case with DPD. Although this might now be seen as a drawback of digital predistortion for a fixed operation power, its benefits will be further discussed in Section 7.5.3 in the context of a power sweep.

The power spectral density of the output signal is represented in Figure 7.9(a) for the original output of the amplifier and the linearized one. Taking into account the reduction in the output power level in the case with DPD, the input power level of the signal without DPD was reduced to compare both spectra for the same output power. It is evident that the GMP DPD is effective in terms of spectral regrowth reduction. As shown in Table 7.1, the adjacent channel power ratio (ACPR) has been reduced about 29 dB for the first lower and upper adjacent channels, denoted in the Table as −1 and +1, respectively, and a reduction of more than 20 dB has been achieved for the second lower and upper adjacent channels, denoted as −2 and +2, respectively. However, the spectrum of the error signal between the linearized output of the amplifier and the scaled input signal shown in Figure 7.9(b) reveals a certain margin for improvement for the in-band spectral content. The good linearization performance is also confirmed by the normalized mean squared error (NMSE) between the output signal and the scaled input signal—referred to

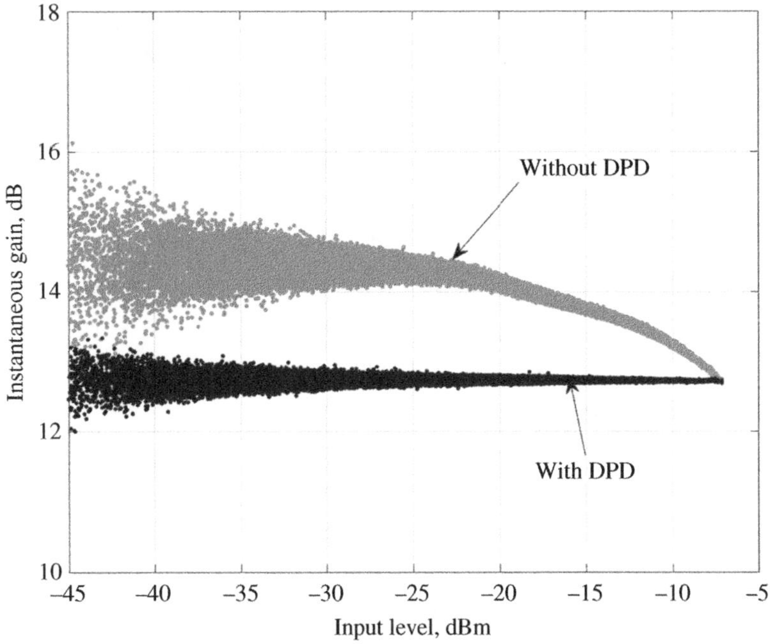

Figure 7.7 Instantaneous gain of a commercial power amplifier versus the input level, with and without linearization following an indirect learning architecture scheme and considering the compressed gain.

as linearization NMSE as explained in Chapter 5—and the error vector magnitude (EVM) provided in Table 7.1. The improvement in NMSE is about 22 dB and the EVM is reduced from 3.8% to 0.5%. Figure 7.10 showing the constellation of the output signal illustrates the decrease in distortion quantified by the EVM.

7.5.2 Case 2: Digital Predistorter with Coefficients Selection and Indirect Learning Architecture

After analyzing the performance provided for a complete set of results for a basic DPD, the second example case is devoted to study the effects of coefficients selection techniques in an ILA scheme with a target gain of the compressed gain value. The same 15-MHz OFDM test signal and power operation point of the previous example are employed in this case. Also, the same GMP model configuration is used for the complete set of regressors, with a total number of 81 regressors before applying components selection.

The OMP (Reina-Tosina et al., 2015), DOMP (Becerra et al., 2018), and PCA (Gilabert et al., 2013) algorithms were run in order to compare their results.

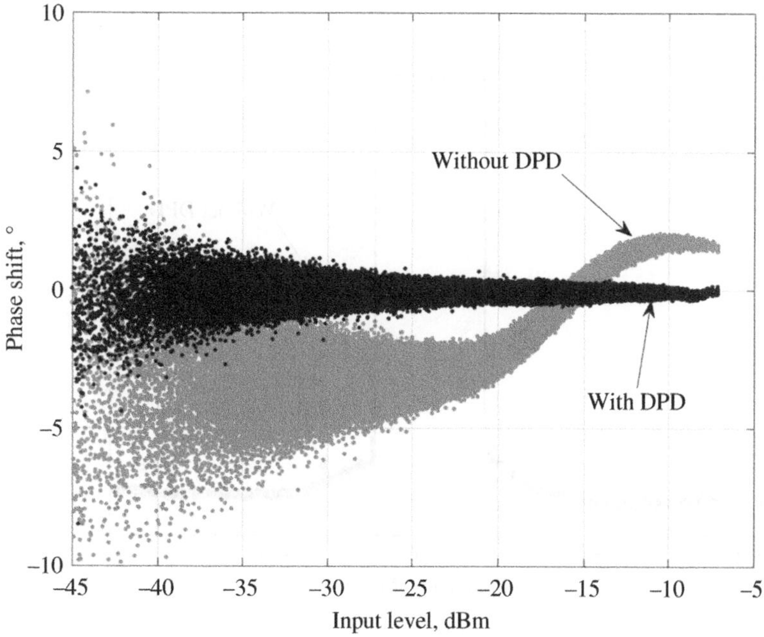

Figure 7.8 AM–PM characteristic of a commercial power amplifier, with and without linearization following an indirect learning architecture scheme.

In each iteration of the algorithms, one regressor was selected in the order determined by each algorithm and a DPD was applied with the reduced-order model structure obtained for all the cumulatively selected components up to this iteration. In this way, it is possible to show the effect of adding new components according to different coefficients selection methods.

The evolution of the linearization NMSE, ACPR and EVM versus the number of selected components that have been added to the active regressors set are shown in Figures 7.11, 7.12, and 7.13, respectively. As it is reasonable to expect, NMSE, ACPR, and EVM follow the same pattern for all the techniques in comparison. Since OMP, DOMP, and PCA select the coefficients incrementally, the evolution of the error metrics is decreasing with the number of components. PCA shows a slow decrease in the first iterations to achieve the best linearization at the end of the span in the number of coefficients. DOMP achieves the fastest pruning, reaching its best linearization performance after just the first few iterations. Regarding the NMSE tolerance, the Pareto front—defined here as the values with lowest tolerance for each number of coefficients (Keßler et al., 2016)—is conformed by the DOMP and by PCA for a number of coefficients greater than 65. Another interesting observation is that the linearization performance of the reduced-order DPDs

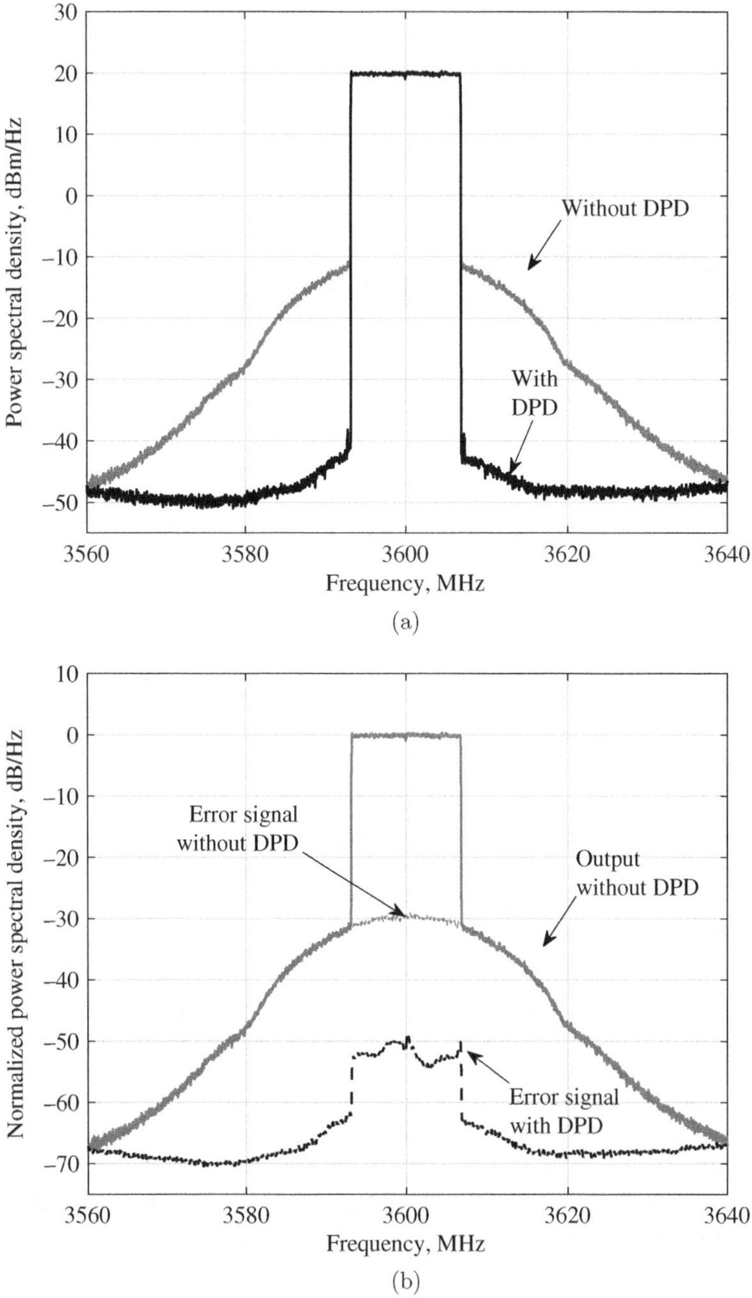

Figure 7.9 Power spectral density of the output signal of a commercial power amplifier (a) and normalized power spectral density of the error signal between the output signal and the ideally linear output (b), with and without linearization following an indirect learning architecture scheme.

Table 7.1 Linearization performance of a basic DPD following an indirect learning architecture scheme.

Case	ACPR (dBc)				NMSE (dB)	EVM (%)	Output power (dBm)
	−2	−1	+1	+2			
w/o DPD	−46.4	−36.7	−36.6	−46.2	−27.9	3.8	28.7
w DPD	−68.3	−65.8	−65.5	−67.2	−50.3	0.5	25.8

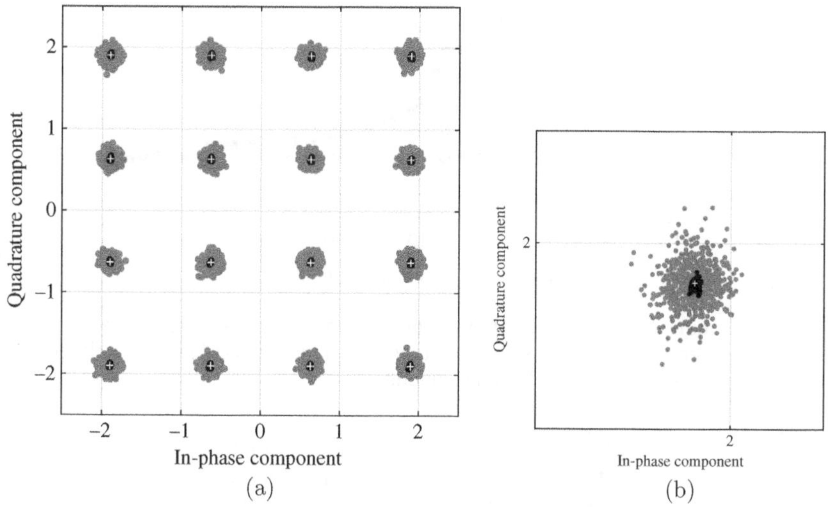

Figure 7.10 Constellation of the output signal of a commercial power amplifier (a) and zoom in one of its symbols (b), with and without linearization following an indirect learning architecture scheme.

can be considered sufficiently good for a certain number of selected components, according to the adopted stopping criterion, so that the computational complexity associated with the implementation of the predistorter can be significantly reduced without degrading the linearization quality.

Regarding the spectral regrowth reduction, the power spectral densities of the linearized signals for the first 20 coefficients selected by each algorithm are shown in Figure 7.14. It is worth mentioning that, for a fair comparison of the spectra in this figure, the input power level of the signal without DPD was reduced so as to produce an output power of 25.8 dBm, matching the achieved output power with DPD when the compressed gain is selected as the target gain, as it is the case. The out-of-band distortion is clearly reduced with respect to the case without digital predistortion. Furthermore, Table 7.2 shows the linearization metrics for a fixed

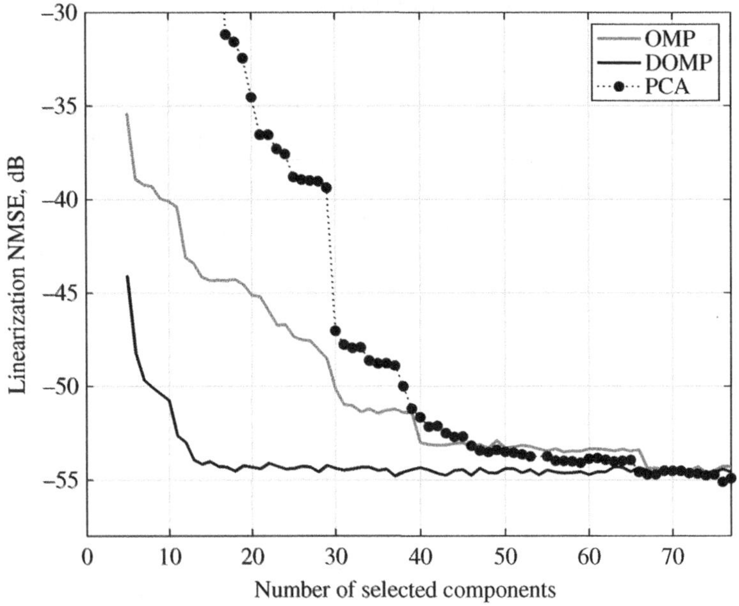

Figure 7.11 Linearization NMSE achieved for a commercial power amplifier by employing different coefficients selection algorithms in an indirect learning architecture scheme versus the number of selected components.

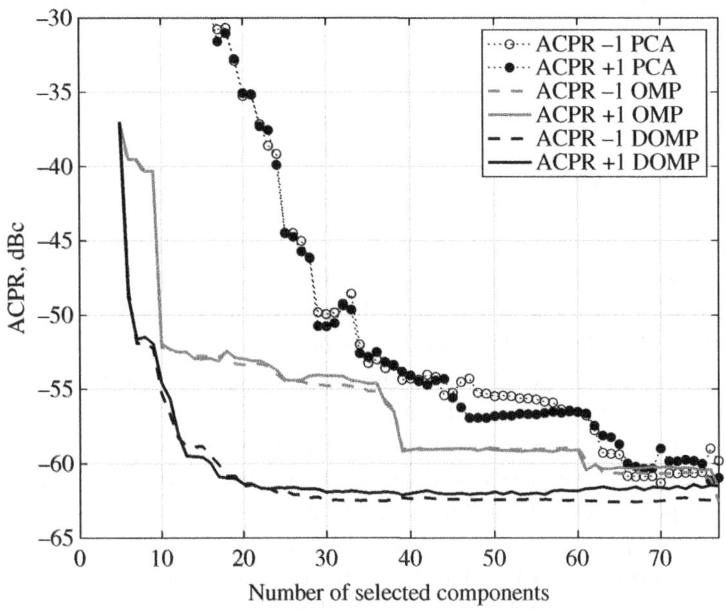

Figure 7.12 Linearization ACPR, denoted as −1 for the lower channel and +1 for the upper channel, achieved for a commercial power amplifier by employing different coefficients selection algorithms in an indirect learning architecture scheme versus the number of selected components.

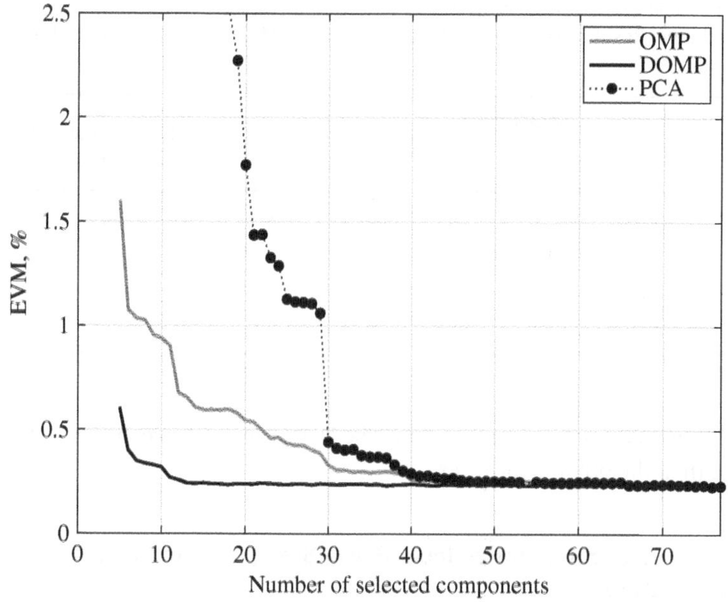

Figure 7.13 Linearization EVM achieved for a commercial power amplifier by employing different coefficients selection algorithms in an indirect learning architecture scheme versus the number of selected components.

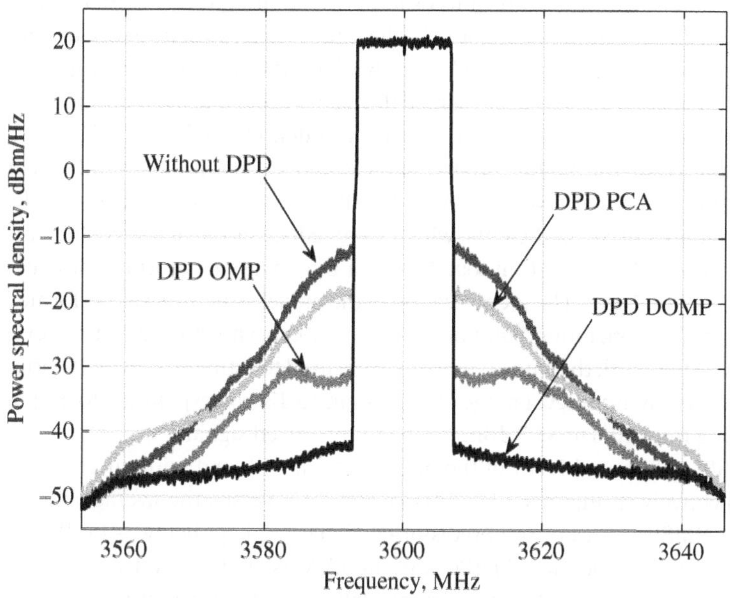

Figure 7.14 Power spectral density of the output signal of a commercial power amplifier, with and without linearization following an indirect learning architecture scheme and different coefficients selection algorithms.

Table 7.2 Linearization performance of a DPD following an indirect learning architecture scheme. Results are provided for the first 20 coefficients selected according to different algorithms.

Case	ACPR (dBc)				NMSE (dB)	EVM (%)	Output power (dBm)
	−2	−1	+1	+2			
w/o DPD	−46.6	−37.1	−36.9	−46.3	−28.3	3.6	28.9
DPD OMP	−54.7	−52.1	−51.6	−53.8	−45.1	0.5	25.8
DPD DOMP	−66.9	−65.1	−65.1	−66.0	−54.3	0.2	25.8
DPD PCA	−49.9	−42.1	−42.4	−49.7	−34.5	1.8	25.8

number of 20 coefficients in each algorithm. The DOMP algorithm exhibits better performance than the other algorithms.

7.5.3 Case 3: Linearization for an Input Power Sweep with Indirect Learning Architecture

In the third example case, the linearization of the power amplifier under test is analyzed for an input power sweep, following an ILA with compressed gain as the target gain (Crespo-Cadenas et al., 2017). In this case, a 30-MHz OFDM test signal was used, differently from the previous examples, exhibiting PAPR = 10.5 dB and acquired with an oversampling factor of 3. The initial model structure was also different in this case, being determined by a univariate GMP model following (3.29) with the following setting: part A presented thirteenth order and a maximum delay $Q = 15$ for the linear term, $Q = 10$ for the third order, $Q = 5$ for the fifth order, and memoryless for the higher orders; parts B and C presented seventh order with maximum values of $L = 1$ and $M = 1$ in both cases. This full basis model had a total number of 59 coefficients, which were reduced to 40 active coefficients selected by the DOMP algorithm together with the Bayesian Information Criterion (Becerra et al., 2020). The identification of the model structure was performed for an intermediate operation point of the power sweep, showing an output level of 27.6 dBm. Afterward, the same regressors were used to estimate the predistorter coefficients over an input level range of more than 13 dB, corresponding to output levels ranging from 20.6 to 35.5 dBm for the unlinearized signal.

The NMSE, ACPR, and EVM of the output signal, with and without DPD, are shown respectively in Figures 7.15, 7.16(a), and 7.17, versus the average output power. The NMSE values of the predistorted signal are kept below −45 dB, the ACPR values are kept below −50 dBc, and the EVM is kept below 1% for all the power levels, with significant improvements with respect to the nonlinear output

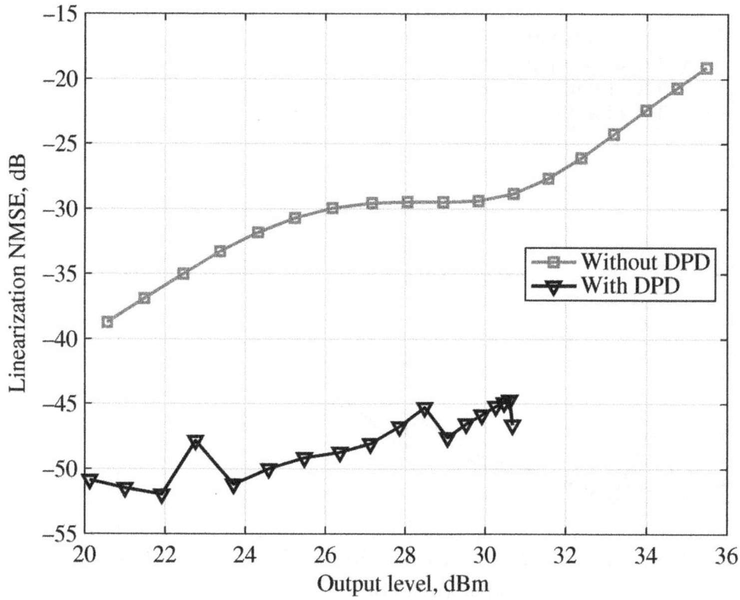

Figure 7.15 Linearization NMSE in a power sweep for a commercial power amplifier, with and without linearization following an indirect learning architecture scheme.

for the highest power levels. As in the previous ILA examples that used the compressed gain of the amplifier as target gain, the price of the good linearization performance is a reduction of the maximum output power available. Despite this reduction, Figure 7.16(b) can help us illustrate that one of the main advantages of applying a linearization technique to a power amplifier consists in being able to employ it with efficiency values that are unfeasible due to the nonlinear effects constraints. In this figure, the ACPR values presented by the signal at the output of the power amplifier are depicted versus the power added efficiency (PAE)[1] provided by the considered power amplifier. The PAE of a power amplifier increases as the output power increases. However, the nonlinear effects of the power amplifier produce also an increasing ACPR with the output power level. Linearity requirements for mobile communication signals usually demand ACPR values lower than −45 dBc in order to avoid that the out-of-band emissions generate interference in the adjacent channels. It can be observed in Figure 7.16(b) that the PAE values are limited below 5% in order to fulfill this requirement without DPD. In contrast, the reduction in the ACPR achieved when a DPD is applied allows us to

1 The power added efficiency of an amplifier is defined as $\mathrm{PAE}(\%) = 100 \cdot \frac{P_{\mathrm{out}} - P_{\mathrm{in}}}{P_{\mathrm{DC}}}$.

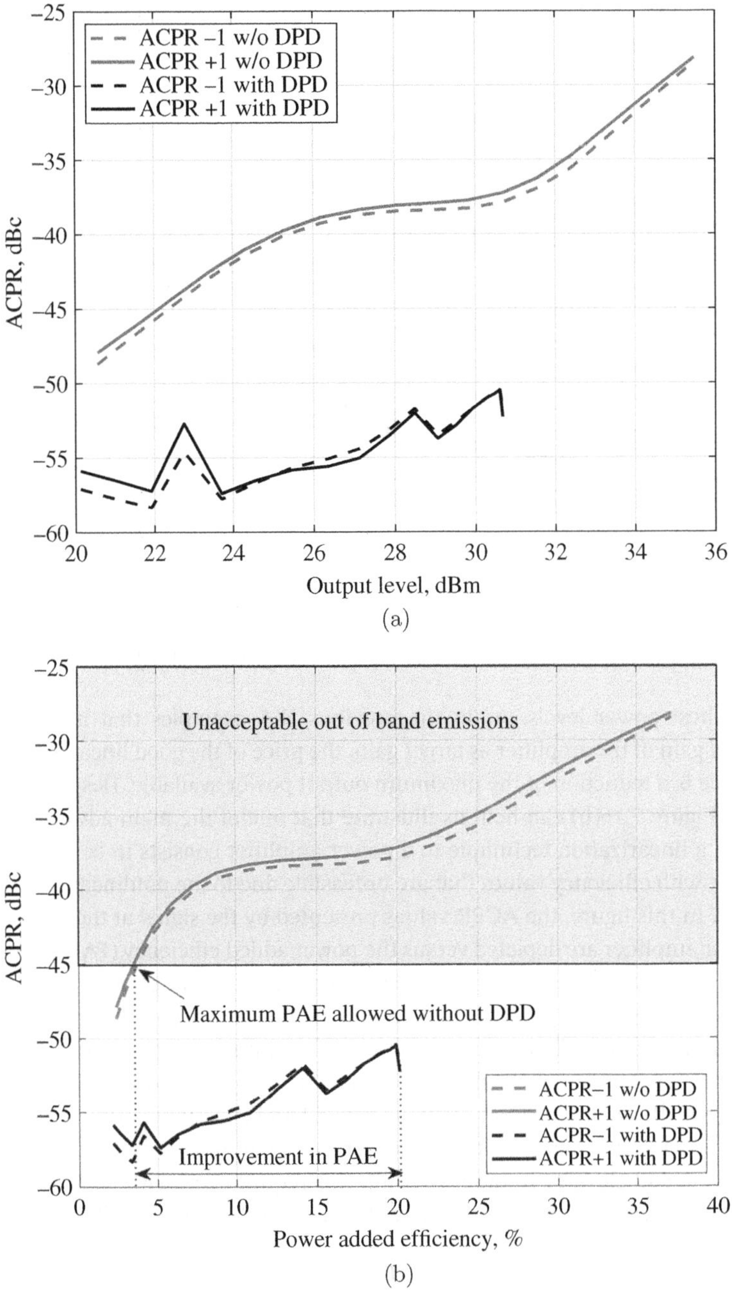

Figure 7.16 Linearization ACPR, denoted as −1 for the lower channel and +1 for the upper channel, in a power sweep for a commercial power amplifier, with and without linearization following an indirect learning architecture scheme. In (a) it is plotted versus the output power level, and in (b) it is plotted versus the power added efficiency.

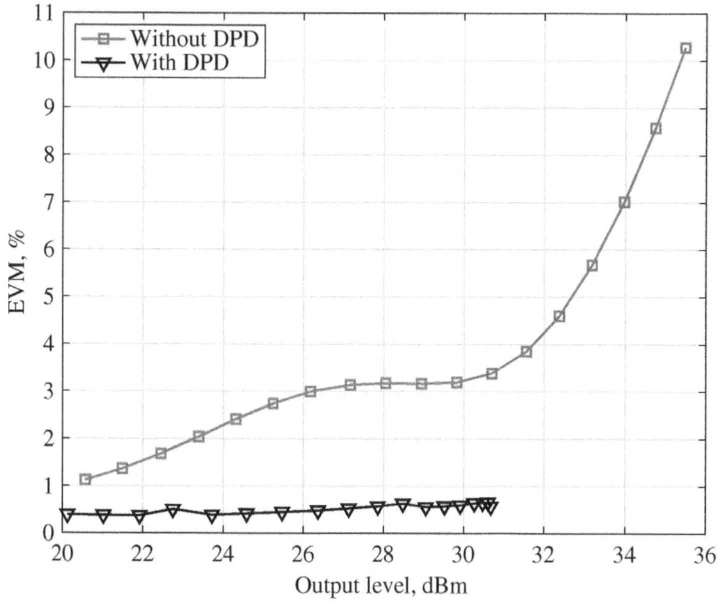

Figure 7.17 Linearization EVM in a power sweep for a commercial power amplifier, with and without linearization following an indirect learning architecture scheme.

reach PAE values of up to 20% since all the measured power levels satisfy the ACPR requirement. Comparable performance gains can be expected for other DUT cases.

7.5.4 Case 4: Basic Digital Predistorter with Direct Learning Architecture

The forth case aims to present a basic DPD implemented following a direct learning scheme (Zhou and DeBrunner, 2007), where the effect of different learning rates will be analyzed. Among other benefits, DLA has shown greater robustness in presence of additive white Gaussian noise (Hussein et al., 2012), a better linearization performance (Paaso and Mammela, 2008; Chani-Cahuana et al., 2018), and a lower impact of feedback noise (Yu and Zhu, 2015) compared to ILA.

In this case, the same test signal than in the previous example is used, a 30-MHz OFDM signal, and the power amplifier exhibited an output power of 31 dBm for the selected operation point, characterized by 2.7 dB of gain compression. It must be noted that in the present example with the direct learning scheme, it was possible to fix the target gain value to the average gain of the power amplifier without significantly degrading the linearization performance, thus obtaining the same output power with the predistorted signal.

The employed model was a thirteenth order GMP model with memory fading from 15 to memoryless in part A, and seventh order with maximum memory depth of 1 in the non-diagonal parts B and C. The identification of the DPD was implemented following a DLA with different values for the learning rate, $\mu = 1/4, 1/3, 1/2$. According to the adaptation parameter selection discussed in Section 7.4, the experimental results for higher values are not shown because they do not perform adequately. Training of the predistorter was executed for 20 iterations. Afterward, the robustness of the identified DPD was tested over 20 more iterations in which the values of the coefficients had been fixed and the sequence of random symbols generated for each iteration was changed so that a new validation signal was set as the input for each iteration.

Figures 7.18–7.20 show the evolution of the linearization metrics for both the training and robustness stages. As expected, the evolution of the error metrics is decreasing with the iterations during the training stage. It can be also observed that the reduction of the error is faster as the learning rate increases. During the robustness stage, the DPD proved its ability to work with signals not used in the training stage. The power spectral density for one of the iterations during the robustness stage is shown in Figure 7.21 with a satisfactory reduction of the spectral regrowth.

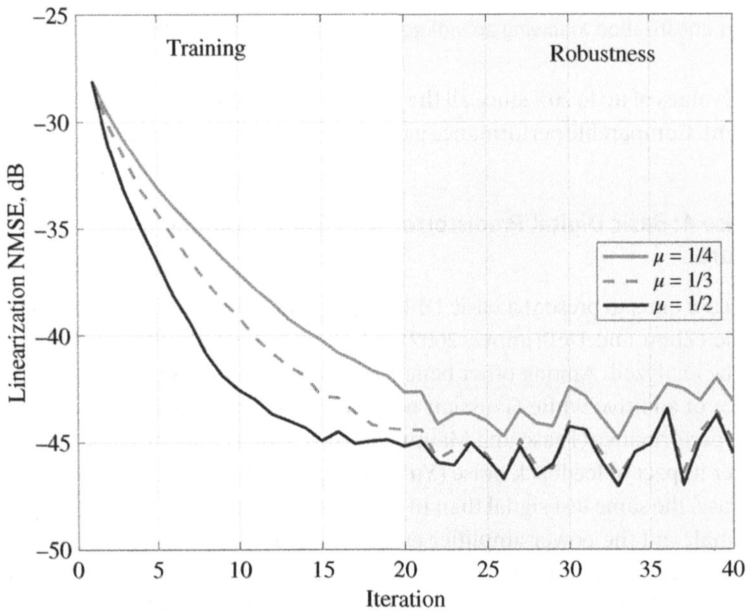

Figure 7.18 Linearization NMSE achieved for a commercial power amplifier by employing different learning rates in a direct learning architecture versus the number of iterations.

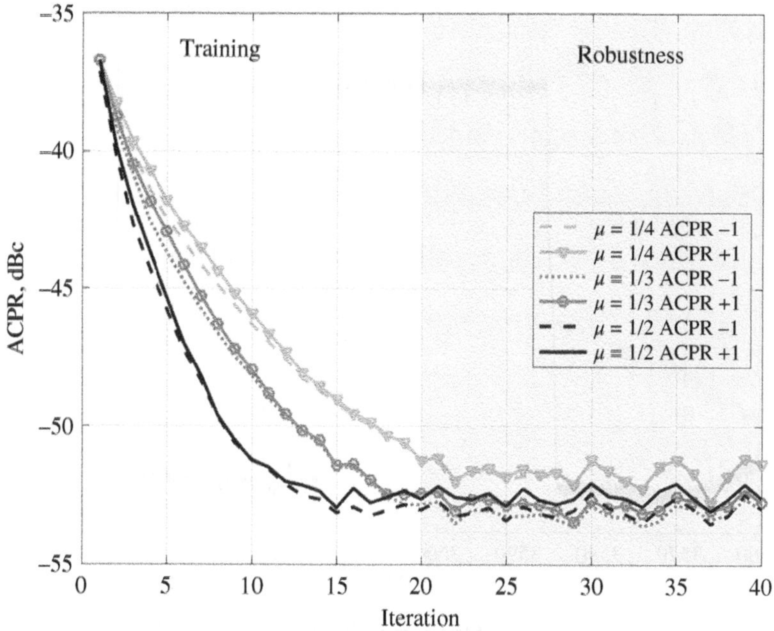

Figure 7.19 Linearization ACPR, denoted as −1 for the lower channel and +1 for the upper channel, achieved for a commercial power amplifier by employing different learning rates in a direct learning architecture scheme versus the number of iterations.

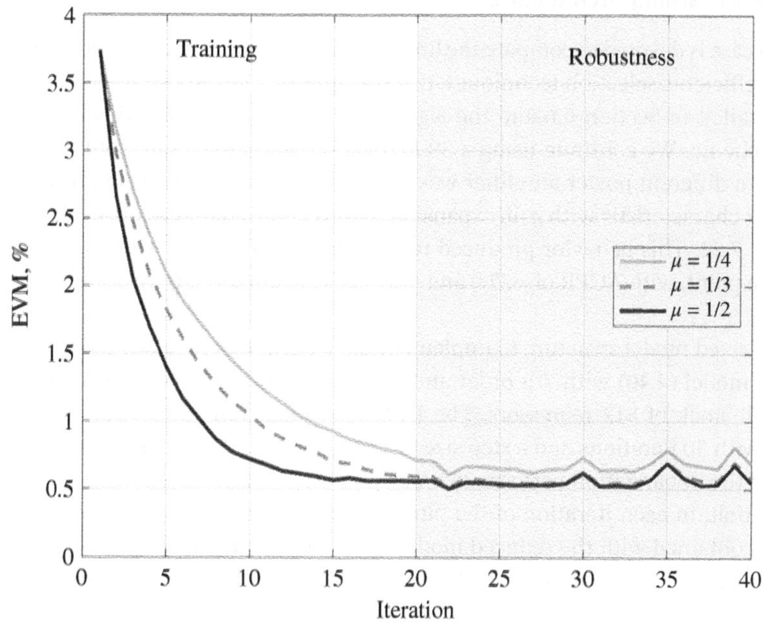

Figure 7.20 Linearization EVM achieved for a commercial power amplifier by employing different learning rates in a direct learning architecture versus the number of iterations.

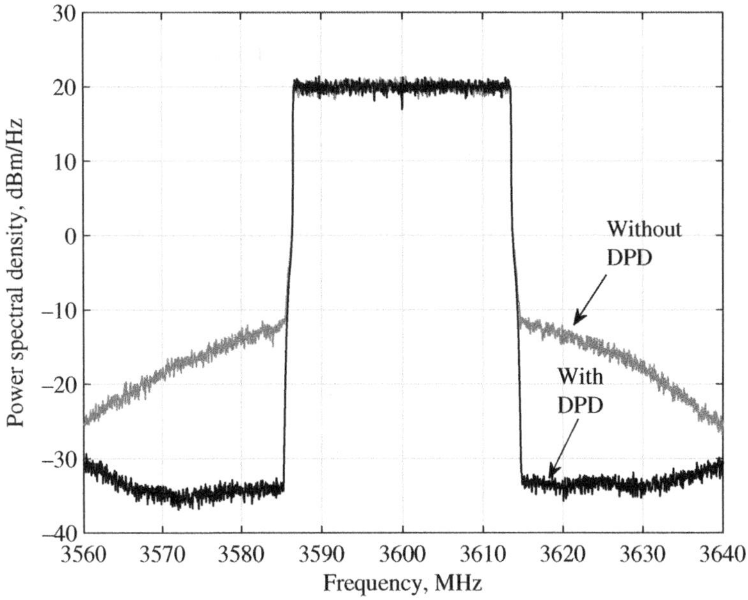

Figure 7.21 Power spectral density of the output signal of a commercial power amplifier, with and without linearization following a direct learning architecture scheme.

7.5.5 Case 5: Digital Predistorter with Coefficients Selection and Direct Learning Architecture

The fifth case is devoted to compare the linearization performance achieved by different coefficients selection techniques, the SBP approach (Crespo-Cadenas et al., 2022) detailed in Section 6.6 and the algorithms OMP and DOMP, this time in a DLA scheme. We continue using a 30-MHz OFDM probing signal, but in this example, a different power amplifier was employed that exhibited a complicated nonlinear characteristic with gain expansion followed by a compression, as shown in Figure 7.22. This behavior produced remarkable nonlinear distortions for the operation point, with ACPR of −27.9 and −28.9 dBc, NMSE of −16.5 dB, and EVM of 14.7%.

The selected model structure to implement the DPD in this case was the bivariate CKV model (4.40) with 7th order and memory depth of 7 taps, providing an initial full stock of 812 regressors. The DPD was implemented through a DLA scheme with 30 iterations and a step size of $1/4$. Then, the DPD coefficients were fixed and the linearization performance was validated with a different signal.

Recall that, in each iteration of the pursuits, one regressor was selected and a DPD was obtained with the reduced model structure accumulated up to this iteration. Therefore, in Figures 7.23, 7.24, and 7.25 we will show the evolution of the

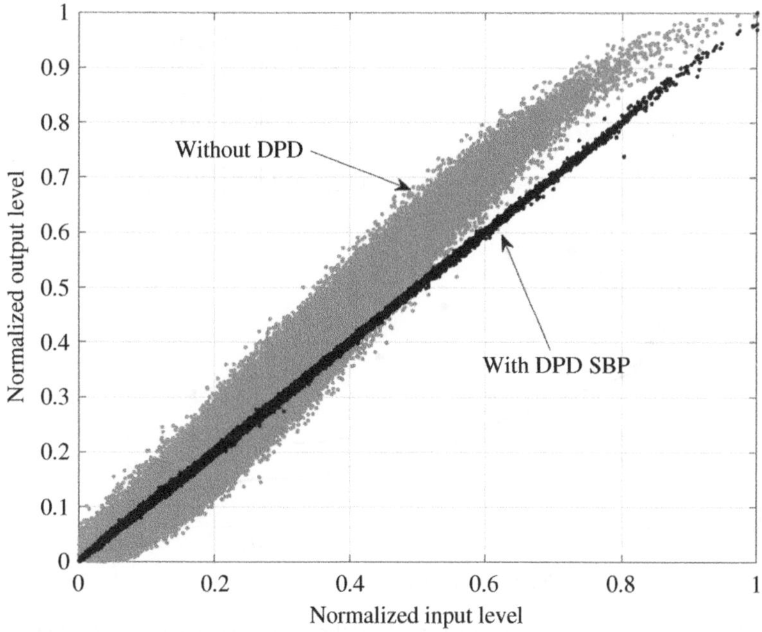

Figure 7.22 Normalized AM–AM characteristic of a commercial power amplifier, with and without linearization following a direct learning architecture scheme and the SBL algorithm for coefficients selection.

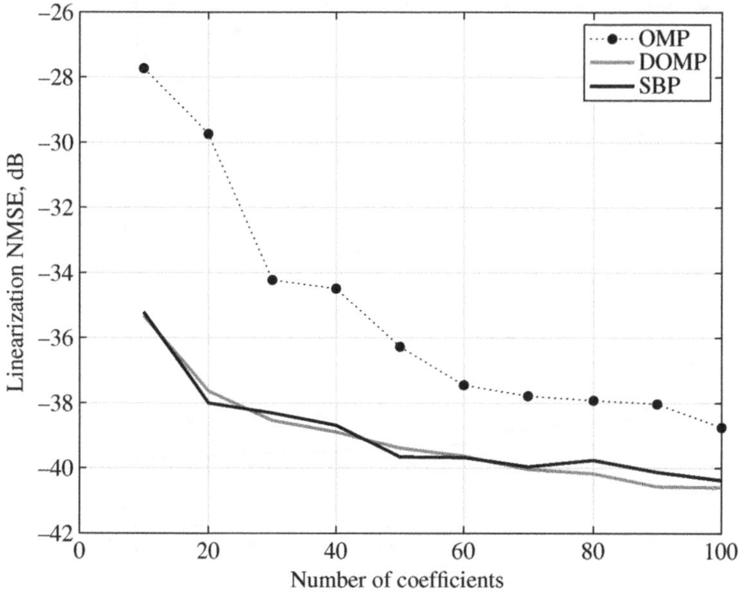

Figure 7.23 Linearization NMSE achieved for a commercial power amplifier by employing different coefficients selection algorithms in a direct learning architecture scheme versus the number of selected components.

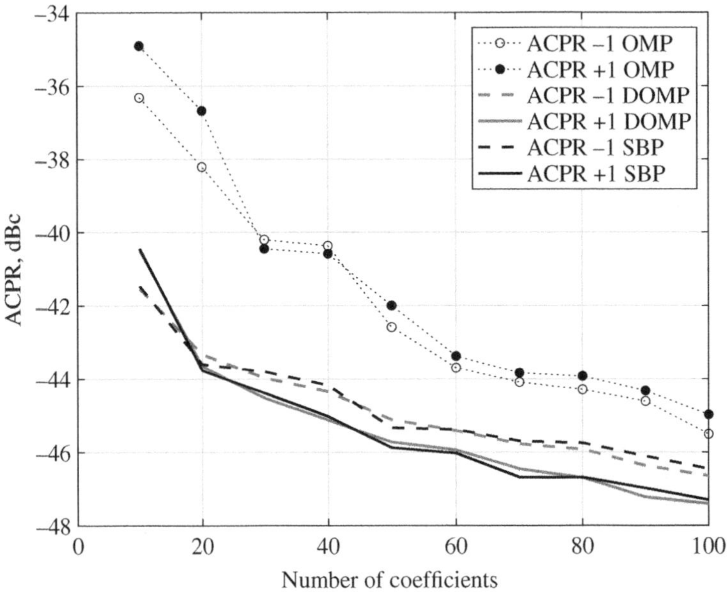

Figure 7.24 Linearization ACPR, denoted as −1 for the lower channel and +1 for the upper channel, achieved for a commercial power amplifier by employing different coefficients selection algorithms in a direct learning architecture scheme versus the number of selected components.

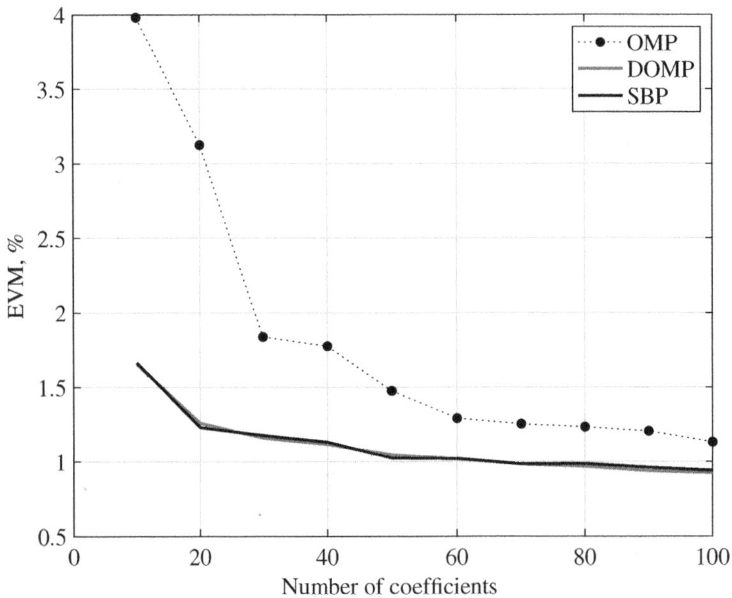

Figure 7.25 Linearization EVM achieved for a commercial power amplifier by employing different coefficients selection algorithms in a direct learning architecture scheme versus the number of selected components.

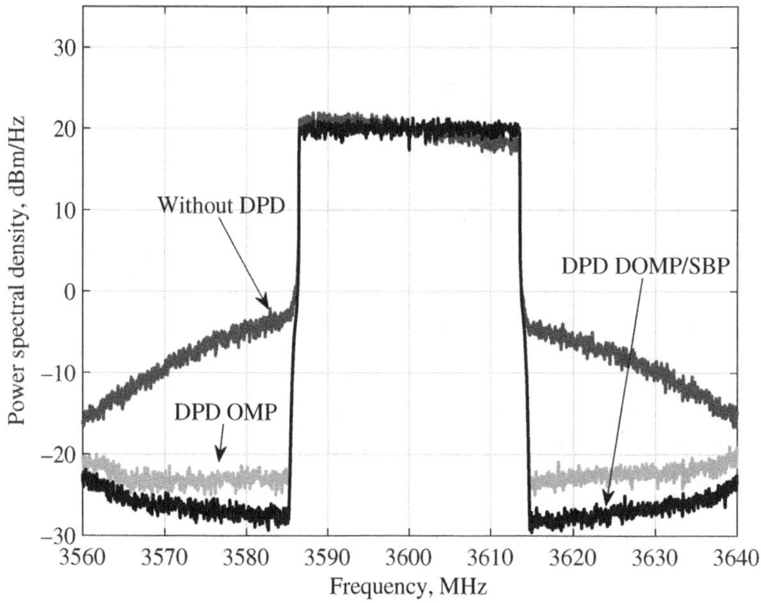

Figure 7.26 Power spectral density of the output signal of a commercial power amplifier, with and without linearization following a direct learning architecture scheme based on a DPD with SBP coefficients selection.

linearization metrics NMSE, ACPR, and EVM, respectively, while new components were incorporated in the order determined by each method.

As it is reasonable to expect, the evolution of NMSE, ACPR, and EVM is similar to that provided in Case 2. Here, the SBP approach achieves almost identical values of NMSE, ACPR, and EVM than DOMP algorithm, with superior performance than the OMP approach. The more complicated nonlinear behavior of the power amplifier without DPD can also be appreciated in the power spectral density at the output of the power amplifier, shown in Figure 7.26, and the output signal constellation provided in Figure 7.27. The signal exhibits a non-flat in-band spectrum together with a notable spectral regrowth, and the corresponding distortion of the constellation diagram of the received 16-QAM symbols. A DPD based on the SBP approach with 100 coefficients can compensate these impairments.

7.5.6 Case 6: Linearization for an Input Power Sweep with Direct Learning Architecture

The sixth case provides an example of application of a DLA DPD with coefficients selection by means of the SBP approach. Moreover, this example is provided in the context of an input power sweep.

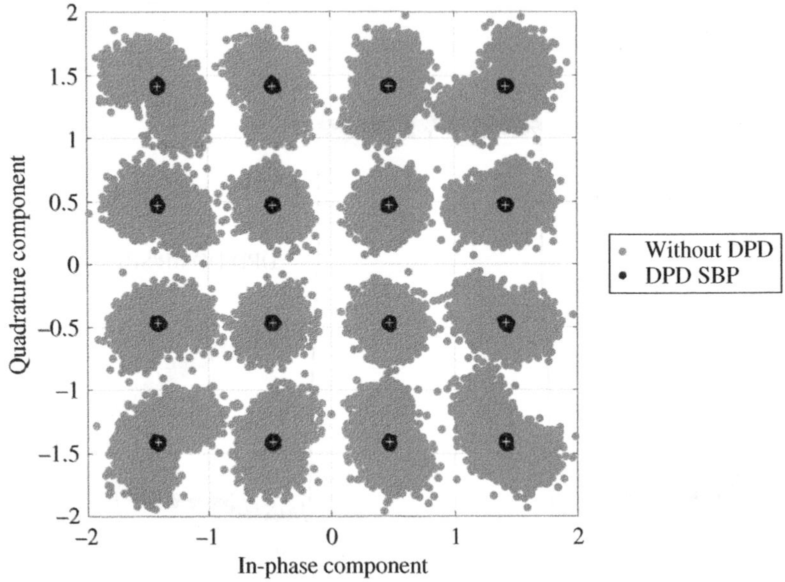

Figure 7.27 Constellation of the output signal of a commercial power amplifier, with and without linearization following a direct learning architecture scheme based on a DPD with SBP coefficients selection.

In this case, the same 30-MHz OFDM test signal was employed and the target gain was also fixed to the value of the average gain of the power amplifier. The model structure employed corresponded to a bivariate CKV model following equation (4.40) with seventh order and maximum memory depth of 7 taps, providing an initial full stock of 812 regressors. The SBP algorithm was executed to select the 30 more likely regressors, using a segment of 1843 samples of the input-output dataset for the identification of the reduced-order model structure at an operation point with the maximum output power level considered in the sweep, in this case 31.9 dBm. These regressors were used to implement a DPD through 10 direct learning iterations with a step size of $1/2$ over a power sweep of 10 dB with a step size of 1 dB, corresponding to an output power level range from 22 to 32 dBm. No new identification of the regressors was necessary after each power change. That is, the 30 more relevant regressors or basis functions were kept unchanged, while their coefficients were updated for each power level through the DLA iterations. Figures 7.28, 7.29, and 7.30 show the linearization NMSE, ACPR, and EVM, respectively, in this power sweep. It can be observed that the performance

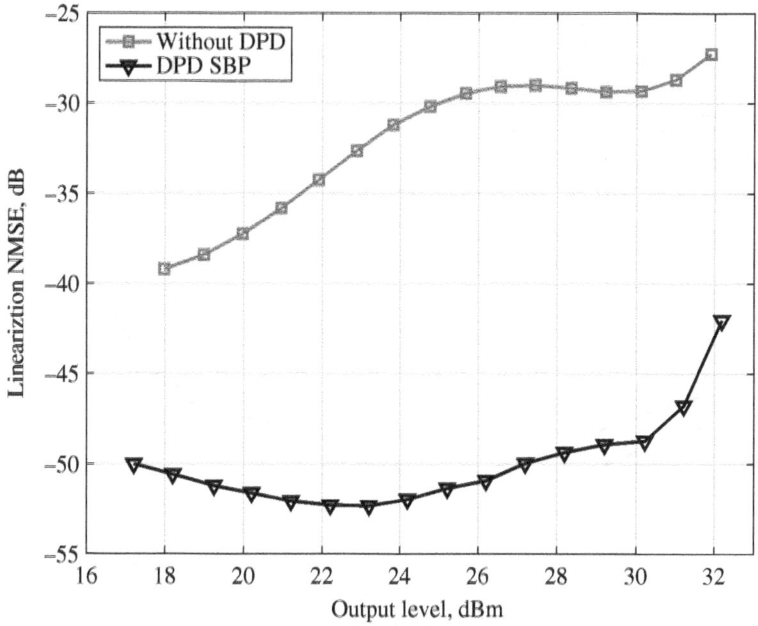

Figure 7.28 Linearization NMSE in a power sweep for a commercial power amplifier, with and without linearization following a direct learning architecture scheme and using the coefficients selection algorithm SBP.

of the linearization is maintained under changes in the power level over a certain range.

Different linearization strategies have been outlined in this chapter, both for ILA and DLA schemes. It is worth noticing the impact that the desired gain G_0 employed for normalization has in the DPD performance. As it can be observed in the provided examples, better linearization metrics can be achieved when using the compressed gain for normalization instead of the average gain. However, it introduces a backoff with an associated reduction in the efficiency of the power amplifier. From the point of view of an RF engineer, who would be mostly interested in maximizing efficiency while maintaining distortions just below the allowed limits, it is interesting to explore trade-off values between the compressed and average gains.

It can be concluded that, to design a high performance low complexity DPD, the best practice is the combination of a model with a sufficiently rich set of basis functions and the application of an effective coefficients selection technique. As

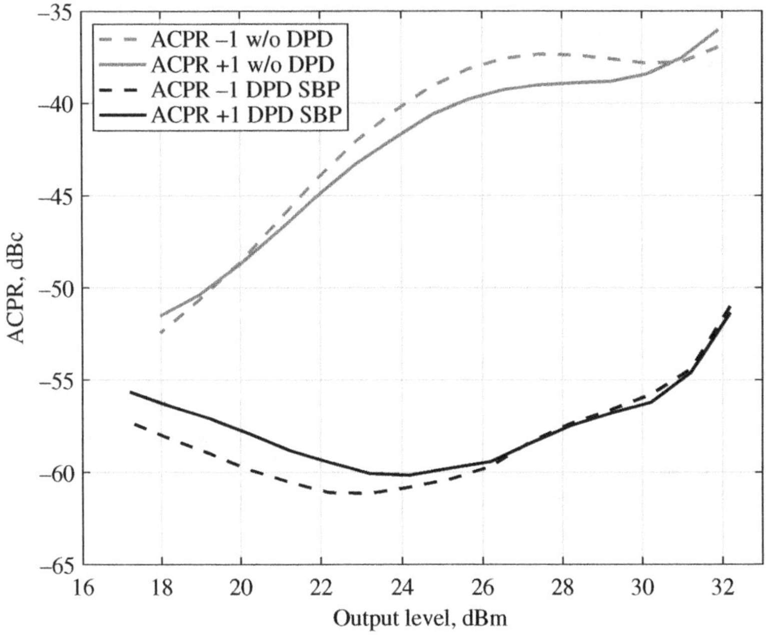

Figure 7.29 Linearization ACPR, denoted as −1 for the lower channel and +1 for the upper channel, in a power sweep for a commercial power amplifier, with and without linearization following a direct learning architecture scheme and using the coefficients selection algorithm SBP.

it has been demonstrated throughout this book, there exists an ample variety of Volterra-based behavioral models whose regressors structure constitutes an appropriate set of basis functions for the design of DPDs. When adopting pruned Volterra models, either by heuristic approaches or based on circuit knowledge, it is important to keep in mind their regressors format and properties in order to select the model structure that is better fitted for each device under test and scenario. For instance, the performance of any general Volterra-based model derived for real-valued RF power amplifiers and represented by its baseband equivalent will be degraded in the presence of I/Q impairments, since a complex-valued system must be considered in this case. Sparse regression based on coefficients selection techniques allows significantly reducing the initial set of basis functions to just the most relevant components while maintaining high-linearization performance. Among the sparse machine learning algorithms that have been reviewed, DOMP and SBL approaches stand out because of their superior performance.

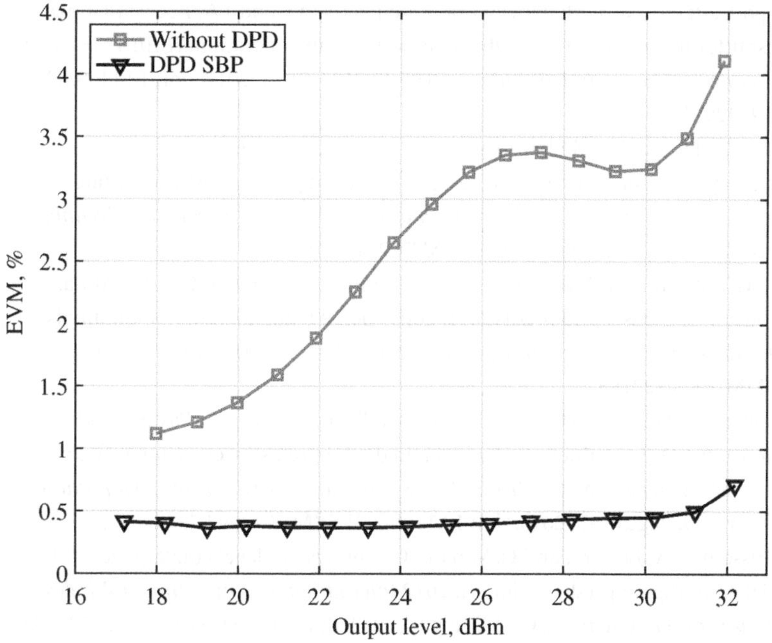

Figure 7.30 Linearization EVM in a power sweep for a commercial power amplifier, with and without linearization following a direct learning architecture scheme and using the coefficients selection algorithm SBP.

Bibliography

J. A. Becerra, M.J. Madero-Ayora, R.G. Noguer, and C. Crespo-Cadenas. On the optimum number of coefficients of sparse digital predistorters: A Bayesian approach. *IEEE Microwave and Wireless Components Letters*, 30(12):1117–1120, 2020. doi: 10.1109/LMWC.2020.3027878.

J.A. Becerra, M.J. Madero-Ayora, J. Reina-Tosina, C. Crespo-Cadenas, J. García-Frías, and G. Arce. A doubly orthogonal matching pursuit algorithm for sparse predistortion of power amplifiers. *IEEE Microwave and Wireless Components Letters*, 28(8):726–728, 2018. doi: 10.1109/LMWC.2018.2845947.

J. Chani-Cahuana, C. Fager, and T. Eriksson. Lower bound for the normalized mean square error in power amplifier linearization. *IEEE Microwave and Wireless Component Letters*, 28(5):425–427, 2018. doi: 10.1109/lmwc.2018.2817021.

C. Crespo-Cadenas, M. J. Madero-Ayora, J. Reina-Tosina, and J.A. Becerra-González. Transmitter linearization adaptable to power-varying operation. *IEEE Transactions on Microwave Theory and Techniques*, 65(10): 3624–3632, 2017. doi: 10.1109/TMTT.2017.2742951.

C. Crespo-Cadenas, M.J. Madero-Ayora, J.A. Becerra, and S. Cruces. A sparse-Bayesian approach for the design of robust digital predistorters under power-varying operation. *IEEE Transactions on Microwave Theory and Techniques*, 70(9):4218–4230, 2022. doi: 10.1109/TMTT.2022.3157586.

L. Ding, G.T. Zhou, D.R. Morgan, Z. Ma, J.S. Kenney, J. Kim, and C.R. Giardina. A robust digital baseband predistorter constructed using memory polynomials. *IEEE Transactions on Communications*, 52(1):159–165, 2004. doi: 10.1109/TCOMM.2003.822188.

P.L. Gilabert, G. Montoro, D. López, N. Bartzoudis, E. Bertran, M. Payaró, and A. Hourtane. Order reduction of wideband digital predistorters using principal component analysis. In *2013 IEEE MTT-S International Microwave Symposium Digest (MTT)*, pages 1–7, 2013. doi: 10.1109/MWSYM.2013.6697687.

M.A. Hussein, V.A. Bohara, and O. Venard. On the system level convergence of ILA and DLA for digital predistortion. In *2012 International Symposium on Wireless Communication Systems (ISWCS)*, August 2012. doi: 10.1109/iswcs.2012.6328492.

A. Katz. Linearization: reducing distortion in power amplifiers. *IEEE Microwave Magazine*, 2(4):37–49, 2001. doi: 10.1109/6668.969934.

T. Keßler, F. Logist, and M. Mangold. Use of predictor corrector methods for multi-objective optimization of dynamic systems. In Z. Kravanja and M. Bogataj, editors, *Computer Aided Chemical Engineering*, pages 313–318. Elsevier, 2016. doi: 10.1016/b978-0-444-63428-3.50057-6.

P.M. Lavrador, T.R. Cunha, P.M. Cabral, and J.C. Pedro. The linearity-efficiency compromise. *IEEE Microwave Magazine*, 11(5):44–58, 2010. doi: 10.1109/MMM.2010.937100.

H. Paaso and A. Mammela. Comparison of direct learning and indirect learning predistortion architectures. In *2008 IEEE International Symposium on Wireless Communication Systems*, October 2008. doi: 10.1109/iswcs.2008.4726067.

J. Reina-Tosina, M. Allegue-Martínez, C. Crespo-Cadenas, C. Yu, and S. Cruces. Behavioral modeling and predistortion of power amplifiers under sparsity hypothesis. *IEEE Transactions on Microwave Theory and Techniques*, 63(2): 745–753, 2015. doi: 10.1109/TMTT.2014.2387852.

M. Schetzen. Theory of pth-order inverses of nonlinear systems. *IEEE Transactions on Circuits and Systems*, 23(5):285–291, 1976. doi: 10.1109/TCS.1976.1084219.

Z. Yu and E. Zhu. A comparative study of learning architecture for digital predistortion. In *2015 Asia-Pacific Microwave Conference (APMC)*, volume 1, pages 1–3, 2015. doi: 10.1109/APMC.2015.7411819.

D. Zhou and V.E. DeBrunner. Novel adaptive nonlinear predistorters based on the direct learning algorithm. *IEEE Transactions on Signal Processing*, 55(1): 120–133, 2007. doi: 10.1109/TSP.2006.882058.

A. Zhu, P.J. Draxler, J.J. Yan, T.J. Brazil, D.F. Kimball, and P.M. Asbeck. Open-loop digital predistorter for RF power amplifiers using dynamic deviation reduction-based Volterra series. *IEEE Transactions on Microwave Theory and Techniques*, 56(7):1524–1534, 2008. doi: 10.1109/TMTT.2008.925211.

Index

A Volterra Approach to Digital Predistortion: Sparse Identification and Estimation, First Edition.
Carlos Crespo-Cadenas, María José Madero-Ayora, and Juan A. Becerra.

m

Printed and bound by CPI Group (UK) Ltd, Croydon, CR0 4YY

17/09/2026

14964329-0001